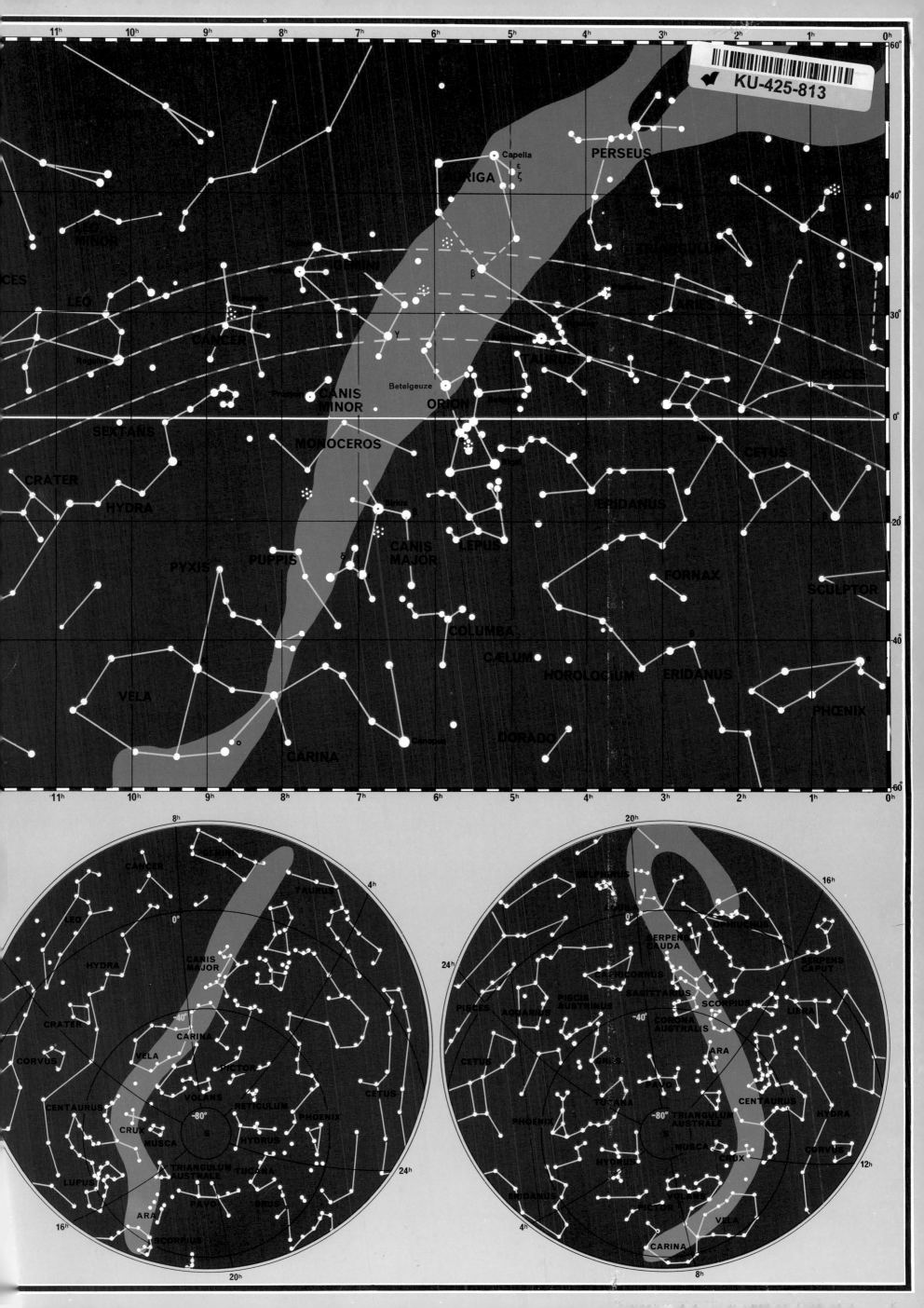

Deryk D
from Jonathan
May 1975

THE MITCHELL BEAZLEY
CONCISE ATLAS
OF THE UNIVERSE

by Patrick Moore

Mitchell Beazley Publishers Limited, London,

The Mitchell Beazley
Concise Atlas of the Universe

Foreword by Professor Sir Bernard Lovell, OBE LLD DSc FRS
Professor of Radio Astronomy and Director of the Experimental Station,
Nuffield Radio Astronomy Laboratories, Jodrell Bank

Consultant on Earth from Space Mapping and Earth Resources
the late Dr Arch C. Gerlach, Chief Geographer, US Geological Survey 1967-73

General Cartographic Consultant
Harold Fullard MSc., Cartographic Director, George Philip & Son Ltd.

The Mitchell Beazley Concise Atlas of the Universe
Edited and designed by Mitchell Beazley Publishers Limited, 14-15 Manette Street, London W1V 5LB
© Mitchell Beazley Publishers Limited, London, 1974
Completely revised from **The Atlas of the Universe** © Mitchell Beazley Ltd. 1970
Moon maps (pages 50-1, 54-5, 58-9) © Rand McNally & Co. 1970
Star maps (pages 148-9, 156-7, 164-5, 172-3) © Hallwag Berne 1970
Relief globe key maps (pages 22, 24, 28, 30) © Geographical Projects 1970

ISBN 0 85533 034 1

Typesetting by James C. Joyce Ltd., London and Tower Typesetting, London.
Printed in the Netherlands by Smeets N.V., Weert
Bound in the Netherlands by Proost en Brandt N.V., Amsterdam

Contents

Foreword

New worlds are charted by new atlases. Each memorable new atlas in the history of the world has been called forth by the discoveries of that era.

About A.D. 150 Ptolemy of Alexandria gave in his *Geographia* instructions for the production of charts of the civilized world which were far in advance of any others because, for the first time, accurate astronomical methods had been used in their compilation.

Ptolemy's world was encompassed by the boundaries of a fraction of the Earth's surface which we know today but his maps represented a vast improvement on anything which had been attempted before. After him mapping techniques of Earth and heavens became more extensive and more accurate. New projections were devised and new observations improved the accuracy.

Some 1450 years later—about a century after the great voyages of Columbus—the first scientific book of maps covering the known world appeared. Drawn by the Flemish geographer Gerhardus Mercator (1512-1594) it gave us a new word "atlas"—from the illustration on the cover which showed the mythological giant Atlas supporting the globe of the Earth. Mercator's great Renaissance Atlas was produced in response to the urgent contemporary need to redefine the boundaries of the old world by including the recent discoveries of the new—America.

Now, four centuries later, comes this *Atlas of the Universe,* a new record of the modern age of discovery. The publication of this *Atlas* marks a significant moment in human history when men recognized their frail status on this Earth in relation to the vastness of space. For everywhere on this Earth people have been fascinated and excited by the space exploits of our age. From the first Sputnik in 1957 to the landing of man on the Moon in 1969 the drama of exploring outer space has been cease

Never before has it been possible to write an Atlas of this kind because the revolutionary strides in our knowledge which have made its publication possible have almost all come about in the last ten years and some of them – the rediscovery by man of Earth from space, the landing on the Moon and the charting of the planet Mars – only at the end of that decade.

But this revolution in man's ability to explore the solar system is only a part of the unfolding of astronomical knowledge during this time. The great radio telescopes and the optical telescopes have discovered objects of a type unknown even a few years ago. Some of these lie in the remote parts of space and time and convey to us on Earth records of the early history of the universe. The *Atlas of the Universe* is original in that it is the first Atlas to draw together astronomical and terrestrial maps of the Earth as seen from Space, the Moon, the Solar System, the Stars, indeed the whole known universe into one volume. It presents all the most exciting discoveries in a new and ambitious form. The beauty of its design and execution is matched by its encyclopedic collection of information to provide a book of great contemporary importance.

I commend the *Atlas of the Universe* to young and old as a guide to the universe as we understand it today. It is reliable, lucid and authoritative: based on the most up-to-date official sources it provides a work of reference of equal value for the home and the expert. As men probe deeper and deeper into the secrets of the universe in coming years the Atlas will retain its value as a guide to the new discoveries.

Professor Sir Bernard Lovell, OBE, LLD, DSc, FRS

Introduction

Are we alone in the universe? How far away are the stars? Does space extend outward for ever? These are only a few of the questions which spring to the mind of anyone who looks up into the night sky. Surely there is no onlooker who can fail to be conscious of a feeling of wonderment and awe; and indeed the science of astronomy must be almost as old as humanity itself.

Early stargazers had to depend upon their eyes alone. Even so, they realized that the Earth is a globe, and gradually it became clear that our world is a normal planet, moving round a normal star – the Sun. Yet before the age of telescopes, astronomy was bound to be very limited. All that could really be done was to measure the positions and the movements of the various bodies in the sky.

The greatest star-catalogue of pre-telescopic times was drawn up by the Danish astronomer Tycho Brahe, whose main work was carried out between 1576 and 1596. Tycho's quadrants depended upon naked-eye estimations, and under the circumstances the accuracy of his measurements was truly amazing. He also paid close attention to the movements of the planets, particularly Mars; and it was his data which enabled his last assistant, Johannes Kepler, to prove that the Sun rather than the Earth is the centre of the planetary system.

Telescopes were invented in the first decade of the 17th century. By far the greatest of the pioneer observers was Galileo, who made a whole series of spectacular discoveries – including the phases of Venus, the four main satellites of Jupiter, spots on the Sun and the countless stars in the Milky Way. All this was done with instruments which were poor by modern standards; the most powerful of them magnified only 30 times!

The invention of the telescope caused a complete revolution in astronomy. With the naked eye alone, observations were restricted to studies of the positions and movements of the various celestial objects – in fact, to what is now called "positional astronomy"; and even here the accuracy left much to be desired, even in the work of a genius such as Tycho Brahe. Nothing whatsoever could be discovered about the nature of the stars or planets, which meant in effect that the universe itself remained a mystery. The telescope altered all this in a few years. Galileo, the first great telescopic observer, made a series of discoveries in 1609-10 which altered the entire situation.

The first such telescopes were refractors, using glass lenses to collect their light; then came reflectors, which used mirrors. The world's most famous telescope today, the "giant eye" of Palomar, has a mirror 200 inches in diameter.

Yet even the Palomar colossus cannot show a star as anything but a point of light. The distances involved in astronomy are vast; even light, moving at 186,000 miles per second, takes over 4 years to reach us from the nearest star. Most of our information comes from the spectroscope, which splits up starlight and gives us at least a partial answer to one of the age-old questions: "What is a star made of?" During the 19th century, too, photography was developed, and it became possible to study pictures taken with large telescopes. The more Man learned, the less important he seemed to become.

However, the human eye and the ordinary photographic plate are limited. They are sensitive only to visible light, which makes up a small part of what is called the electromagnetic spectrum. Very short-wavelength radiations cannot be seen; neither can the long radiations which are known as radio waves. In 1931 it was found that radio waves were coming from the sky, and could be collected by special instruments which are known as radio telescopes, but which are more in the nature of wireless aerials. Nowadays, radio astronomy has become a vitally important branch of science.

Lastly there came the dawn of space research. It is probably true to say that the Space Age began in October 1957, with the launching of the first man-made moon or artificial satellite. By now, men have stepped out on to the lunar rocks, and probes have been sent out to the planets. We are living in the most exciting age in the history of mankind. Yet even so, the story of astronomy may be only just beginning, as you will find as you read through the pages of this Atlas.

1902 : Tsiolkovskii
K. E. Tsiolkovskii is known as "the father of space research". In his papers on the subject of rocket travel, the first of which appeared in 1902 (though actually written about 1895), he laid down the scientific principles of astronautics.

1926 : Goddard
R. H. Goddard fired the first liquid-propellant rocket in 1926 from Auburn, Massachusetts. He undertook fundamental research in rocketry, but was often disinclined to lay his results before the public.

1957 : Sputnik I
The launching of Sputnik I by the Russians on 4 October, 1957, opened the Space Age, and paved the way for manned flight. Sputnik carried a radio transmitter, signals from which were picked up all over the world.

1959: Luna 1
In 1959 the Russian vehicle Lunik I (January) bypassed the Moon at 4000 miles; Lunik 2 (September) crash-landed there; Lunik 3 (October) went round the Moon and sent back the first photographs of the far side.

1961 : Gagarin
The first manned space-flight was achieved on 12 April, 1961 by Yuri Gagarin, who completed a full circuit of the Earth in Vostok 1. He reached an altitude of 203 miles; the journey took 1 hour 48 minutes.

1961 : Shepard
On 5 May 1961 Alan Shepard became the first American in space. He achieved a sub-orbital flight lasting for a quarter of an hour. On 20 February 1962, John Glenn completed a full orbit of the Earth.

1965 : Space-Docking
A successful space-docking operation was carried out in December 1965 by Gemini 6 (Stafford and Schirra) and Gemini 7 (Lovell and Borman). This was an essential preliminary to the successes of the Apollo programme.

1967 : Orbiter
During 1967 the programme of lunar mapping with the U.S. Orbiter vehicles was continued. By the end of the year the entire surface of the Moon had been charted satisfactorily in this way, enabling possible landing sites to be chosen.

1969 : Apollo 11
The final triumph of the Apollo programme came in July 1969, when Neil Armstrong and Edwin Aldrin in Eagle, the lunar module of Apollo 11, made a successful landing on the Mare Tranquillitatis and walked on the Moon.

1970 : Lunokhod 1
In 1970 the Russians sent up Luna 17, carrying the 'crawler' Lunokhod 1. This could move about the Moon, controlled from the Earth, and sent back a great deal of information. A second Lunokhod followed in 1972.

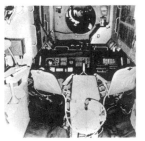
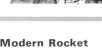

1973 : Skylab
Skylab, America's first space-station, was launched in 1973. It suffered damage during the ascent, and was repaired in orbit by the first crew. A second crew went up to Skylab in July 1973.

Modern Rocket

A modern rocket is a very complicated structure, in which the step-principle is used, as shown in this diagram. An Apollo rocket stands between 360 and 365 feet in height on its launching-pad, but only the relatively small command module survives to bring the astronauts home.

Two and one-half minutes after blast-off, the first stage separates. Later, the second stage separates. The third stage carries the astronauts to a height of 115 miles. After injection into the transfer orbit the command, service and lunar modules separate from the third stage and continue their journey. For a lunar landing, the lunar module is detached, and, after the journey to and from the Moon's surface, is jettisoned For final re-entry into the Earth's atmosphere, the cone-shaped command module comes down blunt-end first, so that the enormous heat caused by friction can be absorbed by the shield.

At present, intensive research is in progress with the aim of developing nuclear-powered rocket motors. Once this has been achieved, the step-method can be dispensed with, and travel times can be shortened drastically. It is likely that nuclear rockets will have to be used for all manned flights beyond the Earth–Moon system.

(A) Launch escape system
(B) Command module
(C) Service module
(D) Lunar module
(E) Third stage
(F) Second stage
(G) First stage

Telescopes

With a refracting telescope, the light from the object to be studied passes through a specially shaped lens known as an object-glass. The light is brought to focus, and an image of the object is produced and enlarged by an eyepiece.

The reflector collects its light by means of a mirror. In the Newtonian form, the light is reflected back up the telescope tube, and strikes a smaller flat mirror; the rays are than brought to a focus near the side of the tube, and the image is magnified by the eyepiece. Radio telescopes may take one of many forms; with the "dish" type the long-wavelength radiations are focused, and the signal is converted to an electrical impulse which is recorded as a trace on a graph.

The Refractor *left.* An image is formed at (A) and magnified at (B).

The Reflector *centre.* In the Newtonian pattern, Light is reflected up the tube to a smaller mirror.

The Radio Telescope *right.* Here radiations are collected by a metal reflecting surface.

The Washington Refractor *left* This 26-inch refractor at the Washington (U.S. Naval) Observatory was installed in 1862.

Schmidt Telescope *centre.* In 1932 Bernhard Schmidt devised this type of telescope which can photograph a large area of sky with a single exposure.

Radio Telescope *right.* An example of the dish type *in situ.*

Electromagnetic Spectrum *below*

Electromagnetic radiation comprises a continuum from the low frequency long-wavelength radio waves to the high frequency short-wavelength gamma-rays. While radiations of all forms originate in the universe, only two limited ranges of wavelength are admitted to the Earthbound astronomer by the visible and radio "windows" in the Earth's atmosphere. To utilize the rest we depend on high flying balloons, aircraft, and spacecraft.

X-rays were once surmised to be energetic particles, but their wave nature has now been established. The wave motion of all electromagnetic radiation is in fact the oscillation of electric and magnetic fields throughout the space occupied by the propagating radiation.

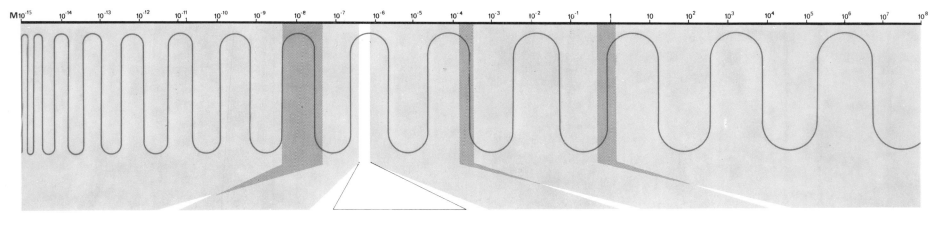

M 10^{-15} 10^{-14} 10^{-13} 10^{-12} 10^{-11} 10^{-10} 10^{-9} 10^{-8} 10^{-7} 10^{-6} 10^{-5} 10^{-4} 10^{-3} 10^{-2} 10^{-1} 1 10 10^2 10^3 10^4 10^5 10^6 10^7 10^8

Gamma and X-rays

Very short, highly penetrating radiations are known as gamma-rays and X-rays. Gamma-ray sources have been identified in the sky, though their nature is still very uncertain; there are many X-ray sources, one of which is the Crab Nebula *below.* Some of the X-ray sources are variable over short periods. Since gamma-rays and X-rays from beyond the Earth are absorbed in the upper atmosphere, rocket techniques have to be used to study them.

Ultra-Violet Radiation

Ultra-violet radiations extend between the X-ray and the visible part of the spectrum. They, too, are absorbed in the upper atmosphere, and for this reason not a great deal was known about ultra-violet sources before the development of rocketry during the past few years. In fact all stars send out ultra-violet radiation, but the hotter stars are the more powerful in this region of the spectrum. Below is Mars in ultra-violet.

Visible Light

Visible light is of limited range. The "optical window" is very restricted, and so long as the astronomer was in a position to study only visible radiations his knowledge was bound to be fragmentary. He was, in fact, in much the position of a musician who is trying to play a composition on a piano which lacks all its notes except those of the middle octave. Some extension of this range was afforded by photography. Below is Mars in visible light.

Infra-Red Radiation

Beyond the visible range comes the infra-red part of the electromagnetic spectrum, extending to the centimetre range. Again there is absorption in the upper atmosphere (apart from a few "windows"). Though all stars are infra-red sources, the late-type cool stars are the most powerful; examples of this are Betelgeux in Orion and the long-period red variable Chi Cygni, which is never brilliant visually. Below is Mars in infra-red.

Microwave Radiation

Microwave radiation can reach the Earth's surface through the "radio window". It is collected and focused by radio telescopes.

Among sources of microwave radiation are the Sun, the planet Jupiter, supernova remnants in our Galaxy, pulsars, some of the external galaxies, and quasars. By now very detailed maps of the "radio sky" have been drawn up. Below is a contour map of the centimetric emission of the Sun.

Long-Wave Radio

Beyond the radio window there is once again total absorption, this time by the ionosphere, so that very long-wavelength radiations can never reach ground level. In the future, installations on the surface of the Moon will no doubt be set up to study them, and Earth satellites have already been used. The wavelengths extend from 100 centimetres up to many kilometres. Below is a contour map of the powerful radio source Centaurus A.

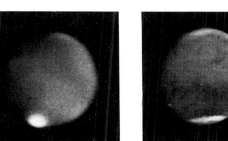

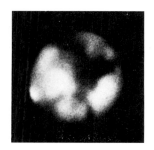
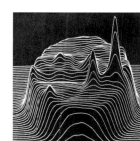
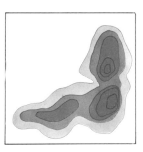

Transparency of our Atmosphere *right*

Atmosphere obscuration plotted against wavelength in Ångströms. Our eyes are only adapted to the region between 3900Å to 7600Å.

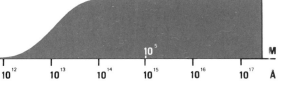

3900 Å 7600 Å

10^{-8} 10^{-5} 1 10^5 M

10^2 10^3 10^4 10^5 10^6 10^7 10^8 10^9 10^{10} 10^{11} 10^{12} 10^{13} 10^{14} 10^{15} 10^{16} 10^{17} Å

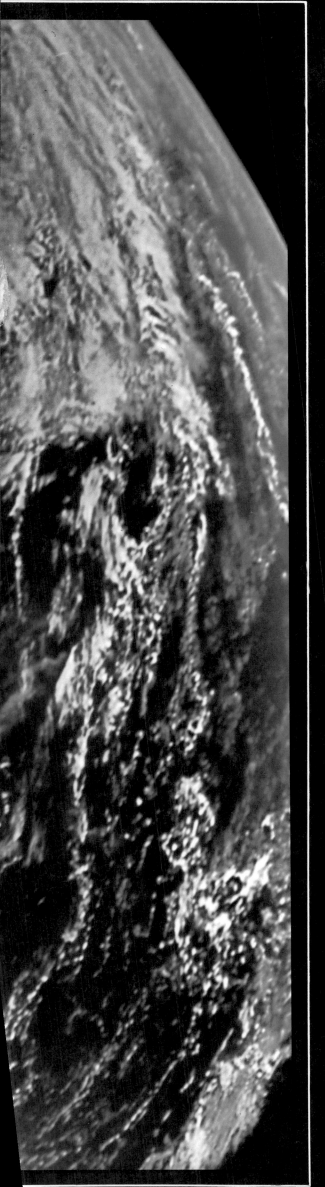

Atlas of the Earth from Space

The Earth, our home in space, is a member of the Sun's family. Originally it was thought to be of supreme importance, but during the last few years we have been able to see it as it really is : an ordinary planet, the third in the Solar System. Photographs taken from space have given us a new view of the Earth, and have shown us details which were previously unknown.

Ancient peoples believed the Earth to be a flat plane in the centre of the universe, so that the entire heavens revolved round it once a day. Some of the old theories sound strange today ; for instance, the Hindus taught that the Earth is supported on the back of four elephants, which are in turn resting on the shell of a tortoise which is swimming in a boundless ocean. However, Aristotle (384–325 B.C.) attacked the flat-earth theory on scientific grounds, and in or about 270 B.C. Eratosthenes of Cyrene was able to make a surprisingly accurate measurement of the Earth's circumference. (His value was better than that used by Christopher Columbus on the famous voyage of exploration seventeen centuries later.)

Earth from Space
An historic photograph taken 3,500 miles from Earth as Apollo 8 sped towards the Moon, December 1968. Florida and the West Indies can be seen to the south, while a huge storm-system swirls over the ocean to the north.

The Earth from Space

Ptolemaic and Copernican Theories

Ptolemy of Alexandria, last of the great astronomers of the ancient world (*c.* A.D. 120–180), perfected a theory of the universe in which all the celestial bodies moved round the Earth in a complicated system of circles and epicycles; an epicycle was a small circle, the centre of which itself moved round the Earth in a perfect circle. The Ptolemaic theory remained officially unchallenged until 1543, when Copernicus, a Polish churchman, published his famous book *De*

Size of Earth *left*
Relative sizes of Jupiter, the largest planet, the Earth and Mercury, the smallest planet; equatorial diameters respectively 88,700 miles, 7927 miles and 3000 miles. Earth is the fifth largest of the nine Solar System planets, discounting Pluto whose size is uncertain. Venus, Mars and Mercury are smaller, though Venus only slightly so (equatorial diameter 7700 miles).

Revolutionibus Orbium Cœlestium, in which he proposed that the Sun, not the Earth, must be the centre of the planetary system. The Copernican theory met with violent opposition, particularly from the Church; Galileo, the great Italian scientist, was brought to trial in 1633 for claiming that he had observational proof of the movement of the Earth and was sentenced to remain under surveillance until the end of his life. In 1687, when Isaac Newton published his classic *Principia*, the Earth was finally relegated to the status of a mere planet.

A Watery World

It is customary to say that the Earth is a typical planet, but this is not entirely correct. It has a high density (its specific gravity, or density on a scale taking water as unity, is 5·41), and it is the only planet which possesses a dense atmosphere rich in oxygen and water-vapour. Also, it alone has wide expanses of ocean on its surface, and it has been appropriately nicknamed "the Watery World". So far as is known, there is no other planet in the Sun's system which is capable of supporting advanced life-forms.

The Ecosphere

Quite apart from the composition of its atmosphere, the Earth's suitability for supporting life is concerned with its distance from the Sun. It lies in the middle of the "ecosphere", or region in which the temperature is neither too hot nor too cold for life to evolve upon a planet with the right sort of atmosphere and of the right size. The orbit does not depart much from the circular form; the eccentricity is 0·017. A more eccentric orbit would have marked effects upon the climate.

Climate

The seasons are due mainly not to the relatively slight changes in the Earth -Sun distance, but to the inclination of the Earth's axis, which is tilted by $23\frac{1}{2}°$ to the perpendicular to the plane of the orbit. In northern summer, the northern hemisphere is tilted toward the Sun; in northern winter, the southern hemisphere is tilted sunward. In fact, the Earth is at perihelion in December and at aphelion in early July, so that in theory the southern hemisphere should have shorter, hotter summers but longer and colder winters than the northern hemisphere. This effect is detectable, but it is largely masked by the differences in the distribution of land and water over the Earth's surface. Water acts as a moderating influence on the climate, and the greatest oceans lie to the south.

Though the Earth is the largest of the four inner planets, it is also the only one with wide areas of ocean. Its land surface is therefore much less than that of Venus, and equal to that of Mars.

Axial Inclination *left*
The Earth's axis is tilted by 23½° to the perpendicular to the plane of its orbit round the Sun. It is this inclination which is the principal cause of the seasons. During a year the solar overhead point traverses 47 degrees of latitude (the Tropics).

The Earth-Moon System

In another way, too, the Earth is unique. It is the only small planet to be attended by a large satellite, and it has been justifiably claimed that the Earth–Moon system should more properly be regarded as a double planet. The two bodies move round their barycentre, or common centre of gravity, which is situated inside the terrestrial globe, since the Earth has 81·3 times the mass of the Moon.

The Moon has profound effects upon the Earth. In particular, it is the main cause of the tidal effects in the water, the land and also the air – though only the oceanic tides are important in everyday life.

To explain how the tides are caused, it is convenient to suppose that the Earth is covered with a shallow, uniform ocean. Immediately under the Moon, where the lunar gravitational pull is strongest, the waters will tend to heap up in a bulge, and there will be another bulge on the far side of the Earth. As the Earth spins on its axis, the water-heaps do not spin with it, but tend to keep lined up with the Moon. The result is

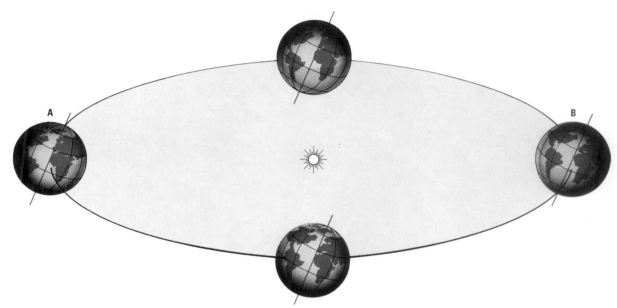

The Orbit of the Earth *above*
The orbit of the Earth does not depart greatly from the circular form, so that its changing distance from the Sun (91½ to 94½ million miles) has only a minor effect on the seasons. The northern hemisphere has its summer with the Earth in position A, and the southern summer occurs with the Earth in position B.

that every point on the oceans' surfaces has two daily high tides. The Moon is moving in its orbit, so that the water-heaps shift slowly as they follow the Moon around; on average, the high tide at any particular place will be 50 minutes later each day.

In practice, the situation is not nearly so simple as this. The oceans are irregular in outline and in depth, so that tidal effects vary considerably from one place to another, particularly on continental shelves with indented and complex coast-lines. The Sun must also be taken into account. As a tide-raising influence the Sun is not so important as the Moon, but at new or full moon (syzygy) the Sun and Moon are pulling in the same sense, producing strong or spring tides; when the Moon is at half-phase (quadrature) the Sun's pull is in opposition to that of the Moon, and the tides are weaker (neap tides). Finally, the Moon's gravitational effects are greatest at lunar perigee, and

the tides are correspondingly higher; the difference amounts to 20 per cent.

Tidal Friction

Because of the friction between the water and the ocean floor, due to the tides, the Earth's rotation is slowing down, so that the "day" is becoming longer. The average increase amounts to only 0·00000002 seconds a day, but over a sufficiently long period of time the effect mounts up. As each day is 0·00000002 seconds longer than the previous day, then a century (36,525 days) ago the length of the day was shorter by 0·00073 seconds. Taking an average between then and now, the length of the day was half of this value, or 0·00036 seconds shorter than at present. But since 36,525 days have passed by, the total error is 36,525 × 0·00036 = 13 seconds. Therefore, the position of the Moon, when "calculated back", will be in error; it will seem to have moved too far, i.e. too fast. This lunar *secular acceleration* must be taken into account in calculations of the times of lunar and solar eclipses which occurred in ancient times.

Another effect of the tidal friction is to make the Moon recede from the Earth at the rate of 4 inches per month. In the early days of the Earth–Moon system, it is thought that the distance between the two bodies must have been much less than at present, so that tidal effects would have been stronger.

Life on Earth

The age of the Earth is known fairly accurately; the most recent estimate is 4700 million years. At first there was no life, and we cannot be confident as to how the first living things appeared, but scientists believe that the original single-celled creatures have developed into the life-forms we know today. This has been possible only because conditions on Earth are suitable for Earth-type life. If there were any other similar planets in the Solar System, it is logical to assume that life would have appeared there also, and few modern astronomers doubt that there must be many inhabited planets moving round other stars. The Earth is unique in the Solar System, but it is not necessarily unique in the universe as a whole.

The Origin of the Earth

Though we may be confident that our estimates of the Earth's age are reasonably accurate, we cannot be so sure that we have a good knowledge of the way in which the Earth and other planets came into being. Many theories have been proposed, though as yet it cannot be said that there is any general agreement among astronomers.

One plausible theory was proposed in the 18th century by the French mathematician Laplace. This was the so-called Nebular Hypothesis, according to which the planets were formed from a shrinking gas

The Position of the Earth in the Solar System *right*
The diagram shows the Earth's distance from the Sun compared with those of the three other planets which lie nearest to the Sun in the Solar System.

Mercury

36,000,000 mi
57,920,000 km

cloud. As the cloud shrank, by virtue of gravitational forces, it threw off "rings", each of which condensed into a planet. It followed from this theory that the outer planets were the oldest members of the Solar System, and that the present-day Sun represented the remains of the original gas-cloud.

Various mathematical objections led to the rejection of the nebular hypothesis, and until fairly recently there was strong support for tidal theories, which involved the gravitational pull of a passing star. The best-known of these was developed by the English astronomer Sir James Jeans. According to Jeans, a passing star drew a cigar-shaped ·tongue of material away from the Sun; after the visitor had passed by, the material left whirling round the Sun broke up into drops, each of which produced a planet. The largest planets, Jupiter and Saturn, were in the middle of the "cigar".

Here, too, mathematical objections were raised, and although Jeans' theory was popular for many years it has now been generally abandoned. An

Earth and Mars Compared *right and below*
It has often been said that the Earth and Mars are similar, but information sent back from the Mariner probes shows that the resemblance is much less close than had been thought. The two are here shown to the correct scale, but since Mars has no oceans the land areas of the two planets are about equal. Mars has a lower density, and the much lower escape velocity has caused it to develop differently. It now seems that Mars is much less like the Earth, and much more like the Moon, than appeared probable before the flights of the Mariner craft.

entirely different concept was proposed by C. von Weizsäcker, in Germany. It was suggested that the planets were formed by accretion, from a cloud of particles collected by the Sun during its passage through an interstellar cloud. This theory sounds plausible, though not all authorities accept it; for instance, it has been suggested by H. Alfvén and others that magnetic forces may have played a vitally important role.

Whichever theory is adopted, it seems very likely that all the planets are of approximately the same age, even though they have developed in different ways. For instance, it is often said that Mars is more advanced in its evolution than is the Earth, and this may be true; but the absolute ages of the two worlds are probably the same. Mars has "aged" more quickly simply because it is smaller and less massive.

On the passing-star theory, planetary systems would be excessively rare. Stellar encounters are very uncommon, and if Jeans' theory were correct there would be only a few Solar Systems in the entire Galaxy, so that we might represent the only intelligent race. Yet if von Weizsäcker's theory is correct – or, for that matter, any hypothesis which does not involve a second star – planet-families are almost certainly very numerous, and this is the present view among astronomers.

Modern Scientific Advances

An important event took place in 1957–8. Scientists of all countries – apart from China – agreed to co-operate in an ambitious programme to find out more about the Earth in all its aspects; the land, sea, interior, and atmosphere. This project became known as the International Geophysical Year, though it lasted for eighteen months. In order to investigate the upper atmosphere and the adjacent space region, it was essential to build vehicles which would stay up

for long periods; and this meant developing artificial satellites which could carry scientific instruments of all kinds. The first satellite, Sputnik 1, was sent up from the U.S.S.R. on 4 October 1957. Others followed both from Russia and from America; on 12 April 1961 Yuri Gagarin of the Soviet Air Force became the first astronaut, when he made a complete circuit of the Earth in his vehicle Vostok 1.

So far as Earth studies are concerned, artificial satellites have been of the utmost value.

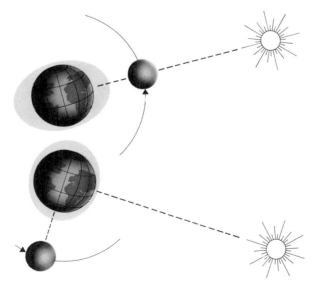

The Tides *above*
The diagrams show the cause of the tides.
upper The Moon and Sun are pulling in the same sense, producing spring tides. *lower* The Moon and Sun are pulling at right angles to each other so that the tides are much weaker (neap tides).

Data
Maximum distance from the Sun 94,537,000 miles
Minimum distance from the Sun 91,377,000 miles
Mean distance from the Sun 92,957,200 miles
Mean orbital velocity 18·5 miles per second (66,000 m.p.h.)
Orbital eccentricity 0·017
Obliquity of the ecliptic (axial inclination to the perpendicular to the orbital plane) 23°27'08".26
Length of the year tropical (equinox to equinox): 365·24 days; sidereal (fixed star to fixed star) 365·26 days
Length of the day mean solar day 24h.03m.56s.555, mean sidereal day 23h.56m.04s.091 mean solar time
Mass 6600 million million million tons
Equatorial diameter 7927 miles
Polar diameter 7900 miles
Oblateness 1/298
Density (specific gravity) 5·41
Mean surface gravitational acceleration (rotating Earth) 32·174 feet/sec. per second
Escape velocity 7 miles per second
Albedo 0·39

Earth

00,000 mi
100,000 km

92,957,200 mi
149,600,000 km

Mars

141.6,000,000 mi
228,000,000 km

11

The Phases of the Earth

Because the Earth is a non-luminous body, and is illuminated by the Sun, half of its surface is in daylight while the other half is dark. From space, the Earth will therefore show phases, basically analogous to those of the Moon; the sunrise and sunset lines mark the Earth's terminator.

These phases were first observed after the development of space-probes. The overall impression is different from the case of the Moon, because the lack of any appreciable lunar atmosphere means that the shadows there are very sharp and black and the terminator is extremely sharp. With Earth, much of the surface may be covered with cloud; the atmosphere diffuses the sunlight, and the terminator is much less sharp and clear-cut. Photographs taken from space, such as those on this page, show the effect well. Also, the albedo of the Earth (39 per cent) is much greater than that of the Moon (7 per cent), though less than that of the permanently cloud-covered Venus. Considerable local colour is seen but the general hue is bluish.

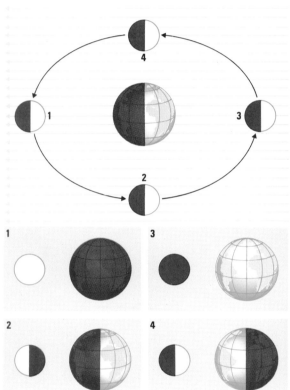

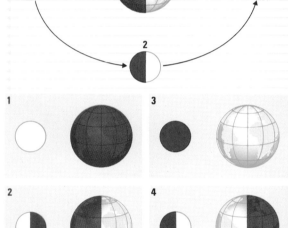

Earth Phases from the Moon *above*
The diagram above shows the phases of the Earth as seen from the Moon. For this purpose, the Earth may be assumed to be stationary, with the Moon moving round it once in 27.3 days. When the Moon is full from Earth (1) an observer on the Moon would have "new Earth"; at our new moon (3), a lunar observer would see full Earth. Below the main diagram the different respective phases of the Moon (left), and the Earth are shown. Neglecting the minor effects of libration, the Earth would appear to stand still in the lunar sky; from the far side of the Moon the Earth would remain below the horizon.

Earth Phases from Orbit *below*
Photographs taken from the ATS-3 satellite, a synchronous vehicle, moving at 22,300 miles above South America. They were taken on 18 November 1967 at 7.30 a.m., 10.30 a.m., noon, 3.30 p.m. and 7.30 p.m. respectively. In the 7.30 a.m. picture, storms are off the African coast; the storms are dissipating by nightfall.

Earth Seen from Lunar Orbit *above*
These photographs, taken from the Command Module of Apollo 10 in May 1969, show the Earth coming into view as the space-craft comes round from behind the far side of the Moon. In the first photograph, it is worth noting the sharpness of the lunar horizon; there is no atmosphere to cause blurring or distortion.

Crescent Earth *right*
Shown here is a view of the crescent Earth, seen from a distance of 10,000 miles. This photograph was taken from Apollo 4 and looks southwest over Africa and South America. Despite considerable cloud cover, the Earth's bluish colour is well in evidence, as is the line of the terminator, separating the sunlit and dark halves.

Th Earth's Magn tosph re

A new view of the Earth – no longer a small, isolated body in space, but a world which is the centre of an extensive region in which its magnetic forces are dominant. This magnetosphere reaches far into space. It forms a barrier against charged particles coming from beyond, and it traps particles so that they cannot escape, producing what is known as the Van Allen radiation zone.

The modern concept of the "magnetosphere" dates only from 1958, when the United States satellite Explorer 1 was launched (1 February). This was the first successful American space-vehicle, and although it weighed only 31 pounds, it carried a full load of equipment in miniature. In particular, it carried a Geiger-counter, which is an instrument used to detect charged particles and high-energy radiation.

The initial orbit of Explorer 1 was elliptical, with a perigee at 224 miles and an apogee of 1585 miles. One of its main functions was to study the intensity of cosmic radiation at various heights above the ground. Cosmic rays are not true rays; they are atomic nuclei coming from space. Although the Sun is one source, most of the cosmic radiation comes from beyond the Solar System, and it is highly penetrative, though the heavy cosmic ray primaries cannot reach the Earth's surface without being broken up by collisions with the particles of the upper atmosphere.

The cosmic ray counts were expected to increase with altitude, but at 600 miles the Geiger-counter on Explorer 1 ceased to function. This happened at each revolution of the satellite. It was then found that the intensity of charged particles had increased so much that the instruments had become saturated, and subsequent investigations showed that there is a zone of intense radiation surrounding the Earth. This is known as the Van Allen zone, after James van Allen, a distinguished scientist and the leading American investigator of the phenomenon.

Later satellites have carried out studies of the magnetic effects beyond the Van Allen region, and it has now become evident that the Earth's magnetosphere is very extensive.

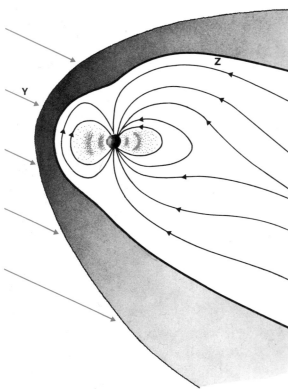

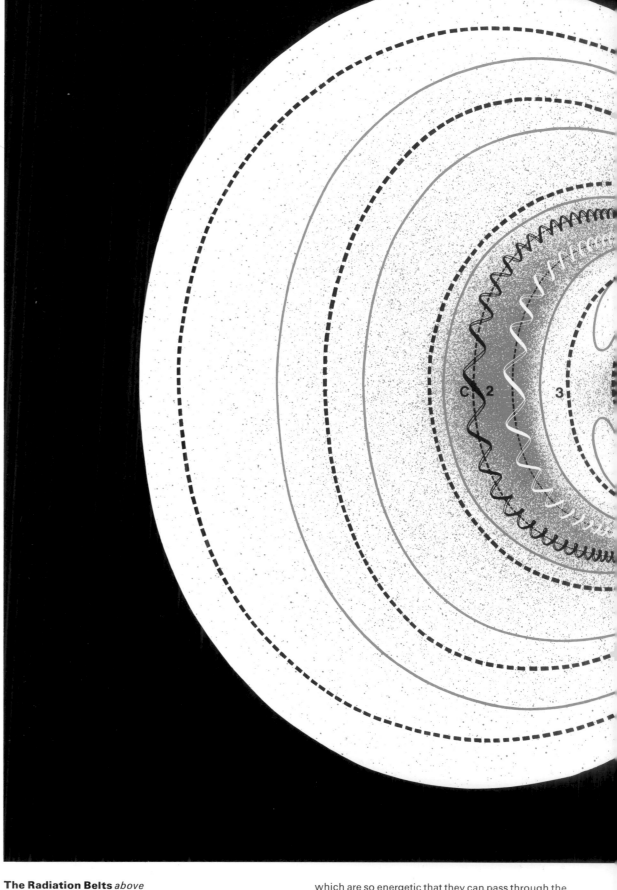

Shape of the Magnetosphere *above*
The magnetosphere resembles a teardrop in shape, with the tail pointing away from the Sun. On the sunward side of the Earth the magnetosphere extends to only 40,000 miles, but on the dark side of the Earth it reaches out to a much greater distance. As the solar wind (Y) comes toward the Earth, it meets the magnetic field, and a shock-wave forms. Inside the shock-wave is a turbulent region, inside which again is a definite boundary, the magnetopause (z). Solar wind is made up of streams of atomic particles, sent out by the sun in all directions. The magnetosphere proper lies on the Earthward side of the magnetopause. On the dark side, away from the Sun, the shock-wave gradually weakens until it has ceased to be detectable. As can be seen, the Van Allen radiation belts lie entirely within the magnetosphere.

The Radiation Belts *above*
The diagram shows the general aspect of the main part of the Earth's magnetosphere, including the Van Allen region. The north and south magnetic poles (X and X) are shown; they are not coincident with the geographical poles (N and S).

The Inner Zone (1)
It seems that the Van Allen region is divided into two distinct zones, separated by what is often called the "slot". The inner zone, whose highest-energy particles are chiefly protons, reaches its maximum intensity at a height of 3000 miles. The flux of high-energy protons is relatively steady, and the inner zone is much more stable than the outer zone. According to a theory developed by C. Christofilos, V. Vernov and S. F. Singer, the protons in the inner zone are caused by cosmic rays (A),

which are so energetic that they can pass through the Earth's magnetic barrier and collide with particles in the upper air, producing various other particles – including neutrons. Since a neutron has no electrical charge, its motion is unaffected by the presence of the magnetic field. It is not a permanent particle, and decays spontaneously into a proton and an electron; the half-life, or time taken for half the neutrons present to decay, is only 15 minutes. Since the mass of a proton is about the same as that of a neutron, while the mass of an electron is much less, almost all the kinetic energy of the old neutron will be transferred to the proton produced by its decay. Neutrons which move upward (i.e. away from the Earth) before they decay are called albedo neutrons, and it is thought that they are responsible for the proton flux in the inner Van Allen belt, since as soon as the neutron decays the charged particles

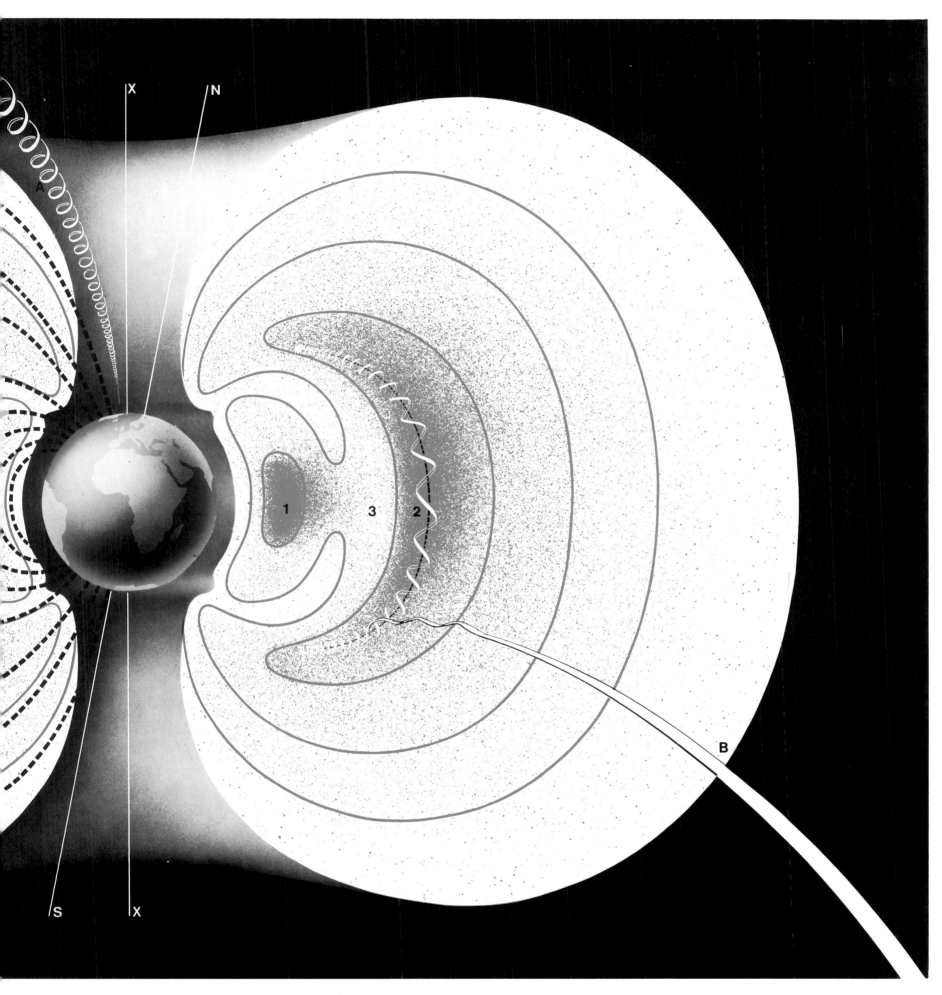

which are produced will be trapped by the Earth's
magnetic field. Beyond 6000 miles, protons sent out in
the solar wind (B) are thought to make a major
contribution to the total. Once a proton has been trapped
in the Earth's magnetic field, it will remain there for
several hundred years.

The Outer Zone (2)

The outer zone contains many high-energy electrons,
and is strongly affected by events taking place in the Sun.
The manner in which the outer zone is replenished is not
yet known in detail, but there seems little doubt that
the electrons (C) are due to the solar wind ; it is possible
that they enter the Van Allen zone through the tail of
the geomagnetic field. The maximum intensity is at an
altitude of 10,000 miles. The outer zone is disturbed by
solar flares, and these too are responsible for the brilliant

displays of aurorae (see page 79). If the Van Allen layer
becomes "overloaded" with electrons, some of its
particles will cascade downward into the atmosphere and
produce aurorae.

The "Slot" (3)

Between the two Van Allen zones is what is called the
"slot", at an altitude of 8000 miles. In this region, the
intensity of charged particles is less than at greater
or lower altitudes.

Limits of the Van Allen Zone

The Van Allen region does not envelop the entire Earth.
It extends from latitude 75 °N to 75 °S on the daylight
hemisphere, and from 70 °N to 70 °S on the night
hemisphere. Magnetic particles spiral back and forth.

along the lines of force, from one hemisphere to the other.
Charged particles which encounter the Earth at higher
latitudes will not be trapped, and will simply follow the
magnetic-field lines into the atmosphere in the
polar regions.

Earth's Magnetic Core

So far as is known, the Earth is the only inner planet
to have an extensive magnetosphere. There is no
appreciable magnetic field associated with the Moon ; no
fields have been detected with Venus or Mars, and it is
not likely that Mercury has one.
Undoubtedly the cause of the Earth's magnetism is the
heavy core, made up of magnetic materials. It is
not yet certain whether the relative weakness of
magnetism with Venus, Mars and the Moon indicates
that these bodies lack comparable cores.

The Earth's Atmosphere

We live at the bottom of an ocean of air. The total weight of the atmosphere has been estimated as 5000 million million (5×10^{15}) tons, and the standard atmospheric pressure at sea-level is 1013 millibars (760 millimetres of mercury). Most of the mass of the atmosphere is concentrated at low levels. At a height of 7000 feet above sea-level, one-quarter of the mass of the atmosphere lies below; half the mass lies below 16,500 feet, and three-quarters of the mass lies below 29,000 feet, the height of Everest. Despite the great weight of the atmosphere, it amounts to only one-third of 1 per cent of the weight of the water in the oceans.

Retention of the Atmosphere

The Earth is able to retain a dense atmosphere. The escape velocity (7 miles per second) is relatively high, and since the velocities of the atmospheric atoms and molecules depend upon the escape velocity and the temperature, the atoms and molecules of the type making up the atmospheric mantle do not move about quickly enough to break free. Were the escape velocity lower, or were the temperature higher, no comparable atmosphere could be held; for instance, the Moon (escape velocity 1·5 miles per second) is devoid of atmosphere.

It is impossible to give a definite upper limit to the atmosphere. There is no sharp boundary; the density decreases until it is no greater than that of the interplanetary medium. The falling-off in density is very rapid. At 10 miles, the value is only 1/10 of that at sea-level; at 20 miles, the density is 1/10 that of the value at 10 miles, and so on. Traces of what may be called "atmosphere" extend out to 5000 miles.

Structure of the Atmosphere

The atmosphere is not homogenous. There is a layered structure, as was discovered by T. de Bort in experiments carried out in 1898. Using balloons, de Bort sent recording equipment into the upper air, and found that he was obtaining unexpected results with regard to temperature. The lapse-rate, or rate at which the temperature decreases with increasing altitude above sea-level, is 1·6°C (3°F) per 1000 feet. Although this is not constant, and there may be temporary "inversions" or increases in temperature at low levels, the lapse-rate does not vary much. De Bort found that above 7 miles, the lapse rate dropped, and the value remained constant at −55°C (−67°F). His names for the two layers – troposphere for the lower, stratosphere for the upper – are still in use. Later research has shown that there are various other layers; several systems of nomenclature are in use, but the simplified form is to use the terms ionosphere for the region between 40 and 400 miles, and exosphere for the outermost layers.

Thickness of the Lower Atmosphere *below*
The Earth, showing the relative thickness of the dense part of the atmosphere ; its depth is about the same as the thickness of the skin relative to an apple.
below, right The Earth, photographed from Apollo 7, over the Gulf of Mexico ; the sharp cut-off of the disk shows that the dense atmosphere is, relatively speaking, very shallow (only 10 miles deep, as against a diameter of 7927 miles for the Earth itself).

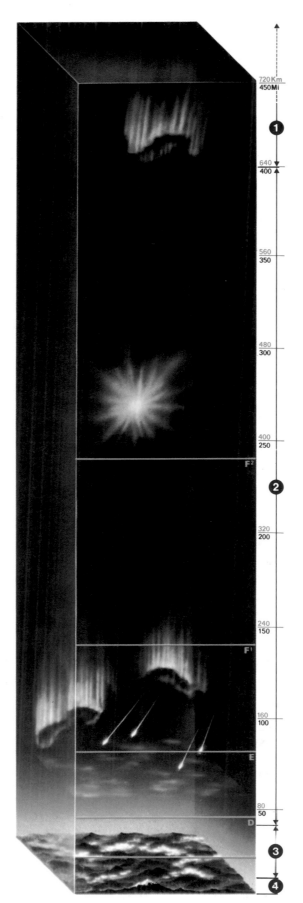

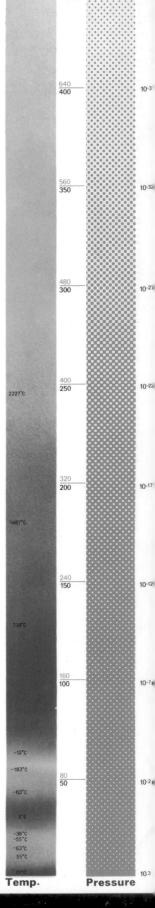

The Structure of the Atmosphere

Exosphere (1)
The exosphere, which merges with the interplanetary medium. Above 1500 miles the main constituent is hydrogen ; between 1500 and 600 miles hydrogen and helium are equally abundant ; below 600 miles, atomic oxygen exists. There is no sharp boundary between the exosphere and the ionosphere.

Ionosphere (2)
The ionosphere is of great importance inasmuch as it contains electrically conducting layers capable of reflecting radio waves. These are the D layer (35/55 miles above sea-level), and E layer (55/90 miles) and the F1 and F2 layers (90 to over 150 miles). The structure of the ionosphere is continually changing, from day to night and also according to events on the Sun. Auroræ appear in the Ionosphere at polar latitudes.

Stratosphere (3)
The stratosphere contains a layer of ozone (triatomic oxygen, O_3) which shields the Earth's surface from some short-wave radiations from space ; without this protection, life could not have developed. The upper part of the stratosphere (above 19 miles) is sometimes known as the mesosphere.

Troposphere 4
The troposphere is the lowest part of the atmosphere, in which all normal clouds and weather occur. It is separated from the stratosphere by the tropopause. The chemical composition of the troposphere is virtually homogeneous, with oxygen and nitrogen predominant.

Temp.   **Pressure**

16

Chemical Structure

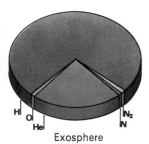

Exosphere

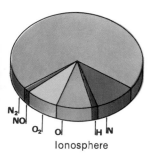

Ionosphere

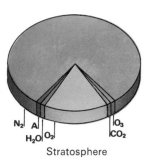

Stratosphere

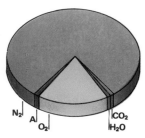

Troposphere

Chemical Composition
above

The diagrams above show the differing chemical composition of the atmosphere.

Key to Diagrams *above*
Nitrogen N_2 (molecular), N (atomic), Oxygen O_2 (molecular) O_3 (ozone), O (atomic) ; Argon A; Carbon Dioxide CO_2 ; Hydrogen H_2 (molecular), H (atomic) ; Helium He ; Water vapour H_2O.

The Protective Atmosphere *right*

The atmosphere acts as a screen against both radiations and small particles from space. It protects the surface of the Earth against bombardment by small meteoritic particles, and normal meteors cannot penetrate below 40 miles above sea level before being destroyed (though meteorites can do so, and land as solid masses).

The ozone layer, with its maximum concentration at 16 miles, is of great importance even though the amount of ozone is only 10 parts per million. It blocks out much of the ultra-violet radiation from space, and by absorbing these radiations it raises the temperature of the atmospheric region above it ; however, "temperature" is defined by the rate of motion of the constituent atoms and molecules, and the high-temperature regions of the upper air have very little "heat".

Of all the radiations from space, only those in the so-called optical and radio windows can penetrate the ground. The optical window extends from 3900 Ångströms to 7500 Ångströms. In the radio range, below about 2 cm. there is an increasing amount of absorption by water vapour in the atmosphere, while above about 10 metres (the exact wavelength depends upon the conditions at the time of observation) the ionosphere gives trouble. For this reason, studies of X-ray, gamma-ray, and ultra-violet sources beyond the Earth must be carried out by means of space-research methods.

Key to Diagram *right*
1 Exosphere ; 2 Ionosphere ; 3 Stratosphere ; 4 Troposphere. A Meteors ; B Radio waves from space ; C Infra-red radiation ; D D-layer ; E E-layer ; F_1 F_1-layer ; F_2 F_2-layer ; G Visible light ; H Ultra-violet radiation penetrating atmosphere ; I Ultra-violet radiation creating ionized layers ; J X-rays absorbed in ionosphere ; K VHF radio transmission ; L Short-wave radio transmission ; M medium-wave radio transmission ; N Long-wave radio transmission.

Weather systems from space

Photographs of the Earth taken from artificial satellites and space-probes have enabled scientific researchers to obtain their first overall view of the weather systems in the Earth's atmosphere. The upper atmosphere (i.e. above the tropopause) is transparent, and the albedo or reflecting power of the Earth is 39 per cent; the general colour is bluish, due to the scattering of sunlight in the air. The extensive cloud coverage in the troposphere often obscures the ground details, and makes the outlines of sea and land difficult to see.

From a meteorological point of view the information is invaluable. In particular, photographs from above the atmosphere can show the early stages in development of dangerous tropical storms. This was done for the first time on 12 July 1961, when the weather satellite Tiros 3 transmitted advance operational data about a hurricane which subsequently struck the American coast. The forewarning enabled people living in the danger-zone to take suitable precautions.

Pacific Ocean *right*
The Pacific Ocean, photographed from Apollo 9 from an altitude of 80 miles. The cloud arrangement is typical of a meteorological depression, and there is considerable cumulus and altocumulus.

Full Earth *below*
A general view, taken from the ATS-3 satellite, showing the cloud distribution over a complete hemisphere of the Earth. There is so much cover that the outlines of the seas and continents are largely obscured.

Clouds over the Equator *below*
Clouds over the Amazon delta photographed from Apollo 9. The cloud cover is not continuous; there is no general layer, and the clouds are chiefly of the cumulus type.

An Atlantic Anticyclone *below*
Another Apollo 9 photograph, showing anticyclonic clouds off the coast of Morocco; note how the cloud coverage does not extend over the African coastline.

A Circular Storm *below*
An Apollo 9 photograph taken over Colombia. A wide area
of cumulonimbus extends up to 22,000 feet above
sea-level, with peripheral cirrus and cumulus.

High-Altitude Clouds *below*
Clouds over the Atlantic, photographed from Apollo 6 ; the
sun glints on these high-altitude clouds which reach 20,000
feet above the ocean.

Mapping the Earth's Weather

All normal "weather" occurs in the troposphere – that is to say, the lowest layer in the atmosphere, extending up to between 7 and 10 miles above sea-level. (The height of the tropopause varies somewhat according to season and to geographical latitude.) The only clouds above the tropopause are the rare mother-of-pearl clouds in the stratosphere, and the ionospheric noctilucent clouds, at altitudes of 50 miles.

Winds are due to the difference in pressure of the atmosphere in different locations. Where the pressure is high, the atmosphere will tend to flow outward across the Earth's surface, so producing a wind. The study of wind systems is of paramount importance in meteorology, and has been greatly helped by observations made from space, since for the first time it is now possible to study large weather systems as a whole. Many "weather satellites" have been launched, beginning with Vanguard 2 on 17 February 1959. It is probably true to say that the meteorological satellites were among the most successful of the space-experiments of the 1960s and early 1970s.

In the near future it is likely that the satellite coverage of weather systems will be extended, and that a continuous record will be kept. Within the next decade a full "meteorological patrol" from space will revolutionize the science of weather forecasting.

South America *below*
There is considerable cloud-cover over South America and the general tendency of the patterns is well defined, following the prevailing wind system.

Southern Africa *below*
Extensive cloud covers the southern part of the African continent, together with Madagascar. June is a winter month in the southern hemisphere.

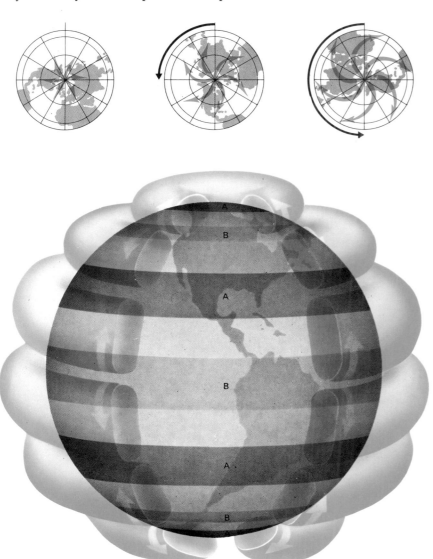

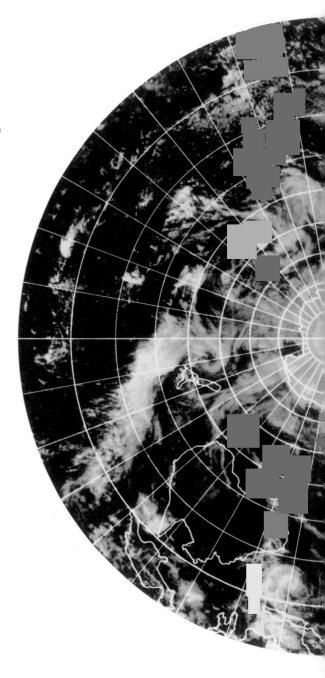

Atmospheric Circulation *left*
The large diagram shows the system of "wind cells" which carry heated air upward from the tropics and cause it to move toward the poles. Where the general circulation is upward, as at latitude 60°, the pressure is low. The pressure is also low at the equator, and the winds tend to flow toward it, but the directions of the prevailing winds are affected by the Earth's rotation (Coriolis effect, shown in the small diagrams). In the northern hemisphere, the winds flowing toward the equator have a north-east direction; in the southern hemisphere the direction is south-east. For the same reason, the movement of the air round a centre of low pressure (a depression) is clockwise in the southern hemisphere, counter-clockwise in the northern. The processes are less simple in operation than they appear, as other influences distort the basic patterns, producing the typical pictures illustrated opposite.

(Diagram key: high pressure A; low pressure B).

Wind Systems *below*
The map below shows the general wind circulation on the Earth's surface. Because of the Coriolis effect, the winds blowing toward the equator are north-easterly in the northern hemisphere, but south-easterly in the southern. The uplifting of warm air at the equator causes a region of light, variable winds known as the Doldrums. Violent disturbances farther from the equator are known variously as tornadoes (inland), hurricanes (in the Caribbean area) and typhoons (in the Pacific area).

The wind circulation in the southern hemisphere is more consistent than in the north, because of the greater amount of ocean, but there are local variations from one region to another. The pattern over the huge land-masses of the northern part of the Earth is less regular.

The general circulation of the atmosphere is very strongly affected, at low levels, by the unequal distribution of land and sea in the two hemispheres, but this does not apply to the circulation of the layers above the troposphere.

Africa and Arabia *below*
Arabia, covered by thinner cloud. The Red Sea area has no cloud-cover, and the outlines of Africa are clearly seen, though a cloud belt extends across the continent south of the equator.

Australia *below*
Australia, well seen because there is no cloud over it. The patterns to either side are comparatively regular, because there are no large land-masses to interrupt the air-flow.

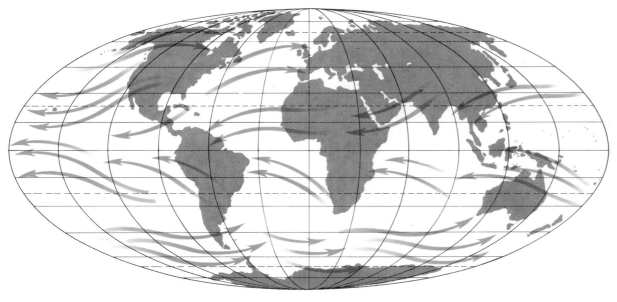

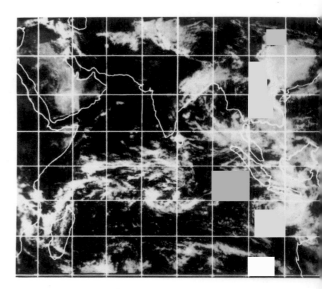

The Pacific *below*
Much of Australia is cloud-free, but again the prevailing winds are shown up by the pattern of cloud formations in the Pacific, beyond New Zealand.

Southern and Northern Hemispheres *left and right*
The illustrations below are digital photographic mosaics of cloud pictures taken by the ESSA-3 weather satellite from polar orbit on 1 June 1967. The outlines of the continents have been superimposed onto the photo-maps. The extreme polar areas are not covered by photographs. Every day the weather data for the entire Earth are processed by high-speed computer into the three maps shown on this page, thus giving each day's weather pattern over the globe.

Sea of Japan *below*
Cloud-systems are clearly visible in the Sea of Japan, an area prone to violent storms and tornadoes which sweep in at speeds of up to 200 mph.

North Africa *below*
Africa north of the Equator is largely free of cloud, but the coastal regions are covered. Cloud cover would normally be expected to the south.

North America *below*
There is cloud coverage over the central part of the United States, but much of the continent is clear, as would be normal during the summer month of June.

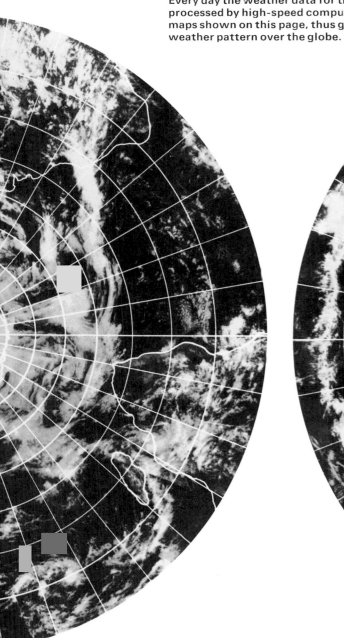

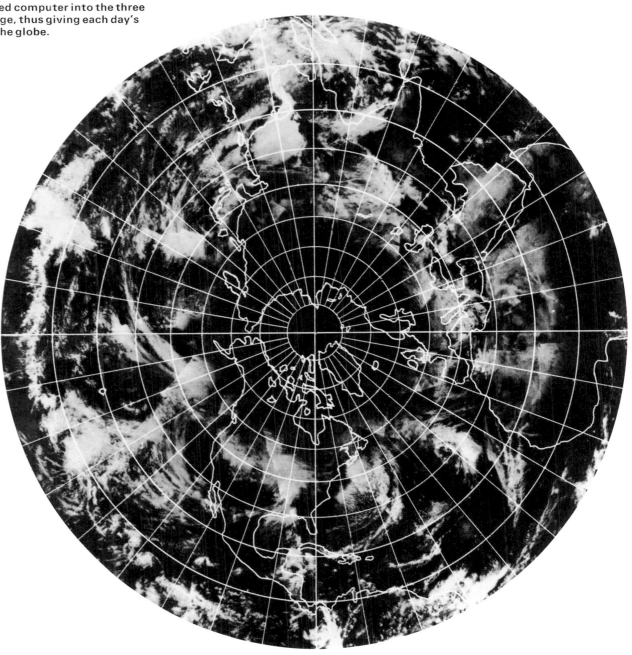

Central America *below*
The Coriolis effect is also well displayed in the region of Central America, where there is a wide area with no cover bounded by extensive cloud systems with north-easterly movement.

The Tropical Belt *below*
Another ESSA-3 photo-mosaic of 1 June 1967, showing the whole equatorial region of the Earth, from 35°N to 35°S, and therefore completing a global map of the weather systems at this time. This Mercator projection may be compared with the two polar stereographic projections given above. In the equatorial weather-map, the Coriolis effect is very noticeable indeed.

South America *below*
Part of the south-east wind system causes the strip of clouds across part of South America, extending across much of Brazil and into the South Atlantic Ocean to the east.

North Africa *below*
The cloud-free strip of Africa extends from the west coast right across to the Arabian Gulf, and can be traced over to the area shown on the extreme left of the map.

Southern Africa *below*
The cloud belt across Africa, south of the equator, again traceable on the far left of the photograph. At this time most of the South Atlantic had comparatively clear skies.

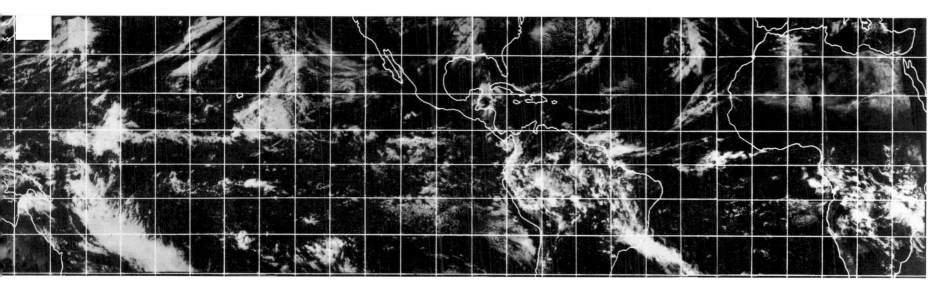

21

Earth Panorama: The Americas

The "New World", made up of North and South America, has been separated from the Eurasian land-mass since the beginning of the Tertiary era, 70 million years ago. Climates range from the Arctic tundra of north Canada to the deserts of the south-western United States, the tropical forests of Brazil and the cold grasslands of Patagonia. The map above gives the locations of the space photographs shown on these pages.

Gulf of California *left*
An Apollo 7 photograph of the Mexican coast and the Gulf of California. The island is Tiburón, with the much smaller island of San Sebastian to the lower left; the long, narrow island is San Lorenzo. The highest point on the photograph is the central peak of Tiburón, which rises steeply to 3986 feet. To the left, the shallows off the coast of Baja California show up well, and details of the seabed can be clearly seen. At this point the Gulf is 60 miles wide. Typically haphazard Mexican field patterns can be seen on the mainland (right) with seasonally dry watercourses to the north.

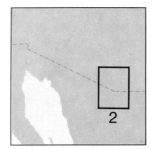

S-W United States and Mexico *below*
The United States/Mexico border-country (Apollo 6). Tucson is centre top; Nogales is centre bottom. Left of the central valley are copper workings. Mt. Wrightson (centre) rises 5000 feet from the valley floor to 9432 feet.

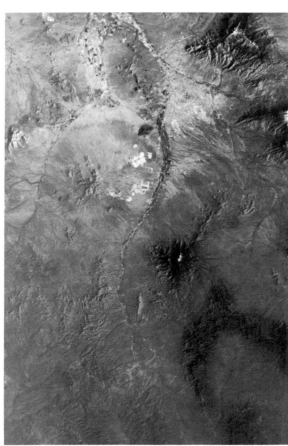

Texas *right*
An Apollo 9 photograph of part of Texas in the United States; Houston, Galveston, Freeport, and Sabine Lake can be seen. The island and lagoon coastline is typical of the north-west shore of the Gulf of Mexico. The port facilities of Galveston can be clearly seen.

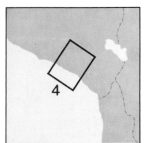

Peru *above*
Part of southern Peru,
photographed from
Gemini II. Low stratus clouds
obscured the coastal
region and cumulus clouds
covered the Andes, 100
miles away. Arequipa,
is to the lower centre
and the mountains to the
north-east rise to 20,000 feet.

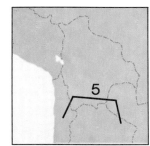

Bolivia *above*
A Gemini 7 photograph of
south-western Bolivia.
Two salt flats and two
small lakes, the Laguna
Pastos Grandes and the
Salina Olaroz, can be
distinguished in the foreground. The Sun was
setting over the Andes when
this picture was taken.

Chile *above*
Another view of South
America : the coast of Chile,
at Antofagasta, photo-
graphed from Apollo 7.
Between the Atacama
Desert and the Andes
(top) is an
interesting area of interior
drainage, apparently of
structural origin.

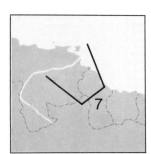

South America *right*
A Gemini 10 photograph,
showing the northern
coast of South America
from Caracas, Venezuela
(left), to Georgetown,
Guyana (right). Landward,
a narrow coastal plain
separates the Guyana
plateau from the sea. The
massive delta of the
Orinoco River is at the
upper centre, with the
mouth of the Essequibo
River to the right ; the larger
tributaries of the
Essequibo River system
are outlined in the cloud
pattern — the clouds are
mainly of the cumulus type.
Sediment is being poured
into the Atlantic from the
rivers and is carried along
the shore by coastal
currents. As can be seen, the
sediment has a discolouring
effect on the water.
Broad parts of this coastal
land are below or barely
above sea-level and are
built up of shales, clays and
lignite originating from
deposits of the river
effluent. The synoptic view
of this process is of
considerable interest.

Earth Panorama: Africa 1

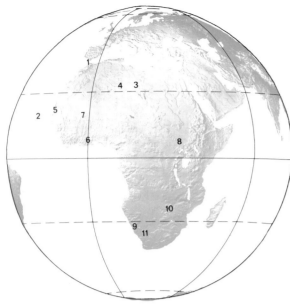

Africa was long known as "the Dark Continent", because the nature of its terrain made it difficult to explore. The climate is essentially equatorial and humid over the central latitudes and tropical or subtropical and dry over northern and southern Africa. This results in frequent clear skies over much of the north and south – ideal conditions for survey by photography from space, and this fortunate fact will provide much information about areas until now little explored or developed.

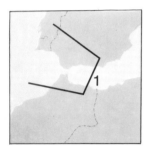

Straits of Gibraltar *right*
The junction of Europe with Africa; the Iberian peninsula and Morocco are separated by the Straits of Gibraltar. There is extensive cloud coverage over the almost landlocked Mediterranean in this photograph taken from Apollo 9.

Cape Verde Islands *above*
Boa Vista and Sal, the two most easterly of the Cape Verde Islands, in the Atlantic Ocean off the coast of West Africa (Apollo 9). Few details can be seen in the ocean itself, which soon reaches a depth of 12,000 feet in this area.

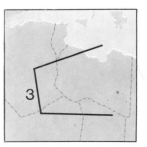

North Africa *right*
Libya, parts of Algeria and Chad, with the Mediterranean to the upper left (Gemini 11). The immense Marzuq Sand Sea is right of centre, with the Tibesti Mountains upper right; the dark irregular area (top left) is the Haruj al Aswad, a Quaternary volcanic field.

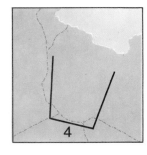

Central Sahara *right*
Idehan Marzuq, a great
sand sea in the Fezzan,
southern Libya, bounded on
the left by the Messak
Mellet range, an eastern
extension of the Hoggar
massif. 600 miles to the
north the Mediterranean Sea
forms the horizon.
(Apollo 9.)

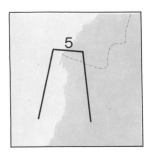

West Africa *below*
The Atlantic coast of
Mauritania looking south to
the Senegal River and a
horizon of clouds over the
Guinea Highlands. Cap
Blanc and Port Étienne are
in the foreground. The edge
of the continental shelf is
about 40 miles off-shore.
(Apollo 9.)

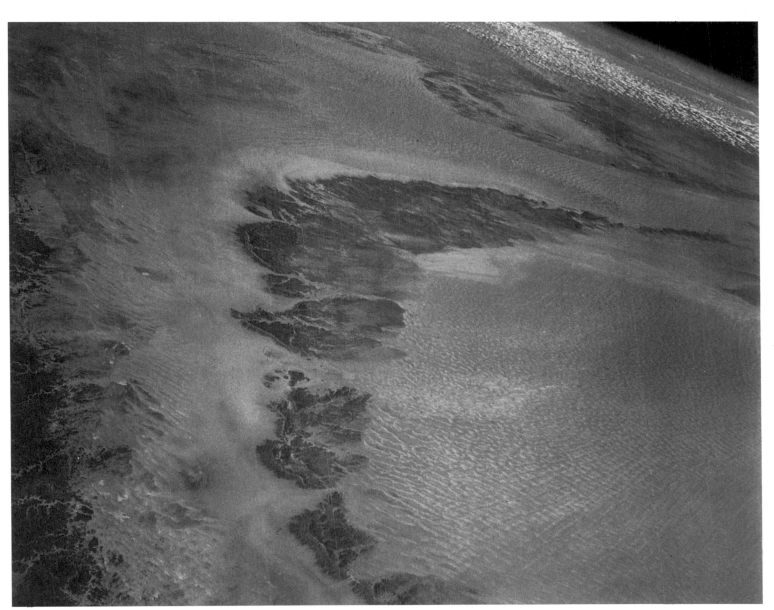

Earth Panorama: Africa 2

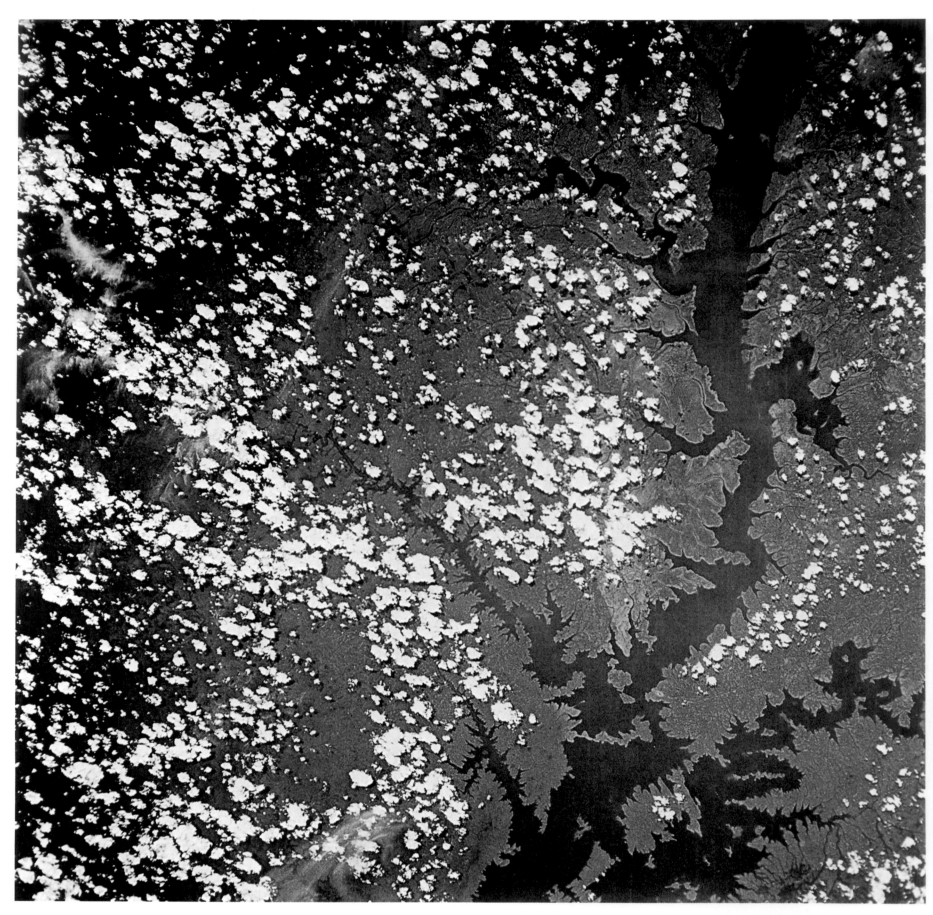

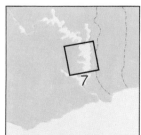

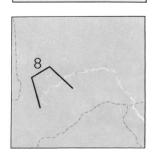

Ghana *above*
Part of Lake Volta, Ghana (Apollo 6), a vast reservoir formed by damming the River Volta at Akosombo, 70 miles upstream from the Gulf of Guinea. This area of Ghana is heavily forested and was typically cloud covered when the photograph was taken.

Mali *right*
An Apollo 9 photograph showing Lake Fagubine and looking south over the inland Niger delta in Mali. The lake, a fault structure, lies west of Timbuktu, on the edge of the Sahara Desert. The Niger forms a vast area of swamp, showing blue-green on the photograph.

Sudan *above*
The White Nile in the Sudan (Gemini 6). The area shown measures 120 miles from east to west. The Nile is flowing out of the Sudd, an immense swamp in whose waters the reflection of the Sun can be seen. The Nile rises 500 miles south in Lake Victoria.

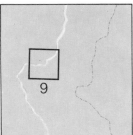

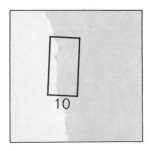

South-West Africa *above*
Walvis Bay is enclosed by the northern of the two capes. 120 miles of coast is covered. The area is part of the barren Namib Desert, underlaid by Pre-Cambrian rocks which have been folded, invaded by granitic magma, and eroded. (Gemini 5.)

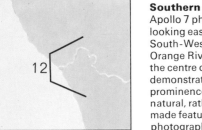

Southern Africa *right*
Apollo 7 photograph, looking east over South and South-West Africa. The Orange River runs across the centre of the photograph, demonstrating the prominence with which natural, rather than man-made features show up when photographed from space.

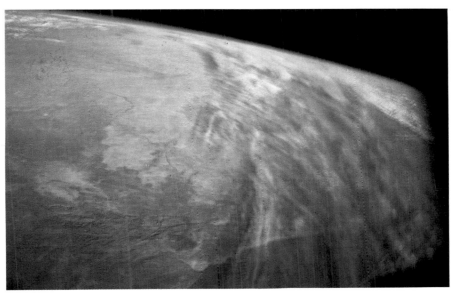

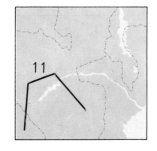

Central Africa *above*
Apollo 7 photograph looking south over Lake Kariba, from Zambia into Rhodesia. Lake Kariba is a huge reservoir. The Victoria Falls, 75 miles west of the lake, are just visible before well-bounded cloud obscures Botswana to the south-west.

Earth Panorama: The Indian Sub-Continent

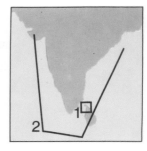

The Indian sub-continent covers an area of 159,500,000 square miles. It is densely inhabited (population 588 million). Northern India lies between 36°N and the Tropic of Cancer, while peninsular India and Ceylon are wholly between the Tropic and the Equator. There are three main geographical regions: the Himalayan mountain region, the great plains of the Ganges, Brahmaputra and Indus rivers, and the Deccan plateau.

India and Ceylon
below
The Indian sub-continent; (Gemini 9, alt. 460 miles). The side-to-side distance at the bottom is over 2000 miles; the distance to the horizon, where the Himalayas are visible, is 2300 miles. The Western Ghats are to the left, while the Deccan Plateau basalts, upper left, are seen as the dark areas.
The large river to the south-east side of the sub-continent is the Coleroon, and Ceylon is to the lower right.

left
A detailed view of northern Ceylon. Again there is considerable cloud cover. (Apollo 7.)

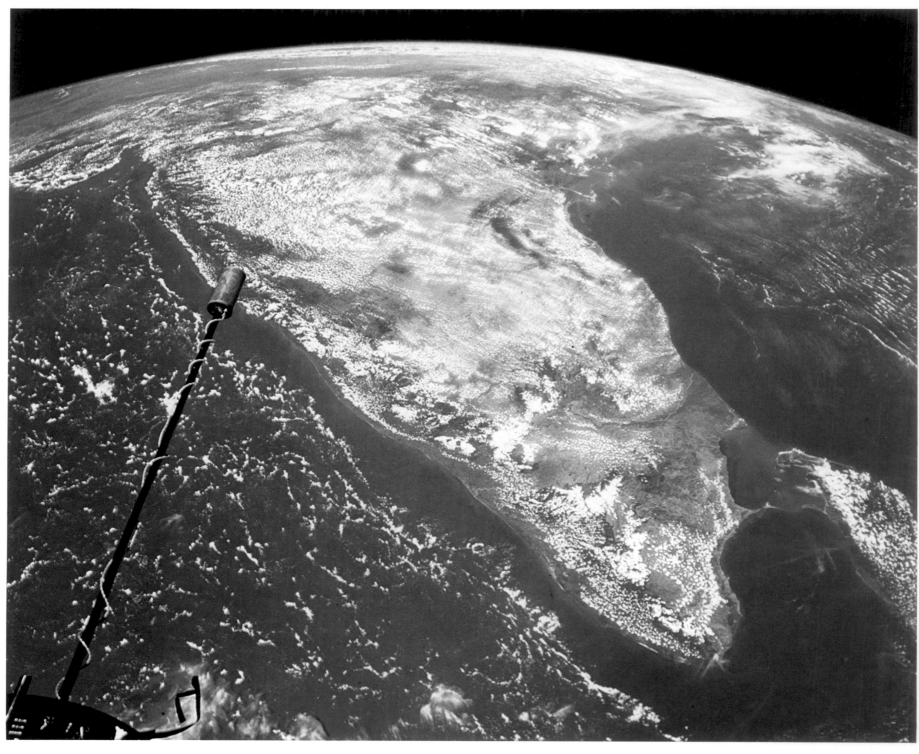

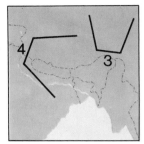

The Himalayas

left Apollo 9 photograph, looking north over the eastern Himalayas in China.

below Apollo 7 photograph, looking east over the central Himalayas, showing the plateau of Tibet left and the Ganges plain right. The Himalayas contain Mount Everest, the world's highest peak (29,029 feet). Geologically they are not an old structure ; they are Miocene, and were formed between 12 and 25 million years ago during the same period of orogenesis that produced the Alps. Because of their youth, they have not yet been markedly reduced by weathering and erosion.

Earth Panorama: The Pacific Area

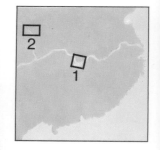

The Pacific is the greatest of all the oceans. It covers an area of 63,855,000 square miles, and averages 14,050 feet in depth. It is larger than all the continents combined, and is circled by what has been called the "belt of fire" – the active volcanoes of the western coasts of South and North America, and the coasts of eastern Asia. It is also one of the few regions of the Earth which has been ocean-covered since early geological time.

China

left The Middle Yangtze Plain (Gemini 5); a large basin, drained by the Yangtze River (left). Lake Tung'ting is 60 miles in length; sedimentation patterns are well shown. The delicate colours and tones shown here would probably not be brought out in air photography.

below The Chialing River, a tributary of the Yangtze, in the provinces of Kansu and Szechwan, winding from the left centre edge to the upper right. To the lower right, the Yung Feng Shan Mountains are clouded. The city of Wutu is in the light-coloured valley to the top left.

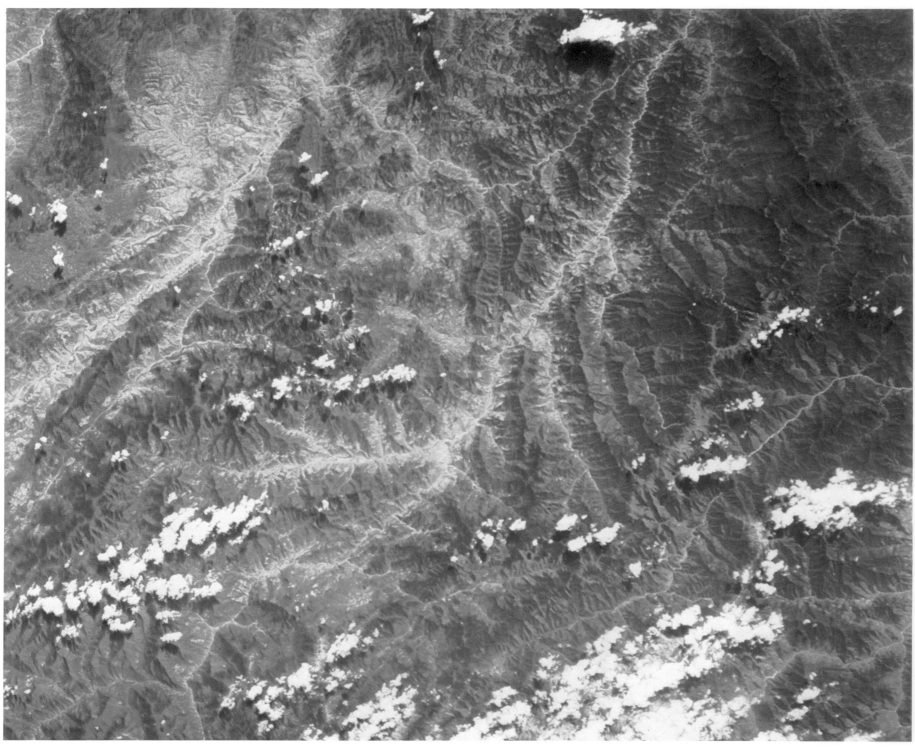

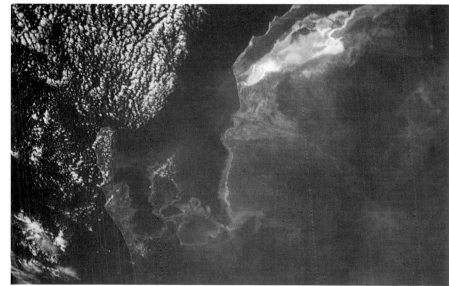

Southern Japan *above*
The western coast of
Kyushu Island (Apollo 7) ;
again there is considerable
cloud cover. Japan is one of
the most unstable areas
of the world, and is
exceptionally subject to
severe earthquake shocks.
The islands form part of
the Pacific volcanic arc.

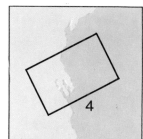

Western Australia *above*
The coast of Western
Australia : Shark Bay, 500
miles north of Perth
(Apollo 7). Inland, the
North-West Basin is
cloud-free, but the ocean
is extensively covered. The
coastal waters are fairly
shallow, sloping gently to
the Indian Ocean basin.

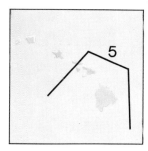

Pacific Ocean *below*
Apollo 9 photograph
looking south over the
Hawaiian Islands in the
North Pacific. Little
surface detail can be seen
except for the volcanoes,
which rise above clouds
generated by warm air
rising over the islands. The
cooler sea is largely clear.

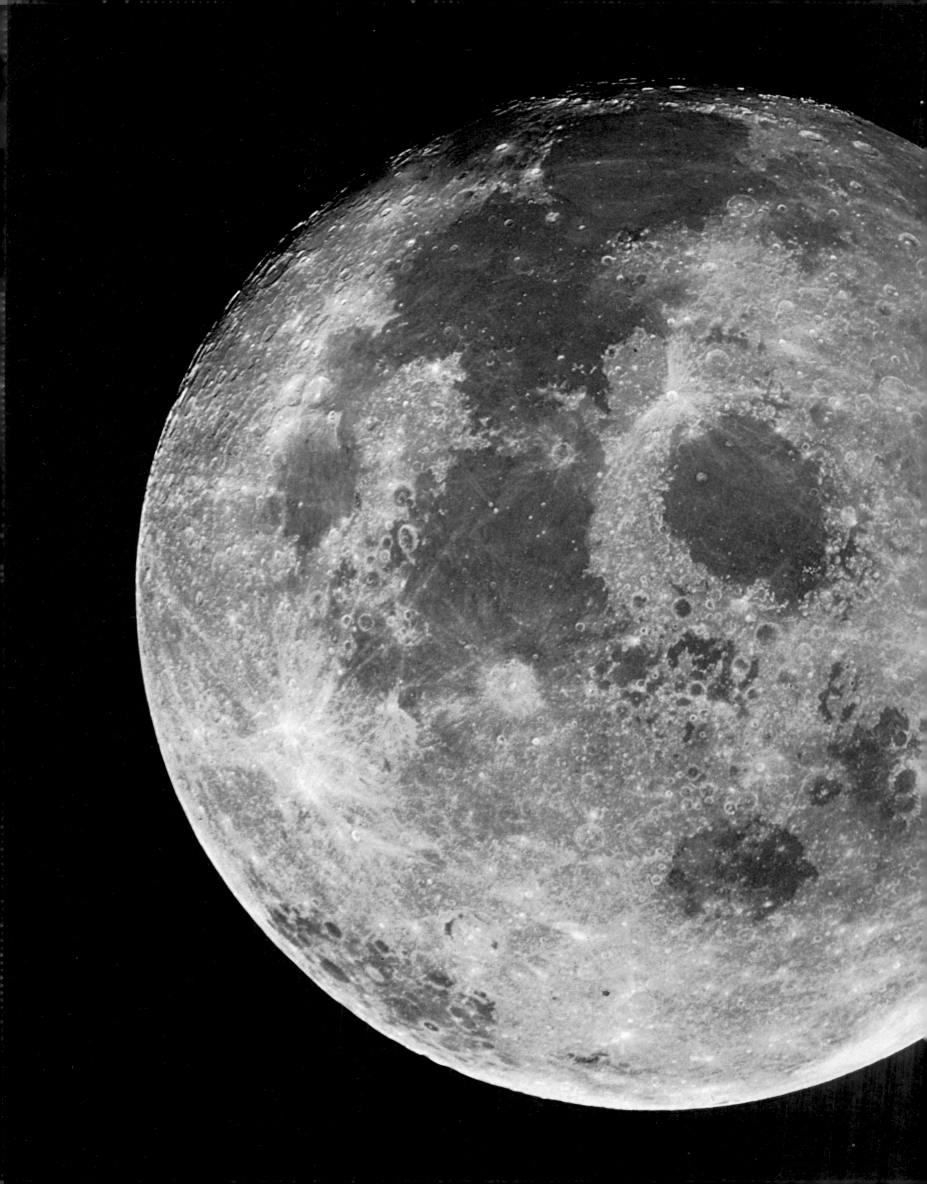

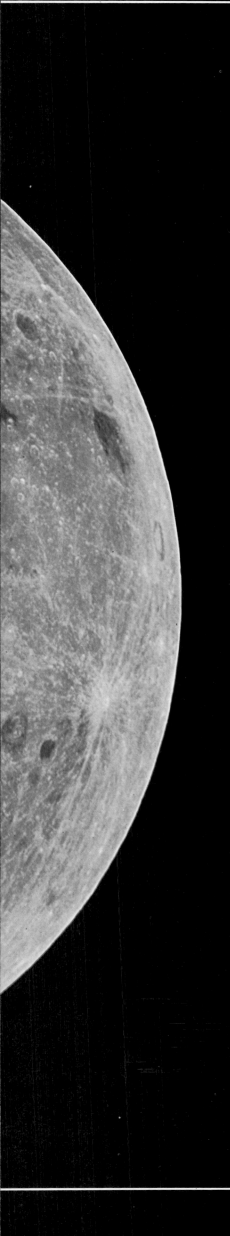

Atlas of the Moon

The Moon is our closest neighbour in space. For this reason—and for this reason alone—it appears as a splendid object in our skies, and it is hardly surprising that ancient peoples worshipped it as a god. Indeed, the Moon-god was often just as important as the god of the Sun, and lunar worship lingered on until a surprisingly late stage. It was always realized that the Moon has a special rôle, and the philosophers of classical Greece knew that it is the Earth's companion. Plutarch, in his essay "On the Face in the Orb of the Moon", described it as an earthy body, with mountains and valleys. Also, the causes of eclipses were known.

With the naked eye, various markings can be seen on the lunar disk; these are the broad dark areas which are still miscalled "seas", together with some of the more delicate features. Yet it was not until the invention of the telescope, in the first decade of the seventeenth century, that serious observation of the Moon could begin.

The Moon seen from Apollo II
The dark plane to the centre is the Mare Crisium. The photograph covers part of the far side which can be seen on the right. To the left is the Mare Tranquillitatis. North is at the top.

The Moon

Previously the main attention had been upon lunar theory – that is to say, the movements of the Moon, which had been worked out with surprising accuracy.

Galileo

It is usually claimed that the first man to look at the Moon through a telescope was Galileo, in the winter of 1609–10. In fact, this is not so. The English scientist, Thomas Harriot (one-time tutor to Sir Walter Raleigh), not only looked at the Moon telescopically several months before Galileo, but even drew up a rough map of the surface features which

Galileo *above*
Galileo, the great astronomer who drew up a map of the Moon. This portrait of him is from an old engraving.

was highly creditable. Another early observer was the Welsh amateur Sir William Lower, who was certainly using his telescope to look at the Moon very soon after Galileo. However, for patience and skill Galileo admittedly stands alone. He made an attempt to measure the heights of lunar mountains, which he thought to be loftier than those of the Earth. Even though he concentrated upon the lunar Apennines, which are by no means the highest of the peaks, though they are the most spectacular, Galileo's results were of the correct order. Moreover, he did not believe the so-called "seas" to be water-filled, so in many ways his views were far ahead of his time.

Galileo's Successors

Galileo was followed by other astronomers, notably Hewelcke (usually known as Hevelius), who compiled a map of the Moon which was better than Galileo's. Hevelius named the lunar features after geographical ones, a system which has not survived; a few years later, in 1651, it was superseded by the system introduced by the Jesuit priest Riccioli.

Riccioli's map, based largely upon the work of his pupil Grimaldi, was not particularly accurate, but it did introduce the lunar nomenclature which is still followed today. Each major crater was named after an eminent person, usually an astronomer or scientist of some other discipline; the seas were given romantic titles such as the Mare Tranquillitatis (Sea of Tranquillity) and the Oceanus Procellarum (Ocean of Storms). It can hardly be said that Riccioli was unprejudiced. He did not believe in the Copernican theory, according to which the Earth moves round the Sun instead of vice versa; and so in selecting a crater for Copernicus, he "flung him into the Ocean of Storms", even though it must be added that the crater is a particularly majestic one. Not unnaturally, Riccioli named prominent craters after Grimaldi and himself. A few of Hevelius' geographical analogies were retained, mainly for mountain ranges.

Schröter

Relatively little mapping was done for the century following the appearance of Riccioli's chart, but in 1775 there came a small but much more accurate chart made by the German astronomer Tobias Mayer. However, the title of the "Father of Selenography" should properly go to Johann Hieronymus Schröter, who was never a professional astronomer, but who was chief magistrate of the town of Lilienthal, near Bremen. Schröter began active lunar work in 1778, and continued patiently for over thirty years; indeed, Lilienthal became very much of a scientific centre.

Schröter has never received the credit which is his due. It has been claimed that his drawings are rough, and even that his telescopes were defective. It may be true that his largest telescope, a 19-in. reflector by Schrader of Kiel, was not of the best quality; but

Riccioli's Map *below*
Riccioli's map (1651) was not a great improvement upon that of Hevelius, but his system of nomenclature was accepted and is still in use.

Hevelius' Map *above*
Hevelius' map; a reproduction from the old engraving which, unfortunately, is no longer in existence.

most of his work was carried out with smaller instruments, one of which was made by no less a person than Sir William Herschel, the best telescope-maker of the time. It is also true that Schröter was not a neat draughtsman. However, his sketches look much rougher than they actually are, and there can be no doubt that he carried out work of immense value. Though he never produced a complete map of the Moon, he made drawings of many areas, and he put selenography upon a really firm footing. His work came to an unhappy end in 1814, during the Franco-Prussian war. Lilienthal was captured by the French, and his telescopes were plundered; all his unpublished observations were destroyed. Unfortunately, Schröter died not long afterwards.

Beer and Mädler

His mantle fell upon three of his countrymen: W. Lohrmann, a land surveyor of Dresden, and two Berliners, Wilhelm Beer and Johann von Mädler. Lohrmann began work upon a large-scale map which was of real accuracy, but unfortunately his health did not permit him to complete it, and the main work was done by Beer and Mädler.

The telescope used was small by modern standards: the $3\frac{1}{2}$-in. Fraunhofer refractor at Beer's private observatory. For many years he and Mädler worked away, and in 1837–8 they published not only a chart of the whole visible surface of the Moon, but also a lengthy text in which every crater was described in detail. It was an amazing achievement, and the map, to a scale of more than 2 feet to the Moon's diameter, is remarkably accurate. It is all the more regrettable that neither of the authors undertook much more lunar work after publication of the book. Beer died in 1850; Mädler went to Tartu, in Estonia (then, as now, part of Russia), to become director of the observatory, devoting himself to somewhat unprofitable cosmological speculations rather than to pioneer work in Moon-mapping.

Ironically, the appearance of Beer and Mädler's *Der Mond* meant that comparatively little progress was made for several decades. Unlike Schröter, Beer and Mädler had concluded that the Moon must be inert as well as devoid of atmosphere; they regarded its surface as essentially changeless. The general view among their contemporaries was that since the surface had now been charted, there was little point in studying it further. Thus it came about that, until the mid-1860s, the Moon was neglected by all professional astronomers and by almost all amateurs. The only exception was Julius Schmidt, a German who spent most of his career in Greece as Director of the Athens Observatory; and it was in 1866 that Schmidt made an announcement that caused attention to swing back to the Moon.

Linné

One of the most prominent features on the lunar disk is the Mare Serenitatis, or Sea of Serenity. There are no really large craters on it; but there is one prominent little formation, the crater known nowadays as Bessel. Also, all the older selenographers had charted another comparable crater, which Mädler had named Linné in honour of the Swedish botanist Carl Linnæus. It was described as being several miles in diameter, and so deep that, when near the terminator, its floor was filled with shadow.

Schmidt had looked at the crater in 1843. In 1866 he checked again, and found that the aspect was different. There was no deep crater; instead there was a white patch, in which later observers found a much smaller crater-pit. Schmidt was convinced that real change had taken place, and his announcement caused tremendous interest. It was this, more than anything else, which ushered in the 'modern phase' of lunar charting. Sir John Herschel, whose opinions carried great weight, suggested that a tremor in the Moon's crust had caused the walls of the old crater Linné to cave in.

It must be admitted that the evidence in favour of real change in Linné is very slight. The pre-1866 observers had generally used small telescopes, and there were not many reliable drawings. Moreover, Mädler, who lived until the 1870s, looked again at Linné and pronounced it to be of the same form as he had seen it in 1837, though admittedly this is rather at variance with the description in *Der Mond*. Lunar photography was then at an early stage; François Arago's claim, made in 1840, that by photographic methods the Moon could be mapped "in a few minutes" was very optimistic. Nowadays it is generally thought that Linné in fact showed no genuine alteration between 1843 and 1866.

It has always been tacitly assumed that the lunar craters must be of internal origin, so that they could be described as "volcanic", though the term must always be used in a broad sense. The general form of a lunar crater is very different from that of a conventional volcano such as Vesuvius, and is more like that of a terrestrial caldera. Franz von Paula Gruithuisen, an enthusiastic but rather imaginative German observer of the mid-nineteenth century, had supported an alternative theory according to which the craters were produced by the impacts of large meteorites. Arguments between the volcanic and impact supporters have been long and sometimes acrimonious; they were still going on when the Apollo programme of manned landings ended in 1972.

Lunar photography had begun some time before, and by the 1860s some reasonable pictures were being produced.

Lunar Charting to the Present Day

Even more important, perhaps, were the measures of the positions on the disk of various lunar features. This work was undertaken by S. A. Saunder, assisted by J. A. Hardcastle, and was of immense value; indeed, the Saunder–Hardcastle measures are still used today, though they have been refined by Schrutka–Reichenstamm and others. In fact, lunar charting had become an exact science – even though it was still mainly an amateur field of research. This situation endured almost until the start of the 1939–45 war. Two more notable lunar maps were produced in England, one by Walter Goodacre and the other by H. P. Wilkins; Wilkins' chart showed the Moon to a scale of 300 inches to the lunar diameter.

In Germany, P. Fauth also produced an elaborate chart. And in the same period Blagg and Müller were the authors of an official lunar map, published by the International Astronomical Union and giving the official nomenclature. It cannot be said that the I.A.U. map was either particularly accurate or particularly clear, but it served its purpose adequately.

In America, the Lunar and Planetary Laboratory at Tucson, Arizona, produced excellent photographs, and a major photographic chart was published under the editorship of G. P. Kuiper. France, too, played a major rôle, and some of the best lunar pictures ever taken from Earth come from the Pic du Midi observatory in the Pyrenees, where Audouin Dollfus and his colleagues have concentrated upon lunar work. There were also various theoretical developments. The volcanic theory of lunar formations was challenged by R. B. Baldwin in 1949, and for the next twenty years it is probably true to say that the impact hypothesis was the more popular of the two, though there were various dissentients. More alarm-

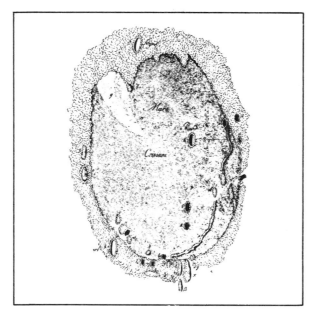

Schröter's Drawing of Mare Crisium *above*
Drawing of the Mare Crisium, made by Schröter. This is typical of his many hundreds of sketches.

ing was the paper published in 1955 by T. Gold, then of the Royal Greenwich Observatory, in which it was suggested that the lunar maria might be covered with soft dust to a depth of many kilometres, so that a space-craft landing there would sink permanently out of sight. This paper gained wide publicity, and the deep-dust theory was not finally disproved until the first soft landings with unmanned probes, though not many practical lunar observers believed that the outcome could be otherwise.

Russian and American Probes

In 1959, less than two years after the launching of Sputnik 1, the Russians opened the era of lunar rocket exploration by dispatching three probes. The first, Luna 1, by-passed the Moon in January and sent back useful information; it confirmed, for instance, that the Moon has no appreciable magnetic field. Luna 2, of the following September, crash-landed on the surface; but it was Luna 3, sent up on 4 October 1959, which was the most important of the three. It went right round the Moon, and on its return to the neighbourhood of Earth it sent back photographs of those areas which are always turned away from us, and which we cannot therefore see.

There had been many discussions about the nature of "the other side of the Moon". A century earlier, the Danish astronomer Hansen had even suggested that all the lunar air and water had been drawn round to the far side, which might well be inhabited. Though Hansen's theory was clearly unsound, and never met with much support, it was true to say that the nature of the far side remained very much of a mystery. Astronomers who supported the volcanic theory of crater formation held the view that it was unlikely to contain many major maria, but that there would be craters in plenty. Yet that was about as much as could be said, though some amateurs had tried to chart the positions of some ray-craters by noting the rays which came round the limb of the

Moon on to the Earth-turned hemisphere. These estimates were later proved to have been quite good.

Luna 3 transmitted photographs which were crude by modern standards, but which showed various features, notably the Mare Moscoviense, as the Russians promptly named it, and the dark-floored crater which the Soviet authorities christened Tsiolkovskii in honour of the great rocket pioneer. But for some time after this there was a lull; and the next important steps came from the United States.

The first American lunar probes, the Pioneers, were uniformly unsuccessful, but then came the Ranger programme, in which vehicles were to be crashed on to the Moon – sending back detailed close-range pictures during the final minutes of their journey. The first triumph came with Ranger 7, which came down in the Mare Nubium (Sea of Clouds) in the general region of the crater Guericke. The pictures obtained were excellent, and showed details so delicate that they could not possibly be recorded from Earth even with large telescopes. Rangers 8 and 9 were even better. The last came down inside the huge crater of Alphonsus, which was already famous for another reason. In 1958 the Russian astronomer N. A. Kozyrev, using the 50-in. telescope at the Crimean Observatory, had recorded a reddish glow inside Alphonsus which indicated a certain amount of surface or sub-surface activity, which Kozyrev interpreted as being of volcanic origin. Ranger 9 confirmed that the whole area is crowded with detail, and includes many of the features known variously as clefts, rills or rilles. Since 1958 other glows have been seen, particularly near the brilliant crater Aristarchus; they are known as Transient Lunar Phenomena, and are associated with minor tremors in the surface, recorded by the instruments left on the Moon by the Apollo astronauts of 1969 and 1971.

The First Moon Landings

Then came another major step. In early 1966 the Russians soft-landed their probe Luna 9 on the Moon's surface, so that its cameras were able to continue transmitting. The dust-drift theory was finally disproved; Luna 9 showed no signs of sinking in to a yielding surface. The pictures sent back showed what looked remarkably like a lava-field. Before long the Americans had similar successes with their Surveyor probes, and it was evident that the age of manned lunar exploration lay close ahead.

Yet no landings could be attempted without detailed charts, and Arago's forecast of 1840 that the Moon could be mapped photographically in a few minutes was still unfulfilled in 1960. This project was tackled by the American team, and with complete success. Rocket vehicles known as Orbiters were put into paths round the Moon, and sent back thousands of photographs from low level. From these pictures it was possible to compile a complete map of the whole Moon – the far side as well as the Earth-turned hemisphere. And it was due to the Orbiter coverage that the Apollo project of 1969 became feasible, even though the site in the Mare Tranquillitatis where Apollo 11 came down had been previously surveyed in detail by Colonel Stafford and Commander Cernan in the lunar module of Apollo 10.

The landing of Apollo 11, with Astronauts Armstrong and Aldrin, was one of the great moments in history. Armstrong's "One small step . . ." will never be forgotten. Other successful Apollo missions followed; by the time that the programme ended, with Apollo 17 in December 1972, our knowledge of the Moon had been increased out of all recognition. At the moment it is hard to say when the next men will go to the Moon, but there is no valid reason to doubt that it will be possible to establish scientific research bases there before the end of the 20th century.

There is one final point to be made. The Moon has no appreciable atmosphere and the temperature conditions, first measured by Lord Rosse a century ago, are hostile to any form of Earth-type life. We may be sure that there is no lunar life now; examination of the rocks brought back from Apollo missions has failed to reveal any trace of past life, and there is every reason to suppose that the Moon has been sterile throughout its long history. It was left to men from Earth to bring life at last to the silent, unfriendly Moon.

The Moon As We See It

The Moon is certainly the most spectacular of all the objects in the sky when viewed through a telescope. It has no atmosphere, so that there is no veiling of the surface features. Up to modern times, all lunar drawings and photographs have been shown with south at the top – because an astronomical telescope gives an inverted image. However, the modern convention (followed throughout this Atlas) is to give north at the top and east to the right.

Seas

The broad dark plains are easily visible to the naked eye – it is hardly surprising that ancient peoples regarded them as seas. Some of them are extremely large. The Oceanus Procellarum (Ocean of Storms) (to the west, or left) covers 2,000,000 square miles, so that it is larger than the Mediterranean. The area of the well-marked Mare Serenitatis (Sea of Serenity) (to the east) is almost exactly equal to that of Great Britain.

Craters

Of course it is the craters which dominate the lunar scene. They are of all sizes, ranging from tiny pits up to vast enclosures well over 150 miles in diameter; some of them are bright-floored, some dark; some have central peaks, others have relatively smooth interiors. There are also many minor features, such as the crack-like clefts or rilles, the domes, and the ridges and faults.

Because terrestrial terms are used to describe lunar features, it is tempting to draw too close an analogy between the Moon's surface and that of the Earth. A lunar mare has never been a sea in the conventional sense of the word; a major lunar crater is unlike that of a volcano such as Vesuvius; and even the mountain ranges are not truly analogous to the Earth's Rockies or Himalayas. Generally speaking, the major ranges of the Moon form the borders to the circular maria; thus the Apennines, Alps and Carpathian Mountains make up part of the boundary to the Mare Imbrium (Sea of Showers). The evolutionary sequence of the Moon has been different from the Earth.

Crescent Moon (2)
Mare Crisium is prominent, between the eastern limb and the terminator. Earthshine is often seen.

Half Moon, First Quarter (3)
The Mare Serenitatis is now in view, with the great chains of craters near the central meridian.

Gibbous Moon (4)
The great ray-craters Tycho and Copernicus are illuminated, though the rays are not yet striking.

Full Moon (5)
There are no shadows, and the rays from Tycho and Copernicus are so prominent that crater identification becomes difficult. The lunar maria take on a decidedly dark hue against the brilliant rays.

The Waning Moon (6)
This is not so brilliant as the waxing gibbous Moon: more dark mare areas are illuminated.

As seen from Earth, the Moon's limb regions appear foreshortened. The well-marked Mare Crisium (Sea of Crises) demonstrates this. Its diameter is 260 miles as measured north–south, 335 miles as measured east–west; yet it appears to be elongated in a north–south direction. Photographs taken from space-probes show it in its correct form

The Full Moon *left*
Oddly enough, full moon is the worst possible time to start trying to identify craters. There are almost no shadows, and the main features are the bright rays associated with a few craters such as Tycho (in the south) and Copernicus (near the centre). The maria are well displayed as dark areas, but even some of the major craters are unrecognizable, though some of the formations, such as the brilliant Aristarchus and the dark-floored Plato, can be identified easily under any conditions of illumination.

Phases of the Moon *below*
The Moon's phases have been known since the dawn of human history. Since the Moon has no light of its own, and shines only by reflected sunlight, it shows apparent, regular changes of shape. When it is approximately between the Sun and the Earth *(1 and 9)*, its dark side is turned towards us, and the Moon cannot be seen; it is then new. When the Earth lies approximately between the Moon and the Sun, the Moon's sunlit hemisphere is turned Earthward, and the Moon is full *(5)*. At other times the phase may be crescent *(2 and 8)*, half *(3 and 7)*, or gibbous *(4 and 6)* (i.e. between half and full). The boundary between the sunlit and night hemispheres is known as the terminator. Unlike the limb, which is the edge of the visible disk as seen from Earth, the terminator does not appear smooth; the Moon has a rough, uneven surface, and elevations will catch the Sun's rays while the valleys below are still in darkness.

The lunation, or interval between one new moon and the next, is 29·53 days; it is longer than the Moon's revolution period (27·32 days), because the Earth is itself moving round the Sun. It is also important to note that the moon does not actually describe a closed loop in its passage round the Sun.

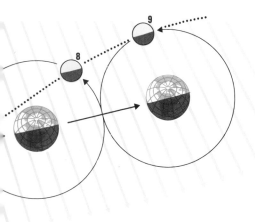

Half Moon, Last Quarter *(7)*
By now the rays are much less striking: the shadows inside the large craters are increasing.

The Old Moon *(8)*
Just before new, visible in the dawn sky. The Earth-shine may often be seen at this phase.

Eclipses

When the Moon passes into the shadow cast by the Earth, its supply of direct sunlight is cut off, and the Moon enters eclipse. (*See diagram below.*) However, it does not disappear. A certain amount of light is refracted on to its surface by way of the Earth's atmosphere, and so the Moon turns a dim, often coppery colour.

Because the Sun is a disk, and not a point source of light, the main shadow, or *umbra*, of the Earth is bordered to either side by what is termed the *penumbra*, or area of partial shadow. The Moon must, of course, pass through the penumbra before entering the umbra. Some eclipses are penumbral only. True eclipses may be either total (when the Moon pasess completely into the *umbra*) or partial, when dimming is noticeable with the naked eye.

A lunar eclipse is reasonably protracted. The average length of the Earth's umbra is 860,000 miles, which is more than three times the distance of the Moon from the Earth, so that at the mean distance of the Moon the cone has a diameter of about 5,700 miles. Totality may last for as much as 1¾ hours, remembering that in the course of an hour the Moon moves across the sky by an amount which is slightly greater than its own apparent diameter. During eclipse the Moon's surface temperature drops sharply, since the lunar surface materials are very poor at retaining heat.

Eclipse in Colour *below*
Lunar eclipse in colour. The bright crescent has necessarily been over-exposed so as to bring out details in the eclipsed portion of the Moon, which takes on a coppery hue. Features can be identified despite the eclipse.

Penumbral Eclipse *below*
Penumbral eclipse of the Moon (1963) seen from Sweden. The northern part of the Moon is covered by the penumbra. Note the irregularities of shadows in the seas on the lunar surface.

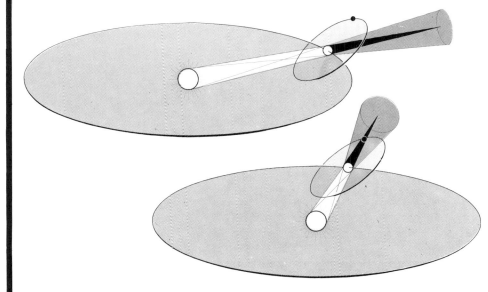

Lunar Eclipse *left*
Since the Moon's orbit is tilted to that of the Earth, the Moon is not eclipsed every month. In this drawing, full moon occurs with the Moon above the plane of the ecliptic, and there is no eclipse.

below When the alignment is exact – that is to say, the Sun, Earth and Moon are in a direct line, with the Earth in the mid position – the Moon enters the cone of shadow cast by the Earth, and is eclipsed.

Sequence of Eclipse
The lunar eclipse of 24–25 June, 1964, as seen from London. The first five photographs were taken before totality; it is seen that the edge of the Earth's shadow is not sharp, and that the border of the shadow is curved. The last photograph was taken after the end of totality, when the shadow of the Earth was starting to pass off the Moon.

The Movements of the Moon

The Moon's distance from the centre of the Earth ranges between 221,460 miles (perigee) (1 in diagram on the right) and 252,700 miles (apogee) (3). The orbital eccentricity is 0·0549; it therefore follows that the Moon's apparent diameter is not constant. When at perigee, the angular diameter is 33′ 31″, but when the Moon reaches apogee, the diameter is 29′ 22″. The difference is not striking, but is easy to measure.

The Moon's apparent magnitude, on the stellar scale, is −12·5. On a dark night, when the full moon rides high in the sky, the illumination is brilliant – and yet it is vastly less than that of the Sun. The albedo, or reflecting power of the lunar rocks, is low; on average, only 7 per cent of the sunlight is reflected. There are local variations, and with Aristarchus, the brightest point on the Moon, the albedo reaches about 20 per cent; but as a general rule the rocks are dark. It is not true that the full moon appears larger when low over the horizon than when high in the sky; this is a well-known illusion.

Relative Size of Earth and Moon *above*
The diameter of the Moon is 2158 miles; the diameter of Earth is 7927 miles. If the Sun is scaled down to a globe 2 feet in diameter, the Earth will be a pea at a distance of 215 feet; the Moon will be a seed 6½ inches from the Earth.

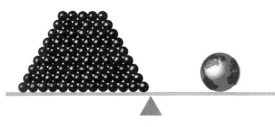

Relative Mass of Earth and Moon *above*
It would take 81·3 Moons to equal the mass of the Earth.

The Moon's Internal Structure

The Moon has a mass of 1/81·3 that of the Earth. The escape velocity is only 1·5 miles per second, and the Moon has no atmosphere; it is this which is the main factor in making the surface conditions on the two worlds so very different. It is not impossible that thousands of millions of years ago the Moon may have had an appreciable atmosphere, but the feeble pull of lunar gravity meant that this atmosphere escaped into space fairly rapidly. The mean density of the Moon is considerably less than that of the Earth; the specific gravity is 3·3, or approximately 0·6 that of Earth.

Accurate information about the Moon's interior was difficult to obtain before the introduction of space-research methods. Most of our knowledge of the internal structure of the Earth is based upon studies of earthquake waves, and there was no way of telling whether the Moon showed any seismic activity. However, in 1969 the first seismometers were set up there, and sent back invaluable information. It may be that the Moon has a layered structure, although investigations are still at an early stage.

Equally important was the detection of a feeble lunar magnetic field. Earlier probes (beginning with Russia's Luna 1, in 1959) had failed to detect any field, but it now seems that lunar magnetism does exist, though it is much weaker than that of the Earth. There may even be a core which contains magnetic materials such as iron sulphide.

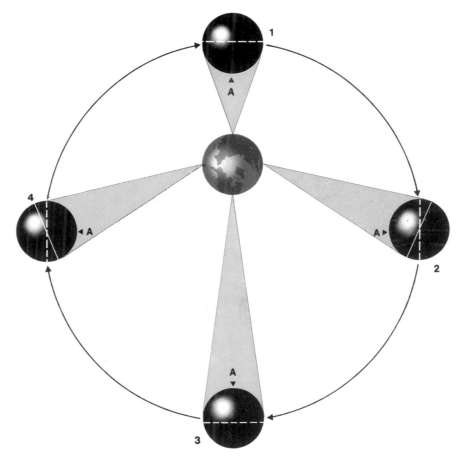

Libration in Longitude
above and right
From Earth, we can study a total of 59 per cent of the Moon, because of effects known as librations. Libration in longitude is due to the fact that although the Moon spins on its axis at a constant rate, its orbital velocity is not constant; it moves quickest when near perigee (1) *above*.
In every lunation, therefore, the orbital position and the amount of spin become "out of step", and the Moon seems to rock very slowly to and fro; first we can see for some way round one limb, then for some way round the other. **A** = the mean apparent centre of the Moon; because of the orbital eccentricity, it is sometimes displaced. The effect of libration in longitude shows up for features near the Moon's limb; Mare Crisium, for example, is better displayed in the upper photograph *right*, position 4, than in the lower, position 2.

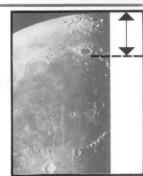

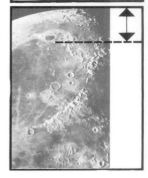

Libration in Latitude
above and right
The second important libration is in latitude. The lunar pole is not perpendicular to the plane of the Moon's orbit (diagram above). The result is that we can see for some way alternately beyond the northern (1) and the southern (2) limbs, producing an effect which is again quite noticeable. Some of the polar formations (such as the craters Challis and Main, not far from the north pole) can be well seen only when the libration in latitude is favourable. The effect is not so striking as the libration in longitude, because the shadow effects are less in evidence. The positions of the crater Plato in the photographs on the right reveal the effects of libration near the northern limb, but libration effects become far less noticeable for features well away from the limb. Lunar observers have always been used to seeing the limb formations as very foreshortened. Photographs from the Orbiter probes, taken from more favourable angles, show them correctly. For instance, Plato is almost perfectly circular, with a diameter of 60 miles.

Diurnal Libration *left*
When the Moon is low, an observer can see for some extra distance round the Moon's mean limb since he is "elevated" by the amount of the Earth's radius. This we call diurnal libration. Observationally, it is not so important as librations in latitude and longitude, because when the Moon is low viewing conditions are always poor.

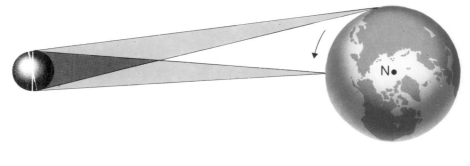

Origin of the Moon

There has been a great deal of discussion about the Moon's origin, and even now it cannot be said that the problem has been cleared up.

Two Theories

Basically, there are two theories. Either the Moon and the Earth used to be one body, and divided in the distant past, or else they have always been separate. Until a few years ago the fission theory was widely accepted, but by now it has fallen into disfavour. The mathematical objections to it seem to be insuperable, and almost all astronomers now agree that the Earth and Moon never formed a single body.

Tidal Theory *above*
The tidal theory, proposed in the latter part of the 19th century by George Darwin (second son of Charles Darwin), assumed that the Earth and Moon originally formed one body, and that the Moon broke away. It was assumed that the Earth was rotating too rapidly to be stable ; the body became first pear-shaped and then dumbbell-shaped, after which the Moon broke away altogether.

Earth/Moon/Mars Theory *above*
Though Darwin's original tidal theory was found to be untenable, it has been revived recently in modified form. The suggestion is that both Mars and the Moon once formed part of the Earth, and that the Moon was left behind as the "débris" after Mars separated from the main body and moved into independent orbit.

A Secondary Planet

If the Moon never formed part of the Earth, as is now generally believed, then what was its origin ? Here we come back to the question of the origin of the Solar System itself, since there are reasons for supposing that the Moon ought to be regarded as a secondary planet rather than as a genuine satellite.

According to one suggestion, the Moon was once a completely independent planet, which approached the Earth so closely that it was "captured". There is nothing impossible in this theory, even though it does involve some rather special circumstances. We can make no real attempt to give a date for the capture, except that it must have been at a relatively early stage of the two worlds.

Alternatively, it may be that the Moon and the Earth were formed at about the same time in the same space region of the Solar System, so that they have been gravitationally linked throughout their careers. This also sounds plausible; the fact that the Moon has developed differently from the Earth can be attributed solely to its initially smaller mass, which led it to lose any atmosphere and water it may originally have had.

Other Theories

There has been much argument as to whether the Moon was very hot in its early period as an independent body. Present evidence seems to indicate that the temperature used to be high, but the problem is very far from being settled. Neither can we entirely discount the idea that the Moon was formed from a cloud of material which once surrounded the Earth.

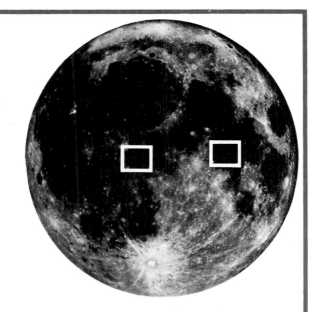

Age of the Moon *above*
The first lunar landings have indicated that the various "seas" are of widely different ages. Rocks brought back from the Mare Tranquillitatis (Apollo 11) *above right* have ages of $3 \cdot 6$ to $4 \cdot 5 \times 10^9$ years, but the age of the rocks from the Oceanus Procellarum (Apollo 12) *above left* is only about $2 \cdot 6 \times 10^9$ years. The diagram below compares the possible time-scales of events in the history of Earth and Moon.

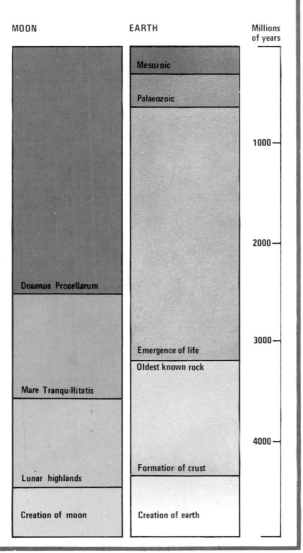

MOON	EARTH	Millions of years
	Mesozoic	
	Palaeozoic	
		1000
		2000
Oceanus Procellarum		
	Emergence of life	3000
	Oldest known rock	
Mare Tranquillitatis		
		4000
Lunar highlands	Formation of crust	
Creation of moon	Creation of earth	

Moon–Earth System

It is incorrect to say simply that the Moon revolves round the Earth. In fact, there can never be a case in which one body revolves round another; with a two-body system, each body moves round the common centre of gravity of the system.

If the two bodies are of equal mass, the centre of gravity of the system will be midway between them. If one body is more massive than the other, then the centre of gravity will be closer to it.

The Earth, as we have seen, is 81 times more massive than the Moon; and therefore the centre of gravity of the system – known as the barycentre – is shifted toward the Earth. It lies, in fact, within the Earth, as the diagram below shows.

In view of this, it is loosely correct to say that the Moon revolves round the Earth; but it does not revolve round the centre of the Earth, and the distinction is important in many astronomical calculations.

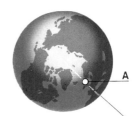

The barycentre
above The barycentre, situated inside the Earth's globe at A.
below Three positions of Earth and Moon in motion round the barycentre.
As the barycentre lies within the Earth's globe, the simple statement that "the Moon revolves round the Earth" is adequate for most purposes.

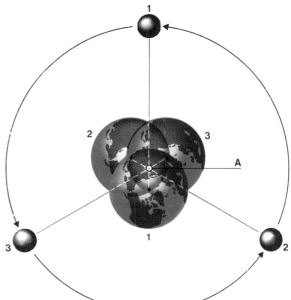

Gravity on the Moon

The surface gravity of a body depends partly on its mass and partly on its size. The Moon is both smaller and less massive than the Earth: on the lunar surface gravity amounts to one sixth of that on the Earth. It has been found that astronauts have no difficulty in adapting to these conditions; there is no problem about walking on the Moon, and of course space suits, which would be very cumbersome on Earth, are also reduced in weight.

Earth–Moon Distance *below*
The mean distance of the Moon from Earth is 238,840 miles, just under thirty times the diameter of the Earth. Venus, closest of the planets, never approaches to within one hundred times the mean distance of the Moon. In the diagram the ovals represent Earth diameters.

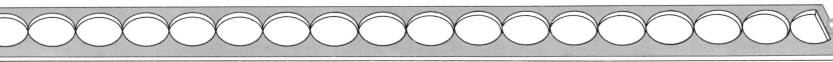

The surface of the Moon

The nature of the Moon's surface has been a subject for discussion over many years. Originally, there was little doubt among astronomers that the craters and maria were of volcanic origin – using the term "volcanic" in a broad sense, since it was quite obvious that a lunar crater, with its massive walls and depressed interior, was very different from a volcanic cone of the Vesuvius or Etna type. Robert Hooke, the contemporary of Isaac Newton, suggested in the late 17th century that craters had been formed by bubbles of gas in the lunar crust, which had burst and had left the features we now see.

Development of Theories

During the 19th century, various astronomers, such as F. Gruithuisen, supported the idea that the craters might have been produced by meteoritic impact. An altogether different view was proposed by James Nasmyth and James Carpenter in their famous book published in 1874; they thought that the craters were due to violent eruptions from central mountains, which showered material in a ring all round the central vent. This "volcanic fountain" idea is hopelessly untenable; the material in the wall of a large crater, such as Clavius, contains too much material to have been formed in such a way. Moreover, there is no known case of a crater whose central peak is as high as the surrounding rampart.

Efforts were made to find terrestrial analogies. Formations of both kinds were discussed; there are volcanic calderas such as Crater Lake (Oregon), and meteorite craters such as that in Arizona. The whole problem led to a great deal of argument, and even after the first lunar landings, in 1969, these arguments continued.

Alternative Theories

There were, of course, other theories. Among the most eccentric were those of the German selenographer, P. Fauth, who believed the Moon to be covered with a thick layer of ice. But, in general, the main problem to be solved was whether the craters were due to endogenic (internal) forces, or whether they were due to impact. This, in turn, was bound up with the question of whether the Moon had been a hot body in its early career.

Seismic Findings

It now seems certain that both volcanic and impact processes have operated on the Moon, so that there are structures of both kinds. The only point at issue is with regard to which of the two processes was the more important – and here, opinions among astronomers are as divided as ever. There is no evidence of any meteoritic material on the Moon, and the rocks are of essentially volcanic type, but they differ in some respects from terrestrial rocks. The link between minor "moonquakes" and the elusive transient phenomena recorded from Earth is established, but this certainly does not support present-day lunar vulcanism.

However, there can no longer be any doubt that the lunar "seas" are lava-plains. Had they been oceans of soft dust, as Gold and others had maintained not so very long ago, the manned exploration of the Moon would have been well-nigh impossible.

Impact Theory *below*
According to the impact theory, illustrated below, the lunar craters are due mainly to meteoritic bombardment. A meteorite will hit the Moon and penetrate the lunar crust, so that its kinetic energy will be transformed into heat and it will act in the manner of a powerful bomb. (In the diagram A = main impact; B = secondary crater produced by ejected material.) Thus the crater which it produces will be essentially circular even if the missile lands at an angle. On the Earth there are various impact craters of this kind, and comparisons can be made between terrestrial meteorite craters, bomb craters and lunar structures. Even the regular maria, such as Imbrium, are attributed to impact. Central peaks are due to a rebound process. The American astronomer, R. B. Baldwin, considers that it is only minor features, such as the crater chains and the obviously volcanic cones, which are of internal origin, though in many cases the impact of a large meteorite may have caused sufficient disturbance to produce extensive lava-flows from below the lunar surface.

Lunar Samples *right*
The analysis of rock samples brought back from the Apollo flights provided some unexpected information. In particular, they contained vast numbers of what have been described as "glass marbles" (see microstructure 5). These have evidently been ejected from the interior of craters, and have impacted upon the surrounding surface. They are very small – the largest is only about half a millimetre in diameter – but they show considerable variation; some of them are elongated, and they are of various colours. Substances identified in the rock samples include plagioclase, ilmenite and pyroxene. Microscopic rubies were also found – but these are much too small to be of any commercial value, so their interest is scientific only.
These five microstructure photographs are of samples brought back by Armstrong and Aldrin from the Mare Tranquillitatis. Specimens collected from the Oceanus Procellarum proved to be somewhat different in structure and content.

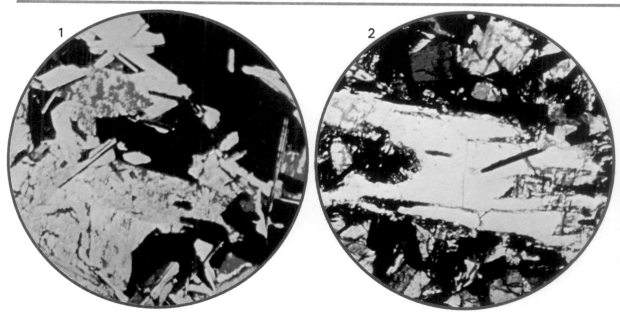

Lunar Surface *left*
Before the landing of the first manned spacecraft, it was essential to ensure that the lunar surface would be sufficiently rigid. The theory that the maria were covered with deep layers of soft dust was current until well into the 1960s. But the first automatic soft landing by Russia's Luna 9 in 1966 showed that a mare surface gives no sign of yielding under the weight of a lunar module. This photograph shows the region of the Mare Tranquillitatis where Apollo 11's module, *Eagle*, landed in July 1969 ; though smooth by lunar standards, the area is still decidedly rough, and there are small crater-pits everywhere, together with larger, shallower depressions. This crater is 35 yards across.

Footprint *top right*
The footprint of an astronaut (Aldrin) on the Mare Tranquillitatis. The depth of penetration is much less than an inch. The astronauts' boots sank in to a slightly greater degree when they were walking near the lips of craters, but there was no suggestion of their sinking dangerously.

Rock Sample *centre right*
One of the rock samples brought back from Apollo 11. It is of basaltic type, and contains many of the interesting "glass marbles". There is very little water content ; the amount of hydrated material on the Moon seems to be very low indeed—and if this applies to the mare regions, it must presumably apply also to the highlands. It now seems most unlikely that there has ever been much surface water on the Moon, and the chances of finding under-ground ice, as had been suggested, are slight.

Detail of Rock *right*
A section of the same rock, considerably enlarged. This photograph was taken at Houston, Texas, during the first analysis of the samples brought back by Armstrong and Aldrin.

Volcanic Theory *left*
Lunar craters are not distributed at random ; even the large walled plains tend to be lined up or to occur in groups and clusters, and on the Earth-turned hemi-sphere there is an obvious alignment with the central meridian. The craters occur along lines of weakness in the Moon's crust, which shows they must be of internal origin. Where one crater intrudes into another, it is almost always the smaller which breaks into the larger, again indicating a non-random distribution ; and the wall of the broken formation is always perfect up to the point of junction, whereas in any violent mode of origin there would be evidence of shaking-down.

Central peaks are only to be expected on a volcanic theory ; regular maria were formed in the same way as the large craters. There is a close analogy between lunar craters and terrestrial calderas, which are formed by a relatively gentle uplift and subsidence process, as shown in the diagram. The rocks brought back from Apollo missions are volcanic. The volcanic and impact theories are not, however, mutually exclusive. It is probable that both processes have operated on the Moon.

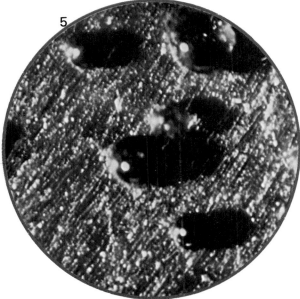

The Lunar Landscape

The Moon's "Seas"

The Moon's seas, or *maria*, occupy a large part of the Earth-turned hemisphere, though there are no major *maria* on the far side. Some, such as the Mare Imbrium and the Mare Crisium, are basically circular; others, such as the Oceanus Procellarum, are irregular. The well-formed seas, such as the Mare Imbrium, have high, mountainous borders. For instance, the lunar Apennines (shown in the photograph below) make up part of the border of the Mare Imbrium, and contain peaks rising to over 15,000 feet above the country below.

The Apennines *below* have peaks of over 15,000 feet.

The Craters of the Moon

The lunar craters are the most striking features of the entire surface. They are of many kinds. Some are well formed, with high, continuous walls; others are broken and ruined, so that in some cases they are difficult to identify at all. On the *maria* there are "ghosts", whose walls appear only as low, discontinuous ridges.

The distribution of the craters is not random. When one crater breaks into another, as frequently happens, it is usually the smaller crater which intrudes into the larger. Moreover, the major craters tend to form groups and chains; Langrenus, for example, is one of a great chain running between Emdymion in the north of the Moon past Janssen in the south.

Each lunar crater has its own individual points of interest. There are some formations with high walls and central mountains, or groups of mountains; others have relatively flat floors with little detail. In general, a crater has walls which rise to a modest height above the outer landscape, but whose floor is sunk below the exterior level. If a large crater is drawn in profile, its form is seen to resemble a saucer more nearly than a deep, steep-sided mine-shaft.

There are many large, regular structures such as Clavius (146 miles in diameter). These are known generally as walled plains. The true craters, with central peaks, may also be of large size – over 100 miles across in some cases – and there are innumerable smaller features, known as craterlets, which are to be found everywhere. No part of the Moon is free from them, as has been shown by the photographs taken from rocket probes and the eye-witness accounts given by the Apollo astronauts.

Near Full Moon, the scene is dominated by the brights streaks or rays which seem to issue from some of the craters, notably Tycho and Copernicus. These rays are surface deposits, and are best seen when the Sun is high over them during the long lunar day.

Mare Crisium *above*
Photographed from Orbiter. It has a relatively smooth floor, with two prominent craters, Picard and Peirce.

Tycho *below*
This is the major ray-centre of the Moon, 54 miles wide. Surveyor 7 landed on the outer slopes to the north.

Craters

Langrenus (*right*) is a typical large crater with a complicated central mountain group. The massive, terraced walls rise to 9000 feet above the deepest part of the floor. There is intricate wall structure, as is seen even when the crater is observed from Earth; this photograph was taken by the crew of Apollo 8 in December 1968, when they were orbiting the Moon at a height of approximately 70 miles.

Langrenus is one of a vast chain of structures in the eastern hemisphere of the Moon. The other members of the chain are Janssen, Vendelinus, Petavius, Cleomedes and Endymion, though they are not all alike – for instance Vendelinus is much less well-preserved than Langrenus or Petavius, with walls which are much lower and less regular. Presumably therefore, Vendelinus is a much older structure than the other members of the chain. It is quite likely that the Mare Crisium should also be regarded as a member, there may be no essential difference between a large crater or walled plain and a circular *mare*.

Walled Plains

Quite different from Langrenus is Plato (*right*), one of the most famous walled structures on the Moon. It lies near the boundary of the vast Mare Imbrium, and is 60 miles in diameter; the walls are relatively low, rising to only 4000 feet above the interior, but Plato is always recognizable at a glance, because of the darkness of its floor. The Danzig astronomer Hevelius, who drew a map of the Moon as long ago as 1645, called it "the Greater Black Lake". (Hevelius's lunar map was probably the best of its time, but his system of nomenclature was soon replaced by that of Riccioli, who drew a new map in 1651 and named the main craters after famous men and women.)

Plato has no central mountain. The floor contains no major structures, though various small craterlets may be seen. There are other walled plains with similar dark floors, but that of Plato is particularly striking. From Earth it appears elliptical, due to foreshortening; in this Orbiter picture it is correctly shown as almost perfectly circular.

Valleys, Rills and Faults

One of the most spectacular objects on the Moon is the great valley which cuts through the Alps on the border of the Mare Imbrium (*right*). It is 80 miles long and is certainly a collapse feature; earlier suggestions that it was produced by a meteorite plunging through the mountain range can now be rejected. Other valleys of the same kind are known, but none is so striking or so regular.

Less prominent, though equally interesting, are the rills, alternatively known as clefts. Superficially they look rather like the cracks in dried mud, though there is no real analogy. Some of them are visible with small telescopes; such is the celebrated Hyginus Rill, which, when photographed from Orbiter 3 (*below*), is seen to be in part a crater-chain. Hyginus itself, the central crater, is 4 miles in diameter. Other rills have no craterlike enlargements, and may form highly complicated networks. Some craters, such as Gassendi, have rill-systems upon their floors. It has been found that some of the rills have convex floors; their walls are comparatively steep.

These great rills are among the most spectacular of all the lunar features. With Apollo 15, the two astronauts Scott and Irwin drove their LRV or Lunar Roving Vehicle to the very edge of the Hadley Rill, in the foothills of the Appennines, and were able to study its structure. Of course, many of the rills shown on Orbiter and Apollo photographs are too delicate to be seen at all from the Earth.

True faults are less common on the Moon, and much the best example is the so-called Straight Wall, in the Mare Nubium (*far right*). It lies between the crater Thebit and the deep, well-marked Birt; its name is inappropriate, since it is neither straight nor a wall! The length is 80 miles, and there is a height difference of 800 feet between the top of the structure and the land to the west. However, the angle of inclination is not so steep as used to be thought, and is only about 40 degrees instead of more than 70. The peaks at the southern end are generally known as the Stag's-Horn Mountains. Before full moon the Straight Wall shows up as a dark line; after full, when the sunlight is falling on to its face, it is seen as a bright line.

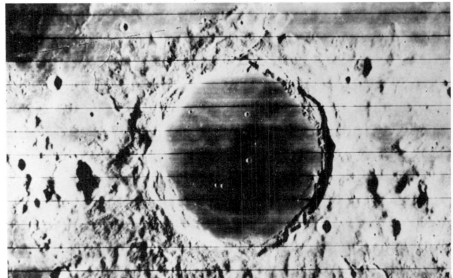

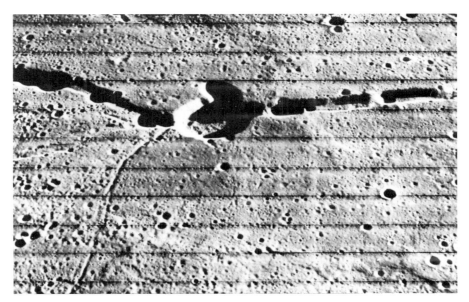

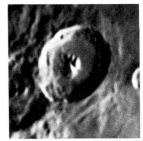

Petavius *above*
Petavius is a member of the great Langrenus chain, and is 100 miles in diameter. It has a prominent central peak, conspicuous rill-valley runs from this to the south-east wall. On the floor are numerous ridges and depressions.

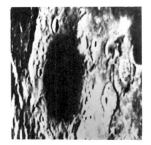

Grimaldi *above*
Grimaldi, near the Moon's west limb as seen from Earth, has low, broken walls; it is 120 miles in diameter, and is noteworthy because of the darkness of its floor. This intense darkness makes Grimaldi recognizable at all times.

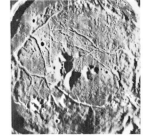

Gassendi *above*
Gassendi, 75 miles in diameter, lies on the border of the Mare Humorum. It has a central mountain group, and its floor is criss-crossed with rills, well shown in this photograph taken by Orbiter 5. Its rill-system is very complex.

Straight Wall *above*
This photograph was taken before full moon, so that the sunlight is coming from the east (*right*) and the Wall is casting shadow. After full moon, it appears as a bright line. It is not a wall but a well-marked fault, over 80 miles long.

Man on the Moon

Manned exploration of the Moon began on 21 December 1968, with the launching of Apollo 8. The three astronauts – Frank Borman, James Lovell and William Anders – completed ten circuits of the Moon on 24 and 25 December, coming down to a height of only 70 miles above the surface; on 27 December they splashed down safely in the Pacific. On 18 May the following year, Apollo 10 tested out the lunar module within ten miles of the Moon. On 21 July 1969, Neil Armstrong and Edwin Aldrin stepped out on to the Mare Tranquillitatis.

On the Moon

The landing of Eagle, the lunar module of Apollo 11, took place almost exactly as planned. After they had arrived, and carried out a thorough check of all the equipment, Armstrong and Aldrin began their E.V.A. (Extra-Vehicular Activity), during which time they collected rock samples and set up various pieces of equipment – including a seismometer, designed to measure any tremors, natural or artificial, in the Moon's crust. The entire E.V.A. was televised.

Apollo 12

Apollo 12 was launched the following November. This time the astronauts were Charles Conrad, Alan Bean and Richard Gordon. Intrepid, the lunar module, carried Conrad and Bean to the landing site in the Oceanus Procellarum; the calculations were so exact that the astronauts were able to walk over to the automatic probe Surveyor 3 and bring back parts of it for analysis. More samples were collected, and the astronauts undertook two "walks". Apart from the failure of the colour television camera, Apollo 12 was a triumphant success.

Apollo 13

So far all had gone well with the NASA programme of manned lunar exploration, and there were no qualms when Apollo 13 was launched on 11 April 1970, carrying astronauts James Lovell, Fred Haise and Jack Swigert. Yet Apollo 13 proved to be the most hazardous of all space ventures. At a distance of 178,000 miles from the Earth there was an explosion which put the service module out of action. The intended lunar landing at Fra Mauro was naturally abandoned, and all efforts were concentrated upon bringing the astronauts home safely. Miracles of improvisation took place; the engine of Aquarius, the lunar module, performed faultlessly when carrying out manoeuvres for which it had never been designed; and on 17 April the command module Odyssey splashed down safely.

From a purely scientific point of view Apollo 13 was a failure; yet it was a human triumph, for the technicians at Houston and above all for the astronauts. Without their superb courage and skill, Lovell, Haise and Swigert could never have survived.

Landing of Apollo 12 *above*
The lunar module Intrepid is seen descending toward the Oceanus Procellarum (19 November 1969). This photograph was taken from the command module. After Intrepid had returned to orbit, and had been abandoned, it was deliberately crashed on to the Moon setting up crustal vibrations which lasted for almost one hour.

Aldrin on the Moon *right*
One of the most dramatic photographs ever taken. Colonel Edwin Aldrin stands on the surface of the Moon, with the whole scene reflected in his helmet. Footprints can be clearly seen; the shadows are jet-black. This photograph was taken by Neil Armstrong on 21 July 1969 – the day when men first reached another world.

The Apollo 12 Mission in Progress *above*
The lunar module Intrepid stands on the Oceanus Procellarum; Commander Alan Bean has finished setting a radio "dish" designed for communication with Earth. The footprints of the astronauts can be seen. This photograph was taken during the first of the EVAs. The only disappointment of the mission was the failure of the colour TV camera.

Antares on the Moon *above*
This spectacular photograph shows the front view of the grounded lunar module, Antares; the reflecting circular flare is caused by the brilliant sun. Shepard and Mitchell said that this unusual sight had a jewel-like appearance. Experiments were set up far enough away from Antares to avoid any adverse effects caused during the blast-off from the Moon.

The Landing Site *above*
As on previous missions, the astronauts set up the U.S. flag. Here James Irwin salutes the flag. The grounded lunar module Falcon is to the centre, and the Rover, which had a range of several miles, on the right. In the background over 20 miles away, looking almost due south, is the mountain known as Hadley Delta. This is one of the most spectacular peaks in the Lunar Apennines.

The Conquest of the Moon

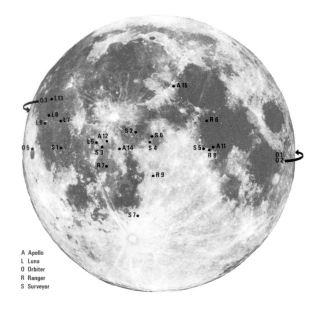

A Apollo
L Luna
O Orbiter
R Ranger
S Surveyor

Luna 3 *above*
In October 1959 the Russian vehicle Luna 3 went round the Moon, sending back the first photographs of the far side. The previously unknown regions were heavily cratered, but there were no major *mare* areas.

Ranger 8 *above*
Nine Ranger probes were launched, between August 1961 and March 1965. Of these, the last three were successful in sending back close-range photographs before they crashed on the lunar surface and were destroyed.

Luna 13 *above*
The first probe to achieve a successful soft landing on the Moon was Luna 9, launched on 31 January 1966. On 21 December 1966 Luna 13 also made a soft landing, and the instruments it carried were capable of analysing the lunar surface material.

Orbiter 5 *above*
Orbiter 1 was launched in 1966. It was followed by four other probes in the series, all of which were successful, and provided information for a very detailed map of almost the whole of the Moon. Fine pictures of the ray-crater Tycho were received.

The Moon from Space

So long as the Moon had to be studied from a distance of a quarter of a million miles, our knowledge of it was bound to be incomplete. The first direct investigations from space were made by the Russians in 1959, when they dispatched three lunar probes. Luna 1 went past the Moon, and sent back valuable information. In October, Luna 3 went round the Moon, sending back photographs of the hidden side.

Landers and Orbiters

For some years after Luna 3 there were no dramatic developments, but then came three successful Ranger probes, sent up from the United States. These crashed on the Moon, destroying themselves, but during the last few minutes of their flights they sent back close-range pictures of the lunar landscape. The next step was to launch probes which would enter paths round the Moon, so that they would be capable of sending back large numbers of photographs. Five Orbiters were sent up and all were successful.

However, there were still doubts as to whether the lunar surface was firm or soft. The question was settled in 1966, when the Russians brought down Luna 9 gently on to the Moon's rocks. The first pictures sent back showed what seemed to be a lava-plain. There was no tendency for the vehicle to sink into dust. It is now known that the lunar surface is firm enough to bear the weight of a large machine.

Luna 9 was followed by several U.S. Surveyor soft-landers, together with the Soviet Luna 13, which repeated the success of Luna 9.

Men on the Moon

Throughout this period the Americans had been concentrating upon the Apollo programme, the aim of which was to send two astronauts to the Moon, and this was finally achieved on 20 July 1969. Another Apollo mission took place in November 1969; and after the near-disaster of Apollo 13, in 1970, another successful mission was achieved with Apollo 14 in 1971.

As well as carrying out reconnaissance, the astronauts left recording instruments on the Moon. Much has been learned from them; for instance, there are perceptible tremors in the crust, due largely to the gravitational strains set up by the Earth, and there are emissions of tenuous gas.

Deliberate impacts of discarded rocket components on to the Moon have shown that the lunar structure is very different from that of the Earth. However, no full explanation of the effects has yet been worked out.

Automatic Probes on the Moon

The Russians at this period were more intent upon exploring the Moon with unmanned vehicles. In 1970 they achieved a major triumph with Luna 16, which landed on the Mare Fecunditatis and then returned, bringing samples of Moon-rock with it. Later in 1970 came Luna 17, which landed in the Mare Imbrium. From it emerged the strange-looking Lunokhod 1, which was a wheeled craft capable of movement; it crawled about the Moon, guided by the controllers on Earth, sending back valuable data.

Lunokhod 1 *above*
Landed in December 1970, it had eight wheels, powered by solar batteries, and could be guided from Earth. It ceased to function on 4 October 1971, having frozen solid after the exhaustion of its atomic heater. Lunokhod 2 followed in 1972.

Track of Lunokhod
right
Lunokhod 1, the first Russian "roving vehicle" operated in the Mare Imbrium, not far from the Sinus Iridum. In this photograph, taken on 7 December 1970, the track made by the vehicle's wheels is clearly shown.

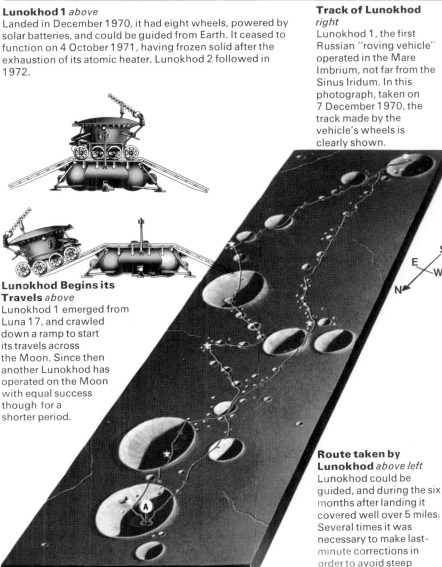

Lunokhod Begins its Travels *above*
Lunokhod 1 emerged from Luna 17, and crawled down a ramp to start its travels across the Moon. Since then another Lunokhod has operated on the Moon with equal success though for a shorter period.

Route taken by Lunokhod *above left*
Lunokhod could be guided, and during the six months after landing it covered well over 5 miles. Several times it was necessary to make last-minute corrections in order to avoid steep slopes.

Apollo 15 mission *below* and *right*

The Apollo 15 flight of 1971 was called "the scientific Apollo", and certainly it provided a vast amount of information. Astronauts David Scott and James Irwin undertook three drives in their LRV (Lunar Roving Vehicle) and went right to the edge of the mile-wide, 1000-feet deep Hadley Rill. They collected many samples, and broadcast detailed and accurate descriptions of the lunar landscape. In the orbiting Command Module, A. Worden, the third member of the team, undertook a long research programme.

A Landing site
B Edge of Hadley Rill
C Apennine foothills
D North crater complex

Luna 16 *above*
Luna 16, launched by the Russians in 1970, was the first automatic probe to go to the Moon and return. It landed in the Mare Fecunditatis, collected samples of lunar material, and took off again. Its lower section was used as a launching platform.

Apollo 11 *above*
Apollo 11, of July 1969, carried two astronauts — Neil Armstrong and Edwin Aldrin — to the Moon. They landed in the Mare Tranquillitatis, and undertook the first direct lunar exploration. Michael Collins remained orbiting in the command module.

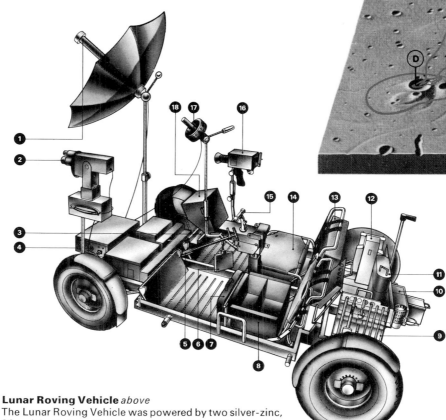

The Lunar Rover *left*
1 High-Gain Antenna
2 Television Camera
3 70 mm. Camera position
4 LCRU
5 Tongs
6 15 Bag Dispenser
7 Lunar Brush Bag
8 Underseat Bag Stowage
9 Low-Gain Antenna
10 C & D Console
11 16 mm. Camera
12 Handhold
13 Hand Controller
14 Buddy Umbilical System Bag
15 Lunar Drill
16 Stereo-camera
17 Magnetometer
18 Stowage Bags
19 Tool Carrier

Lunar Roving Vehicle *above*
The Lunar Roving Vehicle was powered by two silver-zinc, 36-volt batteries with an individual electric motor for each wheel. It was capable of covering considerable distances, at 10 mph on level ground.

Moon-cart *below*
With Apollo 14, of February 1971, the two explorers — Astronauts Alan Shepard and Edgar Mitchell — took a "hand-cart" to help in carrying their equipment. They were scheduled to take the cart up to the rim of Cone Crater, approximately a mile away from the landing site in the Fra Mauro highlands. The cart was very successful, but the journey up Cone Crater proved more exhausting than had been anticipated. Moreover the astronauts temporarily lost track of their exact position. It is now known that they were practically on the rim of Cone Crater when they turned back. However, all the scientific objectives of the walk were carried out successfully. The cart was left on the Moon.

Apollo 14: Antares at Fra Mauro *above*
The landing site of Apollo 14 was rougher than those of Apollos 11 and 12, since it lay in the upland region near Fra Mauro. This photograph shows the lunar module Antares, as seen by the astronauts during their second moon-walk. Tracks left by the moon-cart (known officially as the Modularized Equipment Transporter, MET) can be seen leading away from Antares; the inverted umbrella of the S-band antenna is seen to the left of Antares. The equipment left behind by the astronauts proved equal to expectation and functioned successfully.

Mapping the Moon

Apollo 15 *right*

Apollo 15 proved to be the most spectacular of the lunar missions to date. The landing site was near Mount Hadley, in the Lunar Apennines; the astronauts were David Scott, James Irwin and Alfred Worden. While Scott and Irwin explored the Moon, Worden remained in the command module, carrying out a full programme of scientific experiments. A tremendous amount of valuable information resulted from this orbital programme.

Using the Lunar Rover, Scott and Irwin drove to the region of the great Hadley Rill, a huge crevasse which had been known to astronomers for a long time. The nature of the rill had always been a matter of debate; there seems now no doubt that it is a collapse feature, and by any standards it was spectacular when seen from close range. There was no difficulty in reaching it, and the Rover travelled across the somewhat uneven ground with no trouble at all.

The Apennines themselves were indeed grand, and quite unlike the jagged peaks so often shown in science-fiction novels of past years. Hadley Delta, for instance, was described by Scott as "uniform". Another point of interest was the apparent nearness of the mountain to the astronauts; this was deceptive, since they did not go within about twenty miles of the actual base of the peak.

The photograph shows Hadley Delta in the background, with astronaut Irwin at work beside the Rover. All in all, Apollo 15 was one of the most trouble-free of the missions, and was the first in which a Rover had been used, so that the range of operations was tremendously increased.

Apollo 16 *right*

The landing site for Apollo 16 was in the highlands of Descartes, in one of the rougher parts of the Moon. Again a Rover was taken, and again it performed faultlessly. Astronauts Duke and Young explored a wide area, and set up the usual scientific experiments in the ALSEP (Apollo Lunar Surface Experimental Package). The only failure was with the device intended to measure the heat flowing out from beneath the lunar crust. This was accidentally damaged while being deployed, and could not be repaired. Everything else was set up without incident.

The photograph shows the lunar module in landing configuration, as Duke and Young prepared to come down on to the Moon's surface. Nothing could be less like the sleek, streamlined space-ships of science fiction, and it must be remembered that everything depends upon the motors in the lunar module itself. There is very little margin for error during the landing manoeuvre; and when the astronauts are ready to blast off back into orbit, they have to rely entirely upon the one remaining motor in their craft — the ascent engine. If this does not function flawlessly for the short but vital period of take-off, there can be no second chance, and neither can there be any hope of rescue. Fortunately this did not happen either with Apollo 16 or with the other lunar landing missions; but it may well be true to say that had the Apollo programme continued, there might have been an eventual tragedy. It is quite likely that no more men will be sent to the Moon until there is rescue provision in time of need.

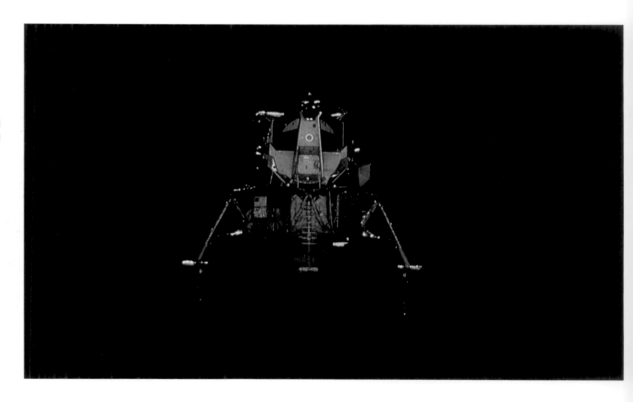

Apollo 17 *right*

The last Apollo mission was that of December 1972. The commander was Eugene Cernan, who had already been round the Moon. With him on the lunar surface was Dr. Harrison H. ('Jack') Schmitt, who was a professional geologist trained as an astronaut. Dr. Schmitt therefore became the first true scientist to set foot on the Moon, though it is fair to say that many of his Apollo predecessors were extremely knowledgeable scientifically.

The landing area was in the volcanic region of Littrow. As before, a Rover was used for extensive surveys, and it was during one of these, close to a small crater nicknamed Shorty, that Schmitt suddenly found what looked like orange soil. Initially he believed it to be evidence of past fumarole activity, but later investigations showed that this was not so.

Partly because of Dr. Schmitt's presence, Apollo 17 was the most valuable, technically, of all the missions. The photograph shows Dr. Schmitt himself working with a lunar scoop and a gnomon, a form of measuring instrument, placed on a large boulder.

The ALSEP of Apollo 17 was deployed without any trouble, and when the astronauts left the Moon everything was working well. Though Apollo was complete, results were still being received from all the ALSEP installations set up there apart from the first. When Cernan stepped off the lunar soil for the last time, it was a dramatic moment; we cannot yet be sure just when the next man from Earth will go there, but in any event the great Apollo triumphs will never be forgotten.

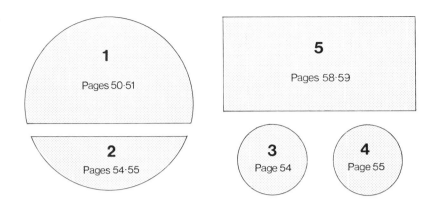

1	5
Pages 50·51	Pages 58·59

2	3	4
Pages 54·55	Page 54	Page 55

Index to Moon Maps

The number in bold type indicates the lunar map on which the formation lies. The page number of the map can be found from the key maps opposite The position is also given latitude (+ = north, − = south), and longitude (+ = east, − = west) According to convention, latitude is given first.

Mean distance from the Earth 238,840 miles maximum 252,700 minimum 221,460
Apparent diameter max. 33' 31", mean 31' 5", min. 29' 22"
Magnitude of full moon at mean distance −12·5
Sidereal period 27d.32166 (27d.7h.43m.11·5s.)
Synodic period 29d.53 (29d.12h.44m.2·8s.)
Inclination of orbit to ecliptic 5°09'
Orbital eccentricity 0·0549
Inclination of equatorial plane to ecliptic 1°32'
Mean orbital velocity 2,287 m.p.h. (0·63 miles per second)
Diameter 2,158 miles Volume 0·0203 that of Earth
Mass 1/81·3 of that of the Earth (=0·0123) : 3·7 x 10^{-8} that of the Sun
Density 3·3 (water=1), 0·60 (Earth=1)
Escape velocity 1·5 miles per second

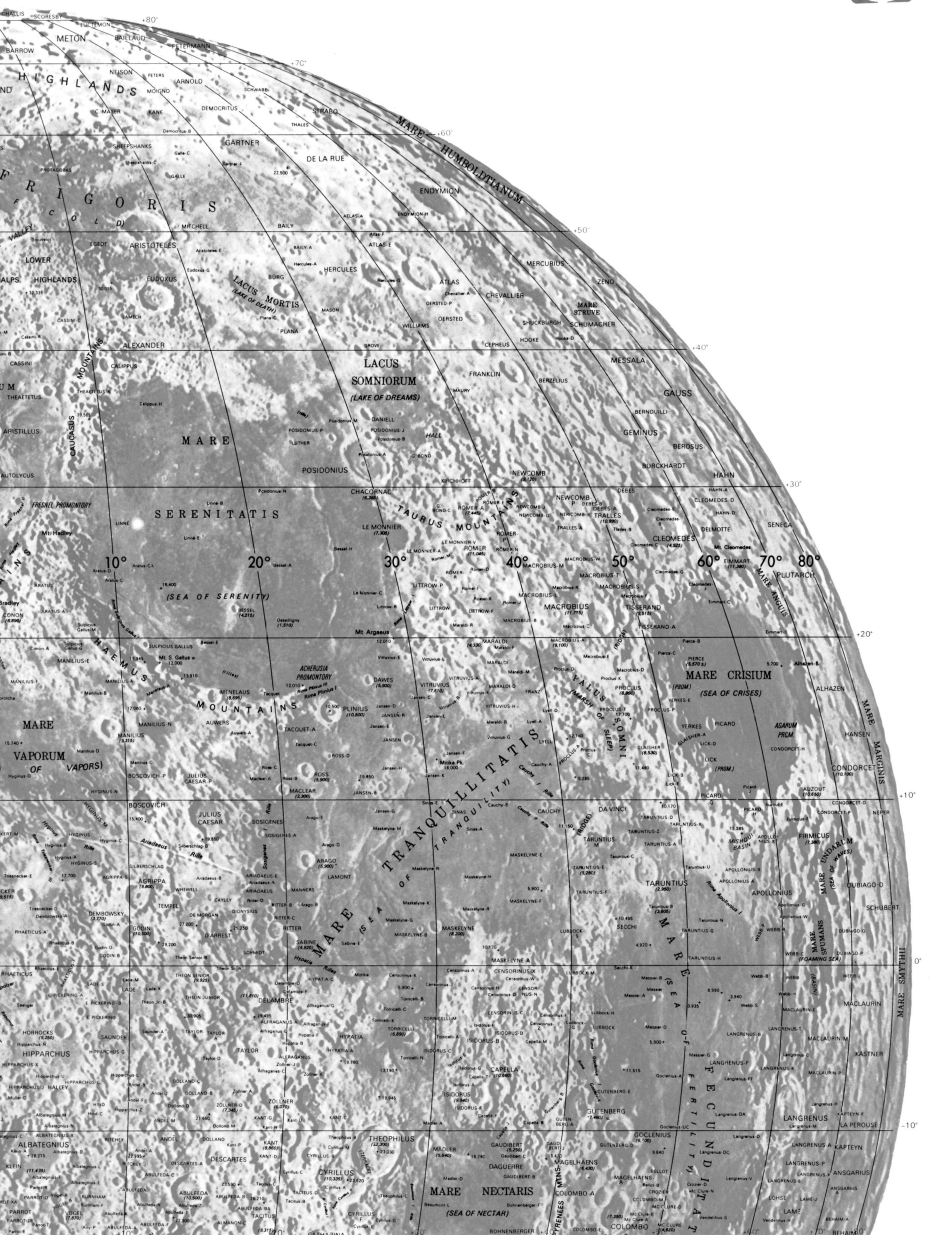

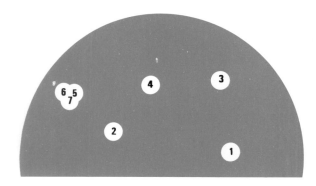

Panorama

The northern hemisphere of the Moon is dominated by the great seas: Mare Imbrium, Mare Serenitatis, Mare Crisium and Mare Tranquillitatis. Of these the first three are basically circular; the Mare Tranquillitatis – containing the point where the Apollo astronauts landed in July 1969 – is much less regular. This hemisphere also contains much of the huge Oceanus Procellarum, which is larger than our own Mediterranean Sea, but has no well-marked borders.

The Mare Imbrium is bordered by some important mountain ranges: the Apennines, Alps and Caucasus. The Apennines are particularly spectacular, and when seen under low lighting cast long shadows across the *mare* surface. The Sinus Iridum, or Bay of Rainbows, leads off the Mare Imbrium; if seen near sunrise, when the solar rays are catching the tops of the peaks to the west, it seems to jut out across the terminator, and has been nicknamed "the Jewelled Handle". It was in the Mare Imbrium, to the south-east of Sinus Iridum, that the Russian vehicle Lunokhod 1 landed in 1970, and carried out its exploration of that part of the lunar surface.

Closer to the equator are the more irregular "seas" such as the Mare Vaporum. The Central Bay, Sinus Medii, contains a complicated system of rills associated with the crater Triesnecker; near here, too, are the conspicuous rills of Hyginus and Ariadaeus.

Of the craters, the most imposing is Copernicus, which is the major ray-centre of the Moon apart from Tycho. On the Mare Imbrium there are Archimedes (50 miles across) and its companions Aristillus and Autolycus; there are also many large structures in the highlands in the north and north-east.

One of the most fascinating regions is that of Aristarchus, in the Oceanus Procellarum, Aristarchus is so bright that it remains visible even when illuminated only by earthlight, and it is in this area that most of the elusive 'red glows' have been seen, probably due to sublurainean emissions of gas.

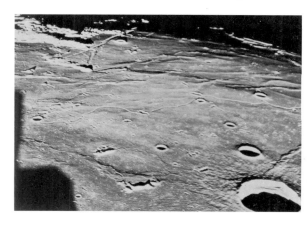

Apollo 11 over the Sea of Tranquillity *above*

The Moon Landing

On 20 July 1969, Neil Armstrong and Edwin Aldrin in the lunar module of Apollo 11 achieved the first landing on the surface of the Moon. The approach to the landing site in the Mare Tranquillitatis is seen in this photograph, taken from the lunar module when it was still docked to the command mondule. The large crater is Maskelyne; the Hypatia Rill is upper left centre, with the crater of Möltke to the right. The landing site is in the centre, close to the edge of the sunlit area. At 02.56 hours GMT on July 21, after a faultless landing, Armstrong stepped out on to the Moon. After completing their explorations, the two astronauts blasted away to rendezvous with the orbiting command module.

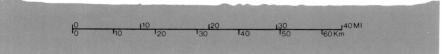

2 Copernicus *above* and *left*

Copernicus, as photographed from Orbiter 2 in 1966 from a height of 28 miles above the Moon. Seen from an oblique angle, the ramparts and the central elevation of Copernicus show up magnificently; the walls are terraced. When the photograph was taken, Orbiter 2 was 149 miles to the south of Copernicus.

(*Left*) An Orbiter photograph of Copernicus, taken almost from above, so that there is no foreshortening. The form of the crater is excellently shown.

(*Upper left*) Cross-section of Copernicus. Like all lunar craters of large size, the depth, though very considerable (well over 10,000 feet), is not great compared with the diameter (56 miles). In profile, a lunar crater resembles a shallow saucer. By lunar standards Copernicus is exceptionally deep for a formation over 50 miles across.

4 Archimedes *above*
The great walled plain Archimedes, 50 miles in diameter, is a prominent feature of the Mare Imbrium.

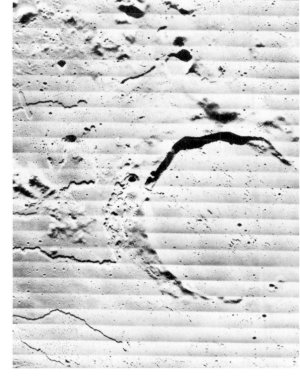

5 Prinz *above*
The incomplete ghost-crater Prinz, over 25 miles in diameter, is seen to be pitted with craterlets.

6 Schroter's Valley *right*
The Orbiter photograph of part of the impressive Schroter's Valley, near Herodotus (the less brilliant companion to Aristarchus), shows the interior detail, too delicate to be seen from Earth. The exact depth of the Valley is still not known with real accuracy, but the cross-section (*below*) is a good approximation.

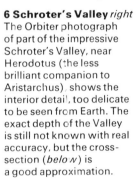

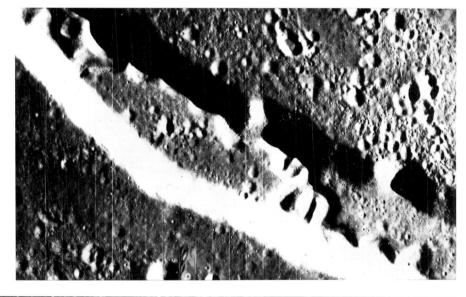

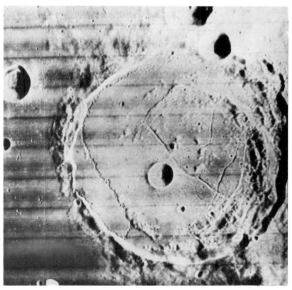

3 Posidonius *above*
Posidonius, on the boundary of the Mare Serenitatis : a plain 62 miles across, with reduced walls, and interior rills.

7 Aristarchus *below* and *right*
Less than 25 miles in diameter, but with well-marked, regular walls, it is the brightest formation on the Moon.

Moon Map: South and Poles

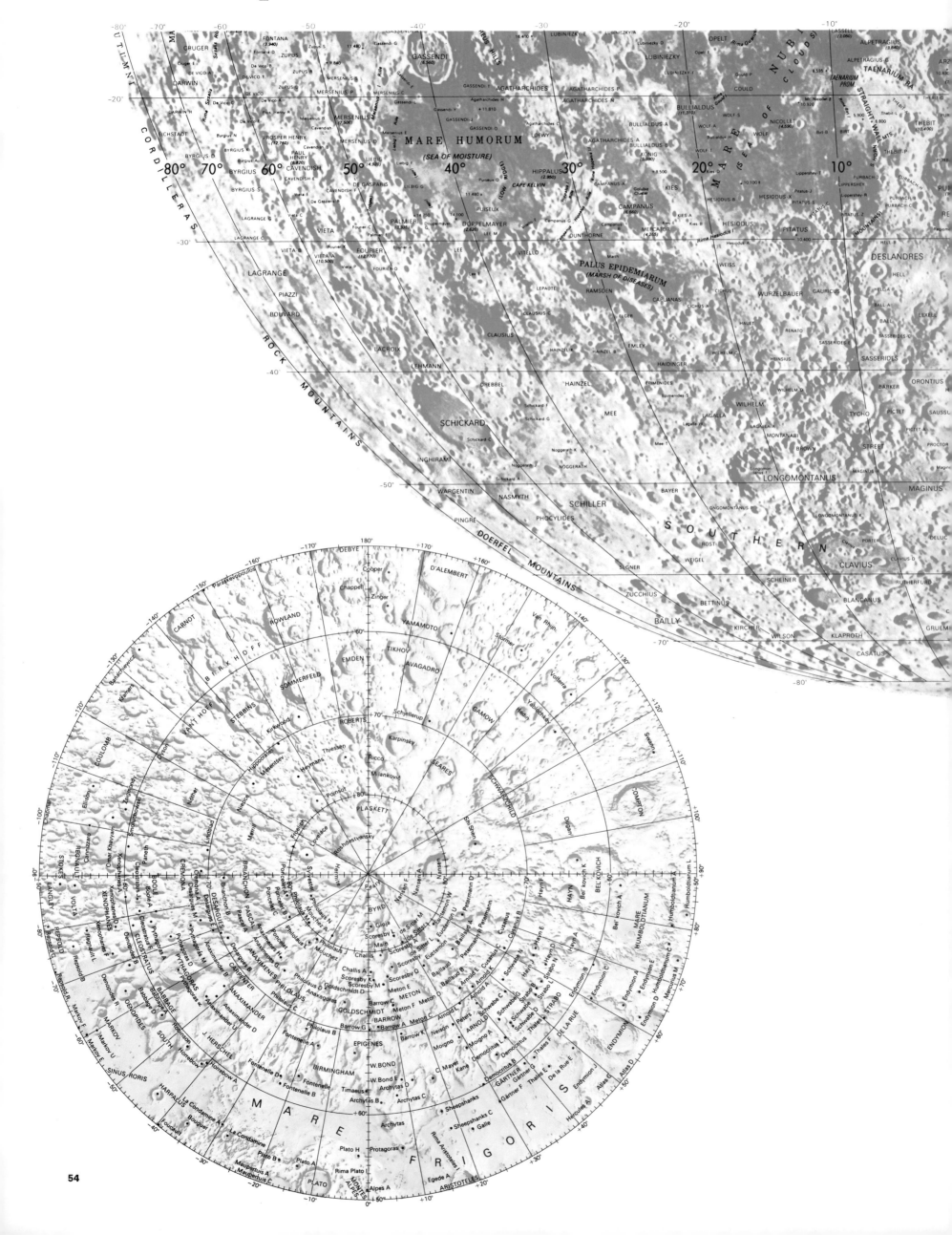

Moon Panorama: South

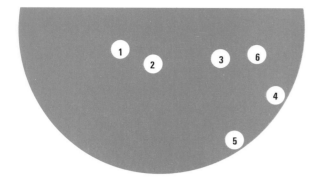

Lunar Landscapes: South

The main seas on the southern hemisphere of the Moon are the Mare Nubium and the Mare Fecunditatis, both of which are rather irregular, but are of great extent. The hemisphere also includes part of the Oceanus Procellarum, together with the much smaller but well-defined Mare Humorum and Mare Nectaris. The south-eastern area is mainly upland, and there are craters in profusion; they crowd against each other, leaving almost no level ground.

Near the centre of the Moon's disk, as seen from Earth, is the great chain of walled plains associated with the 92-mile Ptolemæus. When seen near the terminator Ptolemæus is spectacular, with shadow crossing its floor. Its companions to the south are Alphonsus and Arzachel. Well to the east, on the border of the Mare Nectaris, lie the magnificent Theophilus and its companions, Cyrillus and Catharina.

There are many large, well-formed plains in this hemisphere. Clavius, Maginus, Schickard, and the dark-floored Grimaldi and Riccioli are particularly notable. But under high light the dominant feature is Tycho, 54 miles across, lying in the southern uplands. Its bright rays streak out far across the lunar surface, and drown all other features in their brilliance. They remain striking for some days to either side of full moon, and make up the most splendid of all the ray systems.

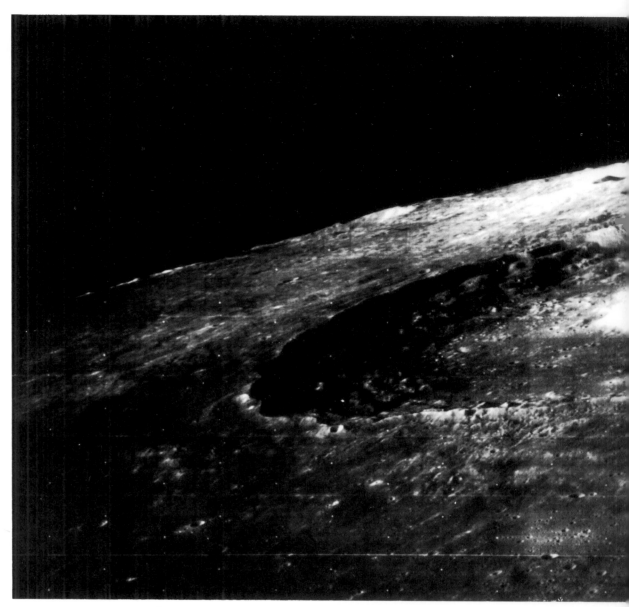

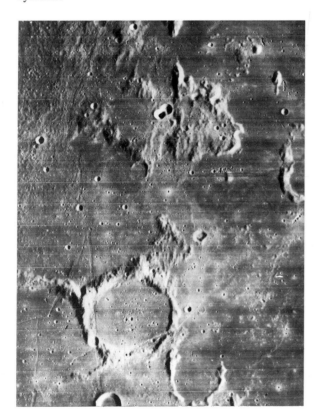

1 Fra Mauro *above*
In 1971 Apollo 14 landed in the highlands just north of Fra Mauro, which is 50 miles across and is very low-walled.

2 Alphonsus *right*
Alphonsus, lying between Ptolemæus to the north and Arzachel to the south, is over 80 miles in diameter. It has a distinct central mountain group, and its floor contains a system of rills, together with other delicate features such as craterlets and peaks. The last successful Ranger vehicle, No. 9, landed here.

Impact of Ranger 9
above Part of the sequence taken as Ranger descended towards the crater Alphonsus. The top one was taken at an altitude of 258 miles, 2 minutes 50 seconds from impact, the second at 115 miles, and the third at 58 miles.

3 Theophilus *left*
Seen from Earth, Theophilus appears almost circular; it is 64 miles across and 18,000 feet deep, so that it is larger and deeper than Copernicus, though it has no comparable system of bright rays. The photograph shown here was taken from Orbiter 3, at an altitude of 34 miles, looking south. Beyond Theophilus, to the upper right, is the companion plain Cyrillus.

4 Wilhelm Humboldt *left and below*
This is a huge walled plain, well over 100 miles across but very foreshortened as seen from Earth (*left*). It has a central mountain group, but was not well mapped before it was photographed from Orbiter probes. It is also seen in profile (*below*).

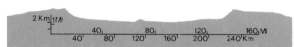

5 Mare Australe *above*
The Mare Australe, near the Moon's east limb as seen from Earth, is not a well-formed sea, and consists mainly of dark patches covered with *mare* material.

6 Goclenius *below*
Goclenius, 32 miles in diameter and 5000 feet deep, photographed from Apollo 8 on Christmas Day 1968.

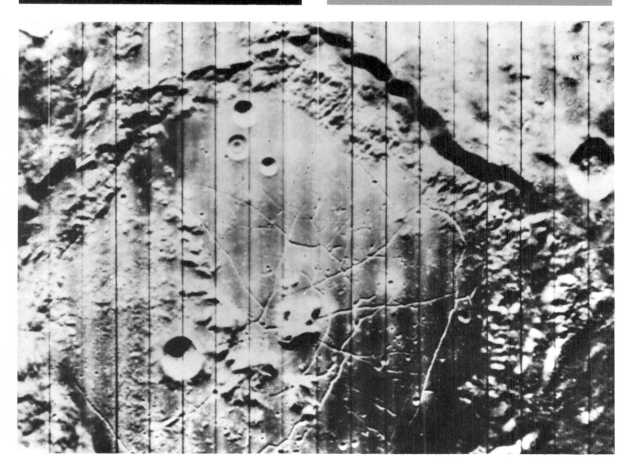

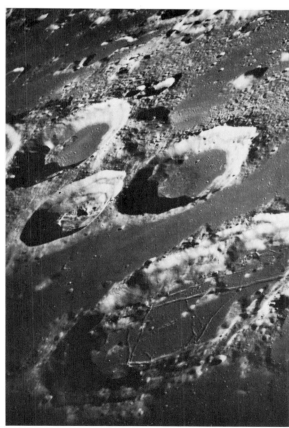

4 Wilhelm Humboldt in close-up *left*
Humboldt proves to have a complex rill system on its floor, together with craterlets and peaks. The walls are not regular, though they contain some high mountains.

The Far Side

When the first maps of the Moon's far side were drawn up, from the Luna 3 photographs of 1959, they showed very little detail – and much of the detail shown was subsequently found to have been wrongly interpreted. This was by no means surprising; it was only with the launching of the later Russian Luna vehicles, and above all with the U.S. Orbiters, that a detailed, correct map of the far side of the Moon could be compiled.

The immediate impression was of a hemisphere more or less devoid of *mare* areas. On the Luna 3 photographs, only two dark regions were shown. One was correctly interpreted as a dark-floored crater with a central peak: Tsiolkovskii. The other, the Mare Moscoviense (Moscow Sea), may be regarded as a true mare, but it is of relatively small dimensions when compared with the great maria of the Earth-turned hemisphere.

Yet the most surprising feature of the far side was, without doubt, the Mare Orientale, which is visible from Earth in very foreshortened form. As this

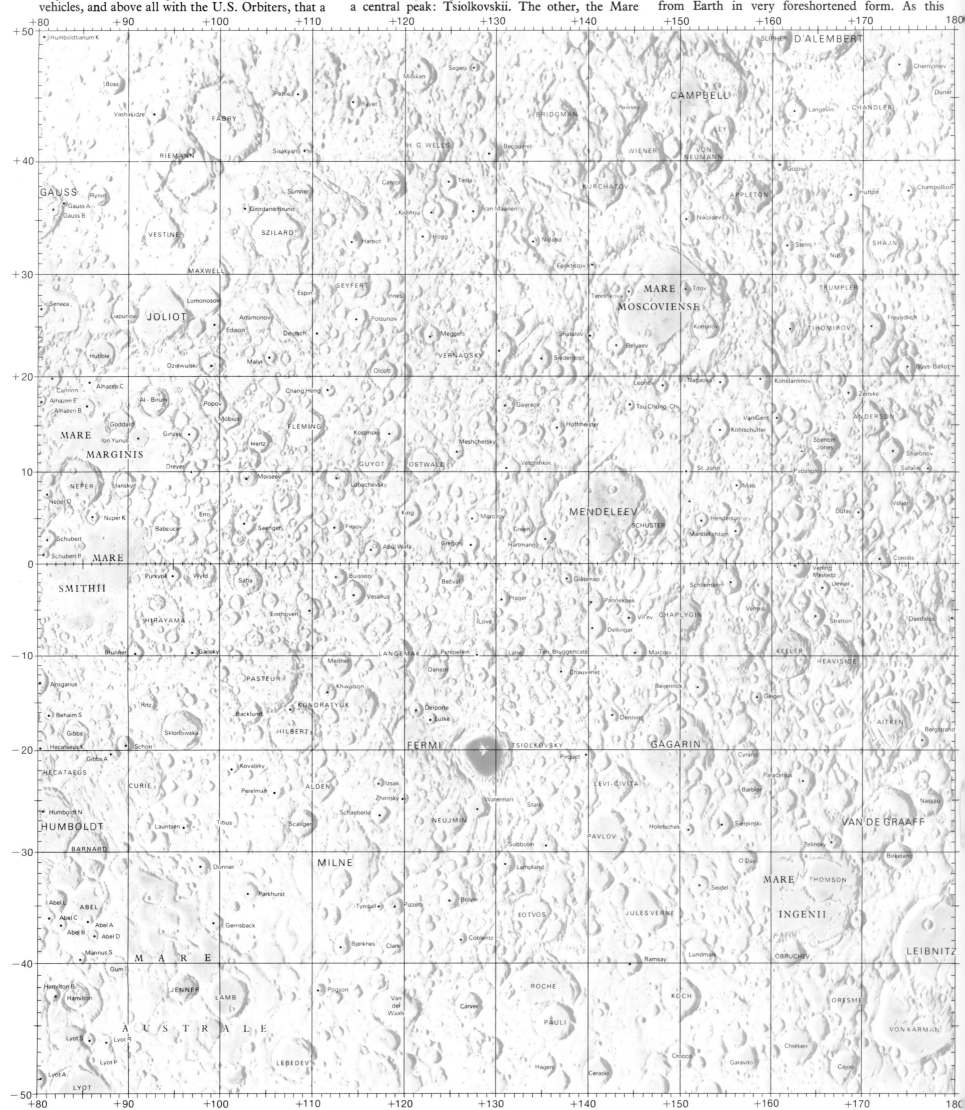

Mercator projection chart shows, it is an immense, complex structure unlike anything on the familiar hemisphere. (Mare Nectaris may be of the same basic type, but the features are much less striking, and much less well-preserved, than with Orientale.) On the opposite limb of the Moon – the eastern limb as seen from Earth – Mare Smythii, Mare Marginis and Mare Australe all proved to be rather more extensive than had been supposed.

On this map, many features are named. These names were allotted in 1970 by the relevant committee of the International Astronomical Union. The familiar system of naming craters after men and women of science has been followed, one notable addition being Jules Verne, the crater to the lower right of Tsiolkovskii.

The Mercator chart brings out the arrangement of the craters; there are lines, chains and pairs. The overall aspect is, of course, strongly affected by the lack of *mare* areas, and this is extra evidence that the distribution of the surface features is not random.

Moon Panorama : The Far Side

Mapping the Far Side

Though a total of 59 per cent of the Moon's surface can be seen from Earth at one time or another, it is not true to say that 59 per cent of it can be well mapped. Near the limb, in the libration zones, the foreshortening is so marked that accurate charting becomes impossible. It was only in the years following 1959, when Luna 3 made its first journey round the Moon, that lunar mapping entered what can be called its modern phase. Not only the libration zones, but also the permanently averted 41 per cent, could be studied, though good photographs were not obtained until the U.S. Orbiters of the mid-1960s.

However, even the first Luna 3 photograph showed that there was an important difference between the familiar and the unseen areas. On the Moon's far side there were no *maria* comparable with Imbrium or even Crisium. The only feature which was regarded as a true *mare* was that which the Russians named Mare Moscoviense; still regarded as a sea, it has a dark floor covered with *mare* material but in size is no larger than the crater Bailly.

With the development of the Orbiters, it was also found that there are large enclosures of mare size which are not dark-floored; they have been called "empty seas", though the term should not be taken at all literally. There are no mountain ranges comparable with the Apennines, which is only to be expected in view of the fact that important ranges are almost always associated with regular maria.

There has been considerable discussion as to the cause of the difference in aspect between the two sides of the Moon. Since it is likely that the lunar rotation has been synchronous from a fairly early stage in the evolution of the Earth–Moon system, the arrangement is not too hard to interpret if the main features (both craters and maria) are endogenic, since the gravitational pull of the Earth must have played a major rôle; note also that, on the visible side, the most important walled formations tend to line up with the Moon's central meridian. If the main cratering process were exogenic, the differences would be much less easy to understand.

Mare Orientale

Though none of the connected maria extend over the limb of the Earth-turned hemisphere, some of the others do. The most striking of these is the Mare Orientale, which has proved to be a vast and complex structure unlike anything else on the Moon; this came as a major surprise, since, from Earth, it gave the impression of being nothing more than a minor feature of Mare Humboltianum or Mare Marginis type. But its effects are widespread, and in lunar geology it may be as important as the Mare Imbrium.

The main features on the far side have now been named. It is true to say that our charts of the areas turned away from the Earth are now as complete and accurate as those of the side of the Moon which we have always known.

General View *above*
The far side of the Moon (Orbiter). Tsiolkovskii is the feature near the limb, to the upper right; it is recognizable because of its dark floor and bright central mountain. The long crater-valley near the centre of the photograph is in the south polar region, near the crater Hale.

Mare Orientale *right*
The Mare Orientale, as photographed from Orbiter 4. The Cordilleras, Rook and d'Alembert Mountains prove to be associated with it. To the right is Grimaldi, easily recognizable because of its dark floor; Riccioli lies immediately upper left of Grimaldi.

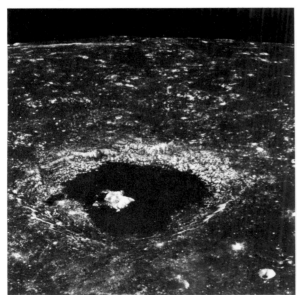

Tsiolkovskii *above*
Though Tsiolkovskii has a prominent central peak, the floor is partly flooded with mare material; it may represent a link between the walled plains and the maria. It is unique on the Moon's far hemisphere.

Another View of Tsiolkovskii *above*
This view of Tsiolkovskii was taken under conditions of higher illumination, but the dark floor is still much in evidence. To the left of Tsiolkovskii is a large formation with a light, rough floor; Tsiolkovskii intrudes into it.

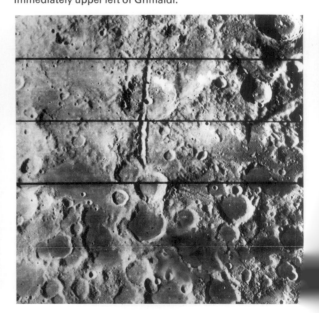

Polar Valley *above*
Enlargement of a part of an Orbiter 4 photograph of the south polar region just beyond the limits of visibility from Earth. The polar valley cuts through several large craters, and is itself cut by a presumably younger crater.

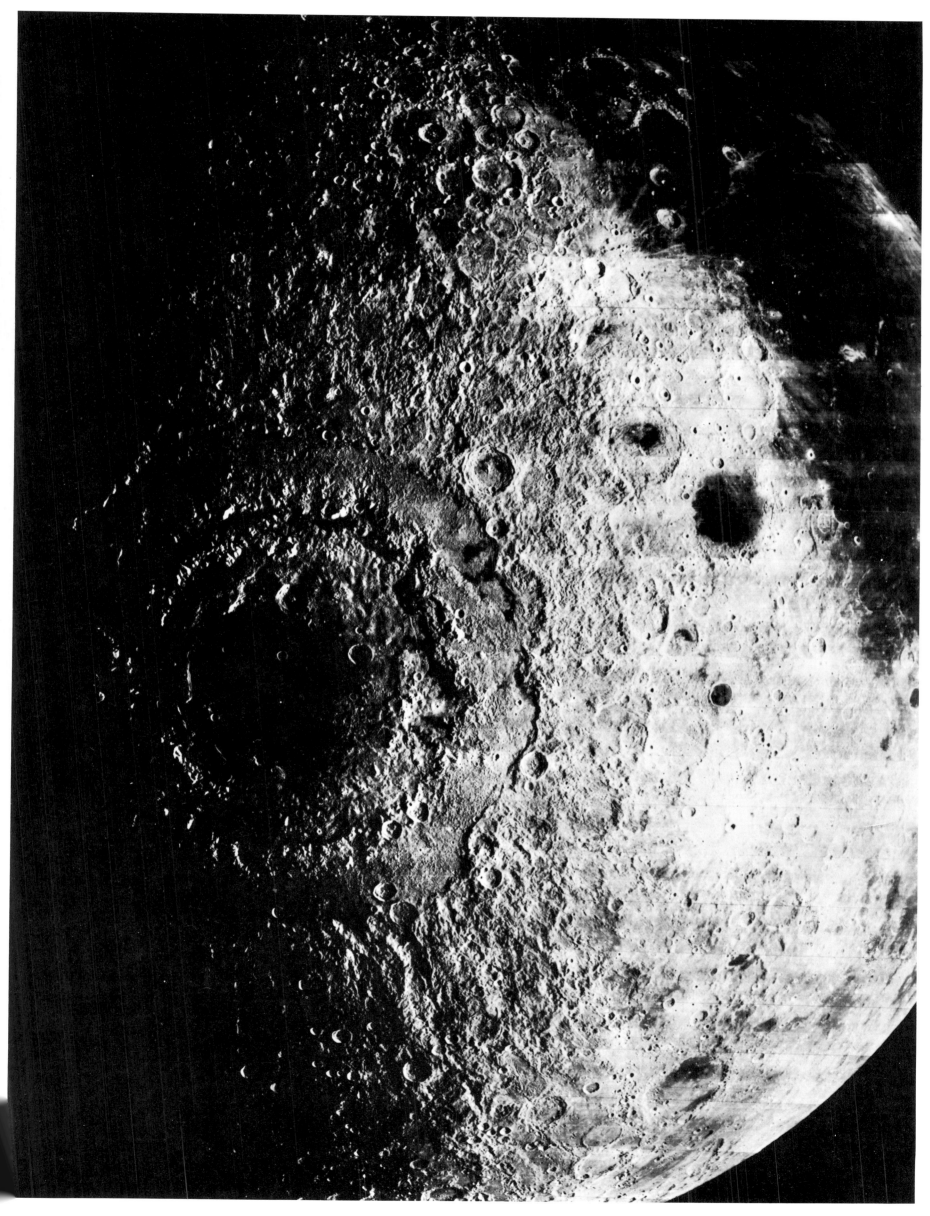

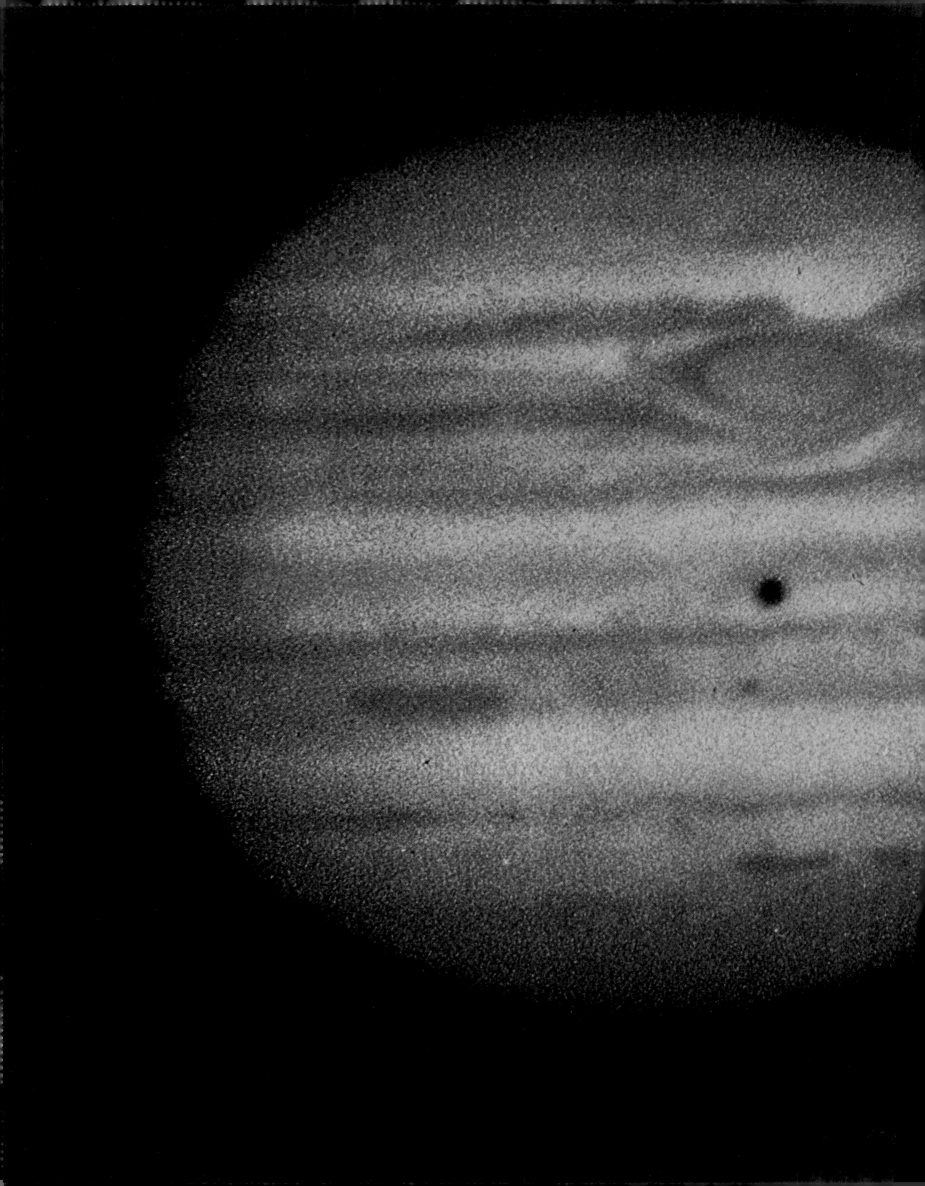

Atlas of the Solar System

It is impossible to say when the planets were first found to be different in nature from the stars. Pythagoras, about 580 B.C., certainly knew it, but the discovery may have been made much earlier. The word "planet" really means "wandering star"; ancient peoples knew that these individual move-ments must mean that the planets are closer to us than the so-called fixed stars. They also deduced that the Moon is much the nearest body in the entire sky.

The nature of the Sun was much more problematical. Anaxagoras, one of the most famous Greek philosophers, was accused of heresy because he taught that the Sun is a red-hot body larger than the Peloponnesus, and was banished from his home city of Athens. Moreover, the ancients could make very little progress in understanding the Solar System, because they believed that the Earth must lie in the centre of the universe, with every-thing else revolving round it once a day.

Jupiter
This colour photograph of the planet Jupiter—probably the best that anyone has been able to obtain—was taken by G. P. Kuiper with the 61-in. reflector at the Catalina Observatory, Texas.

The Copernican Revolution

In 1543 a Polish churchman, Copernicus, published a book called *De Revolutionibus Orbium Coelestium* (Concerning the Revolutions of the Celestial Bodies). It marked the start of a new era, and began a violent controversy which lasted for more than a hundred years. Copernicus rejected the Ptolemaic theory and proposed instead a plan in which the Sun, not the Earth, occupied the centre of the planetary system.

Copernicus was not an observer, but he was a skilled mathematician, and he objected to the artificial and clumsy Ptolemaic scheme. If the Sun lay in the central position, many of the complexities could be removed at one stroke. In this Copernicus was correct; but in his book he retained the concept of perfectly circular orbits, so that he was even reduced to bringing back epicycles.

Copernicus versus Ptolemy

Copernicus was well aware that his ideas would not meet with the approval of the Church. To regard the Earth as anything but supreme would be to invite a charge of heresy. Prudently, Copernicus refused to publish his book until the last part of his life; it is said, possibly with truth, that he received the first printed copy just before he died.

His fears were well-founded. To teach that the Earth moved round the Sun was regarded as heresy of the worst kind, as some luckless Copernicans found out to their cost. One, Giordano Bruno, was burned at the stake in Rome in 1600 (though it must be added that his defence of Copernicanism was only one of his many crimes in the view of the Inquisition). Finally, Copernicus' great work was placed on the Papal Index of forbidden books – and it remained there until 1835!

Tycho Brahe

Ironically, the next main character in the story was no Copernican. He was a Danish nobleman, Tycho Brahe, who refused to believe that the Earth could be anything but pre-eminent. On the other hand he realized that the old Ptolemaic system must be wrong, and he substituted a theory of his own, which was in the nature of a hybrid. He was also a firm believer in astrology, and he was in many ways eccentric as well as hot-tempered. Yet as an observer he was supreme, and between 1575 and 1595 he compiled a star-catalogue which was much more accurate than any previously drawn up. In addition he made excellent measurements of the positions of the planets, particularly Mars. When he died, in 1601, his observations came into the hands of his last assistant, Kepler, who used them well.

Kepler's Laws

Kepler had complete faith in Tycho's observations, and he found that the measurements of Mars could not be made to fit any circular orbit, whether round the Earth or round the Sun. At last, in 1609, he found the answer. The planets move round the Sun in orbits which are not circular, but elliptical – though for all the bright planets the paths do not depart greatly from the circular form; with Mars, the eccentricity is 0·093, while with the Earth it is only 0·017.

From these results, Kepler was able to draw up the three famous Laws of Planetary Motion which bear his name, and upon which all subsequent work has been based. He was not persecuted for his beliefs, though in many ways his life was an unhappy one. He died in 1630.

From Galileo to Newton

The Copernican revolution was not yet over. One of the strongest supporters of Copernicanism was the great Italian mathematician Galileo Galilei, best remembered for his pioneer telescopic observations. He was outspoken and tactless, and, moreover, failed to provide definite proof of the Copernican system; he was summoned to Rome by the Inquisition in 1633, and made to "abjure, curse and detest" the absurd view that the Earth moves round the Sun. He was subsequently kept under surveillance until his death 9 years later.

Galileo was the last scientist to be persecuted for teaching that the Earth is an ordinary planet. Newton (1642–1727) published his book the *Principia* in 1687; in this, he laid down the laws of gravitation, and the revolution in human thought was complete.

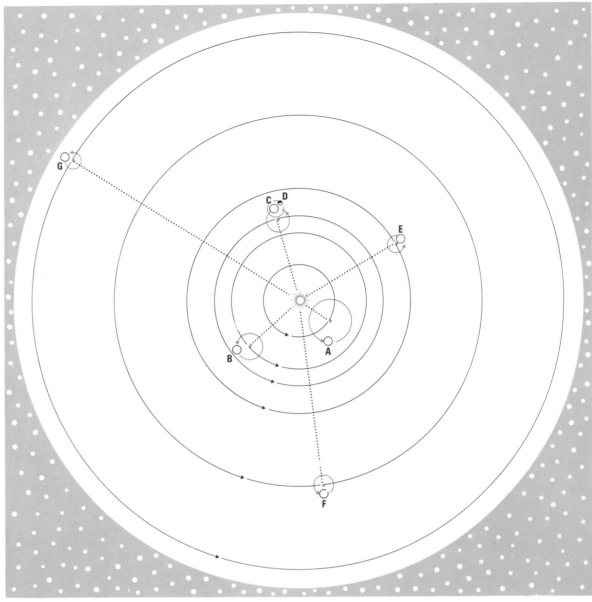

The Copernican Theory *above*
According to the Copernican theory, the Sun lies in the centre of the Solar System, with the planets moving round it – Mercury (A), Venus (B), Earth (C), Mars (E), Jupiter (F), Saturn (G) : only the Moon (D) is in orbit round the Earth. Circular orbits were retained, and the planets' irregularities were accounted for by epicycles.

The theory had probably been drawn up by 1530, but Copernicus refused to publish it, and was finally persuaded to do so only at the insistence of Georg Rhæticus, Professor of Mathematics at Wittenberg. Wisely, Copernicus dedicated the book to the Pope (Paul III), but the publisher, Osiander, added an unauthorized preface to the effect that the Copernican theory was merely a "mathematical fiction" which would be useful in predicting the positions of the planets.

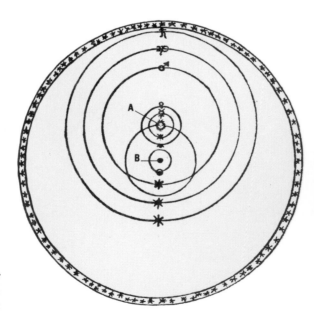

Tycho Brahe *above and right*
Tycho Brahe was born in 1546 and died in 1601. He refused to believe in the Copernican theory, and held that while the planets moved round the Sun (A), the Sun must itself be in orbit round the Earth (B). Tycho was undoubtedly the greatest of all observers of pre-telescopic times. He observed the supernova of 1572, and proved that the brilliant comet of 1577 was much more distant than the Moon.
right A room in his observatory at Uraniborg, provided by the generosity of King Frederick II of Denmark, incorporating Tycho's great quadrant.

Johannes Kepler

Johannes Kepler (1571–1630) was both mathematical genius and astrological mystic. After lecturing at the Universities of Tübingen and Gratz, he joined Tycho Brahe at his observatory.

After Tycho's death, Kepler used his observations to prove that the Sun, not the Earth, is the centre of the planetary system. His first Law (illustrated below) states that the planets move round the Sun in ellipses; the Sun occupies one focus (f1) of the ellipse, while the other focus (f2) is empty.

Yet despite his brilliant and far-sighted mathematical work, Kepler's outlook was far from modern. He believed that the planets emitted musical notes as they moved (the "music of the spheres"), and he linked their paths with five regular solids, which he thought could be fitted between their orbits *right*.

Imagine a cube whose corners touch the inside of a sphere and inscribe another sphere inside this cube. Inside this sphere place a tetrahedron with its corners just touching. Inscribe yet another sphere inside this containing a dodecahedron. This encloses the fourth sphere which contains an icosahedron. The sphere which lies within this encloses an octahedron. The sixth and

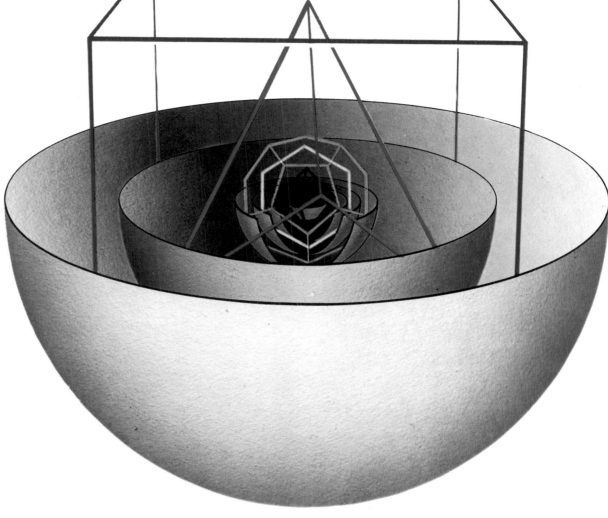

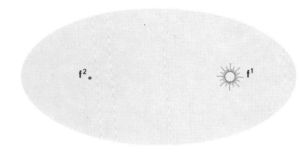

final sphere should lie within this but Tycho was forced to resort to a circle. The radii of the circle and the spheres thus obtained were proportional, he hoped, to the radii of the planets' orbits. He elaborated further variants of this idea without great success.

In 1618 he published his book called *Harmonice Mundi* (The Harmony of the World), which contained the important Third Law of Planetary Motion, but which was mixed up with quite unfounded mystical speculations. This sort of attitude – a cross between the medieval and the modern – was typical of Kepler.

Kepler's Five Solids *right*
The five regular solids, which Kepler associated with the orbits of the planets:
(1) Cube (2) Tetrahedron
(3) Dodecahedron
(4) Icosahedron
(5) Octahedron.

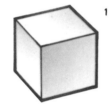
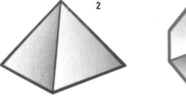

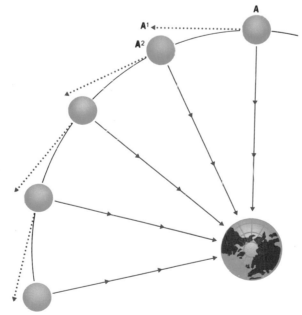

Newton *above*
Isaac Newton was one of the most brilliant mathematicians of all time. In 1687 he published the *Principia*, in which he laid down the laws of gravitation and explained many other phenomena – such as the tides, which had always been regarded as puzzling. Newton's book has been described as the greatest mental effort ever made by one man, and it ushered in the modern era of astronomy.

Newton was also interested in chemistry, biblical chronology and theology, and was elected President of the Royal Society in 1703.

Centripetal Acceleration *above*
Newton showed that the almost uniform circular motion of the Moon is an instance of accelerated motion toward the centre of the Earth. In the diagram above the Moon would move from A to A¹ if undisturbed. In fact it moves to A². There has hence been a change of velocity in the direction of A¹–A² which has a component towards the centre of the Earth. It is the Earth's gravitation which causes this acceleration and keeps the Moon in orbit about the Earth. The theory was first laid down by Newton in his *Principia*.

Greenwich Observatory *above*
Astronomy had always been known to be the basis of all navigation. Sailors found it difficult to find their longitude when out of sight of land. The only way to do so was by astronomical observation, and this meant using a good star catalogue. In 1675 King Charles II authorized the founding of Greenwich Observatory, and an accurate catalogue was compiled by the first Astronomer Royal, John Flamsteed. Old Greenwich is now a museum; in 1957 the scientific equipment was moved to Herstmonceux Castle in Sussex.

Exploring the Sun's Family

Though the movements of the planets could be worked out by naked-eye observations, as had been shown by Tycho Brahe, Kepler and others, some optical aid was essential if anything were to be discovered about the planets themselves. In the early years of the 17th century telescopes were invented, and Galileo and others applied them to astronomy.

Galileo was not the first telescopic observer, but he was by far the most skilful and energetic. He made a telescope which magnified 30 times, and even with his original telescope, which was less powerful, he was able to make a whole series of startling discoveries. In the winter of 1609-10 he observed the mountains and craters of the Moon, the satellites of Jupiter, the stars in the Milky Way, and the phases of Venus.

Galileo's Observations

Galileo was particularly interested in Venus, which he found to show lunar-type phases from new to full. This was particularly significant, as at this time the argument between the Ptolemaic and Copernican

Kepler *right*
Johannes Kepler (1571–1630), the German mathematician who drew up the famous Laws of Planetary Motion. From an old woodcut.

systems was still raging, and Galileo had already come out strongly in favour of the new ideas. On the Ptolemaic theory, Venus always remained close to the line joining the Earth to the Sun, and so could never appear as a three-quarter or full disk. Galileo's work showed at once that the Ptolemaic theory, at least in its original form, could not be correct. Additional evidence came from Jupiter, the giant planet, which was found to be attended by four moons or satellites. This showed that the Earth could not be the only centre of motion in the universe, as had always been maintained. Galileo's observations caused great consternation among churchmen, and one official even refused to look through the telescope, on the grounds that it must be bewitched. Galileo commented that when the official died, he should at least have a good view of Jupiter's moons while on his way to Heaven!

Galileo's telescopic observations were many and varied. He also studied the Sun, and was one of the earliest discoverers of the sunspots, which he knew to be features of the solar disk. But as soon as the first flood of discoveries had been made, there was bound to be something of a lull, and the next major developments did not come until the middle of the 17th century, by which time telescopes had been considerably improved. Excellent observations were carried out by C. Huygens in Holland and by G. D. Cassini in France, resulting in the discovery of features such as Saturn's ring-system (in its true guise) and the polar caps of Mars. Another feat of Cassini's was to record the Zodiacal Light, a cone-shaped glow now believed to be an effect resulting from the presence of interplanetary debris.

Even more important was Cassini's determination of the distance of the Sun. Earlier, Kepler had given 14 million miles; Cassini increased this to 86 million miles, which was of the right order. The scale of the Solar System was becoming known, and the distances of the five naked-eye planets were worked out with reasonable accuracy.

Newton and Halley

The work of Isaac Newton gave the final death-blow to the Ptolemaic theory, and his great work known generally as the *Principia*, published in 1687, ushered in what we may call the modern era. Newton was sensitive and retiring. His work was written at the insistence of his colleague Edmond Halley, later to become Astronomer Royal and to achieve lasting fame because of his successful prediction of the return of the comet which now bears his name. Halley's

Halley *left*
Edmond Halley (1656–1742), the second Astronomer Royal, who was responsible for persuading Isaac Newton to publish the *Principia*. Halley was an expert mathematician and observer in his own right.

reputation is naturally somewhat overshadowed by that of Newton, but there is no doubt that astronomy owes a great deal to him.

William Herschel

Five planets had been known since earliest times, and these, together with the Sun and the Moon, made a grand total of seven. Since seven is the mystic number of the astrologers, no new planet was

expected. It thus came as a great surprise when, in 1781, a then unknown amateur named William Herschel, born in Hanover but an immigrant to England, discovered the planet we now call Uranus, moving round the Sun at a distance much greater than that of Saturn. Herschel did not immediately recognize its nature; he thought that it must be a comet, but its motion against the starry background soon betrayed its true nature.

Herschel's reputation was made, and he spent the rest of his life in astronomical research. His main work was in connection with the stars, and as an observer he has seldom been equalled. Yet his ideas about life in the universe were strange by modern standards: he believed that there must be intelligent

Herschel *left*
Sir William Herschel (1738–1822), who discovered the planet Uranus with a telescope of his own construction.

beings on the Moon, the planets, and even inside the Sun! It is true that these views were not shared by most of his contemporaries, but when we remember that Herschel died as recently as 1822 we can appreciate how great has been the advance in our knowledge since his day.

The Solar System was still not complete. Four small planets, now known as the asteroids were discovered in the first decade of the 19th century. Then, "Herschel's planet", Uranus, was not moving as it had been expected to do; something was pulling

it out of position. In England, J. C. Adams suspected that the perturbations must be due to an unknown planet. In France, Urbain Le Verrier undertook the same investigation, and it was on the basis of Le Verrier's work that J. Galle and H. d'Arrest, in Berlin, identified the planet we now call Neptune.

20th-Century Discoveries

Much later, in 1930, the ninth planet, Pluto, was discovered in the same way, this time as a result of calculations made by Percival Lowell in the United States. Whether there is yet another planet in the depths of the Solar System remains to be seen.

Of course, Man's understanding of the universe is linked with our ideas about the status of the Sun. For some time after the general acceptance of the Copernican system it was thought that the Sun must be of special importance, but the work of Herschel and others showed that this is not so; the Sun is nothing more than an ordinary star. Neither is it "burning" in the accepted meaning of the term. It is

Cassini *left*
G. D. Cassini (1625–1712), the Italian astronomer who made many important discoveries regarding the planets and first accurately determined the distance of the Sun from the Earth.

Le Verrier *below*
Urbain Le Verrier, the great French astronomer whose calculations led to the discovery of Neptune by providing a basis on which later investigators could work.

producing its energy by means of nuclear transformations, though this was not discovered until our own century; the pioneer work was carried out by H. Bethe and C. von Weizsäcker in 1939.

Meantime, there had been important advances in the study of the Moon and planets. Surface details were seen on Mars, and it was even thought that some of the features there might be due to the activity of intelligent beings – a theory which was not finally rejected until after the death of Lowell, its chief protagonist, in 1916. Venus was found to have a dense, cloudy atmosphere which hid the surface permanently, while Jupiter and the other giant planets are gaseous, at least in their outer layers.

Until modern times, all studies of the planets had to be carried out visually. By the end of the 19th century excellent photographic charts of the Moon had been produced, but physical observations of the Moon and planets were left largely in the hands of amateurs, who produced excellent work. It was only after the development of space research methods, after World War II, that the whole situation changed.

Artificial Probes

The first artificial satellite was launched in 1957, and two years later the Russians dispatched their first successful lunar probes, one of which (Luna 3) went round the Moon and sent back photographs of those regions of the Moon which are always turned away from the Earth, and had therefore never been seen before. Next came the first attempts at sending probes to the planets. Here the Americans achieved the first triumph, with Mariner 2 in 1962, when they sent a vehicle past Venus and obtained the first reliable measurements of the temperatures there. More recently, probes have been sent to Mars, Venus and even Jupiter. Vehicles to other planets will certainly be launched in the near future.

Future Prospects

What, then, of the future? Our ideas of the bodies in the Solar System have been altered dramatically since the opening of the Space Age in 1957. It now seems improbable that there is any life in the Sun's system, except for that on Earth; the most we can hope for is a certain amount of primitive organic matter on Mars, and even this now appears unlikely. So far as the Moon is concerned, we have the direct testimony of the astronauts who have been there, and by 1976 we should have information sent back from the surface of Mars by an American automatic probe.

Great care must be taken to avoid carrying terrestrial contamination to other worlds (particularly to Mars, which has an atmosphere and which may not be completely sterile). This is a point which has not always received the attention that it merits. But the most important prospects, perhaps, involve the setting up of research bases upon other worlds, beginning with the Moon and progressing to Mars and then to other planets. When this can be done, the

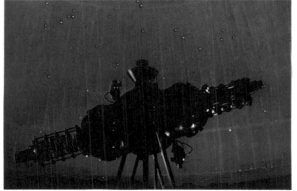

Orrery and Planetarium *above*
top An old orrery, made in 1690, showing the Sun with Mercury, Venus, the Earth and the Moon.
above The projector of the Armagh Planetarium, in Northern Ireland – the modern development of the orrery. In a planetarium, the celestial bodies are represented by lights projected on to the inside of a large dome.

benefits to science will be tremendous, and this in turn will benefit mankind.

Life Elsewhere in the Solar System?

There has been much discussion as to the possibility of life elsewhere in the Solar System. The hostile nature of the Moon is very evident, but until recent times it was still thought likely that there might be living organisms upon Venus and Mars. The modern view is, however, different. Venus has a very high surface temperature, while the conditions even upon Mars, the most earth-like of the planets, do not seem suitable for life in any form. In all probability, there are no living things in the whole of the Solar System except on Earth.

We are still at an early stage in our exploration of the Solar System, and much remains to be learned, but a start has been made. We are gaining knowledge rapidly, and we may hope for striking developments before the 20th century draws to an end.

The Solar System: Introduction

Nature of the Solar System

Look up into the sky on any dark, clear night, and you will see the glorious band of light that we call the Milky Way. The stars in it are not genuinely crowded together, even though they may look as though they are almost touching each other. Our star-system, or Galaxy, is a somewhat flattened system; the Sun, with its family of planets, lies nearly in the main plane – and when we look along the main "thickness" of the system, we are in fact seeing many stars in approximately the same direction.

Altogether the Galaxy contains about 100,000 million stars, of which our Sun is undistinguished. In size, luminosity and in other ways it is quite average. It is important to us only because we happen to live on a planet moving round it.

The Sun is 30,000 light-years from the centre of the Galaxy. (A light-year is the distance travelled by a ray of light in one year; it is equal to 5·886 million million miles.) The whole Galaxy is in a state of rotation, and the Sun is moving round the centre, taking 225 million years to complete one journey – a period which is sometimes known as the "cosmic year". One cosmic year ago, the Earth was going through its Coal Forest period, and the highest forms of life were amphibians; even the great dinosaurs still lay in the future.

At present the Sun is travelling toward a position marked in the sky by the constellation Hercules, but this motion is not dangerous, since the Sun is a long way from its nearest stellar neighbour. The closest star is a red dwarf known as Proxima Centauri, at 4·25 light-years (parallax 0″.765). The brilliant star Sirius, at 8·6 light-years, is one of our near neighbours but the closest stars which are at all similar to our Sun in size and luminosity, Epsilon Eridani and Tau Ceti, lie at 11 light-years from us.

The Sun is the major body of the Solar System, which is made up of the nine planets and their satellites as well as comets, and various minor bodies such as meteorites and meteors. The outermost planet, Pluto, moves at a mean distance of over 3600 million miles from the Sun. Yet how far does the Solar System really extend? In all probability, some of the great comets travel out to distances of at least 2 light-years before starting their return journey toward the Sun; if so, the Solar System is vast indeed, though it makes up a very small part of the Galaxy.

Formation of the Solar System

The Earth is about 4700 million years old. The Sun is obviously at least as old as the Earth, and is probably older; but we have to confess that there is no general agreement as to the way in which the planets were formed. Many theories have been proposed. One, due largely to the late Sir James Jeans, involved a passing star, which was assumed to make a close approach to the Sun and to tear a tongue of material away from the solar surface; after the stranger retreated, the Sun was left with a whirling "cigar" of material, which broke up and condensed into planets. This theory sounds plausible, but mathematical objections have been raised, and the idea has been given up.

According to C. von Weizsäcker, the planets were formed by accretion (that is to say, gradual building up) from material collected by the Sun during its passage through a nebula, or cloud of dust and gas in space. Alternatively, it may be that both the Sun and planets came from what may be called a "solar nebula", and it may be that magnetic effects have been very important, as the Swedish astronomer H. Alfvén believes. But the whole problem is still unsolved. All we can say is that according to the best available evidence, all the planets were formed at about the same time.

Solar systems are probably very common in the universe. What can happen to the Sun can also happen to other stars. It is also logical to assume that many planets of other stars are inhabited. In our Solar System, only the Earth seems to be able to support advanced life-forms; it lies in the centre of the Sun's "ecosphere", or region where the temperature is neither too hot nor too cold for life to flourish. Of the other planets, Mars is on the outer edge of the ecosphere and Venus on the inner edge. Our neighbour worlds are not welcoming, but each has its own points of interest, and there is plenty of variety in the Solar System.

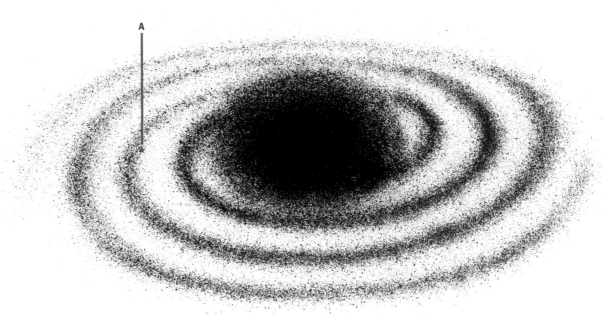

The Galaxy *above*
The drawing above is a representation of our Galaxy as it would appear to anyone observing it from a great distance; it takes the form of a loose spiral.

The Sun (A) lies near the main plane, but 30,000 light-years from the centre of the system. It is in motion round the centre of the Galaxy, and takes 225 million years to complete one journey. The diameter of the Galaxy is about 100,000 light-years.

Our Galaxy is the second largest system in the Local Group; it is surpassed only by the Andromeda Galaxy, M.31.

The Milky Way *left*
This photograph, taken with the 200-in. reflector at Palomar, shows a very small part of our Galaxy. Even so, it contains a very great number of suns, many of which are more luminous than our Sun.

The view of the Galaxy that we have from Earth cannot be complete, owing to the obscuring effect of the interstellar matter. We cannot see to the centre of the Galaxy, though it is possible to study the galactic nucleus by means of its radio emission.

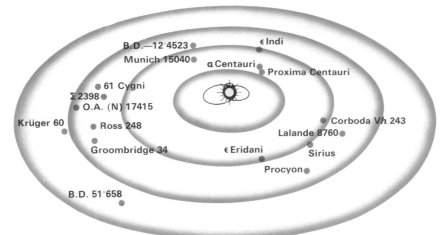

The Sun's Neighbours *above*
The diagram shows a very small area of the Milky Way, and represents the nearest stars to the Solar System. Proxima Centauri, lying at 4·25 light-years, is the nearest, and there are several stars within 10 light-years of us. It is possible that some comets move out to at least 2 light-years from the Sun, as shown in the diagram.

The Sun and the Planets *right*
A part of the Sun, showing its curvature, in relation to the nine planets, drawn to scale. Though the Sun is an ordinary star, it can be seen that it is much larger than any planet.

Mercury   Venus   Earth

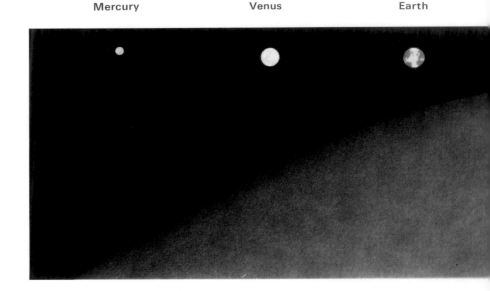

The Ecosphere *above*
Diagram of the ecosphere – the area around the Sun where the temperature is neither too hot nor too cold for life to exist, provided that other conditions are suitable for it. It is seen from the diagram that only the Earth lies near the centre of the ecosphere. Mars is close to the outer limit, so that it is too cold to be suitable; Venus is close to the inner limit, so that it would be too hot even if it had a more welcoming atmosphere.

| Mars | Jupiter | Saturn | Uranus | Neptune | Pluto |

Map of the Solar System

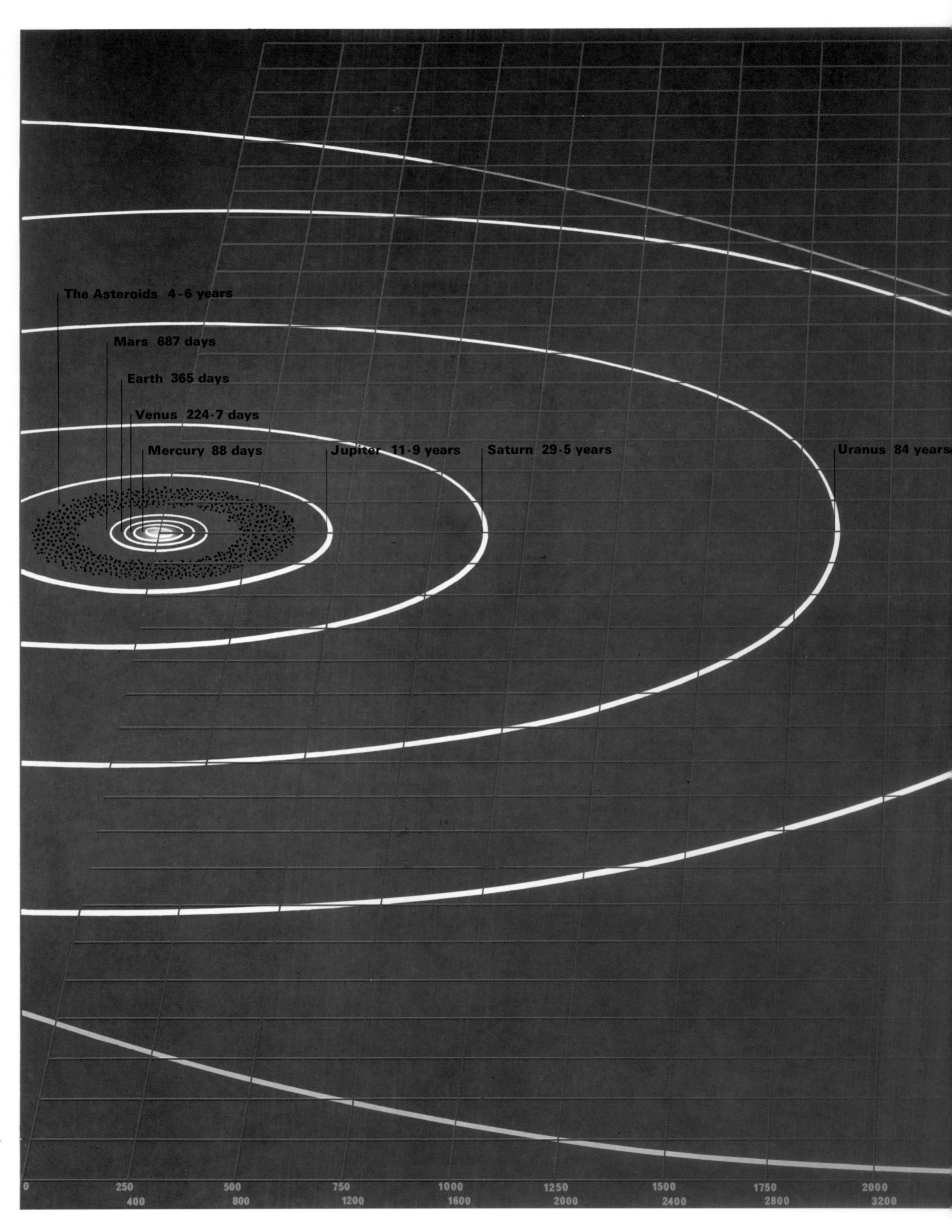

The Asteroids 4-6 years

Mars 687 days

Earth 365 days

Venus 224·7 days

Mercury 88 days   Jupiter 11·9 years   Saturn 29·5 years   Uranus 84 years

0	250		500		750		1000		1250		1500		1750		2000
	400		800		1200		1600		2000		2400		2800		3200

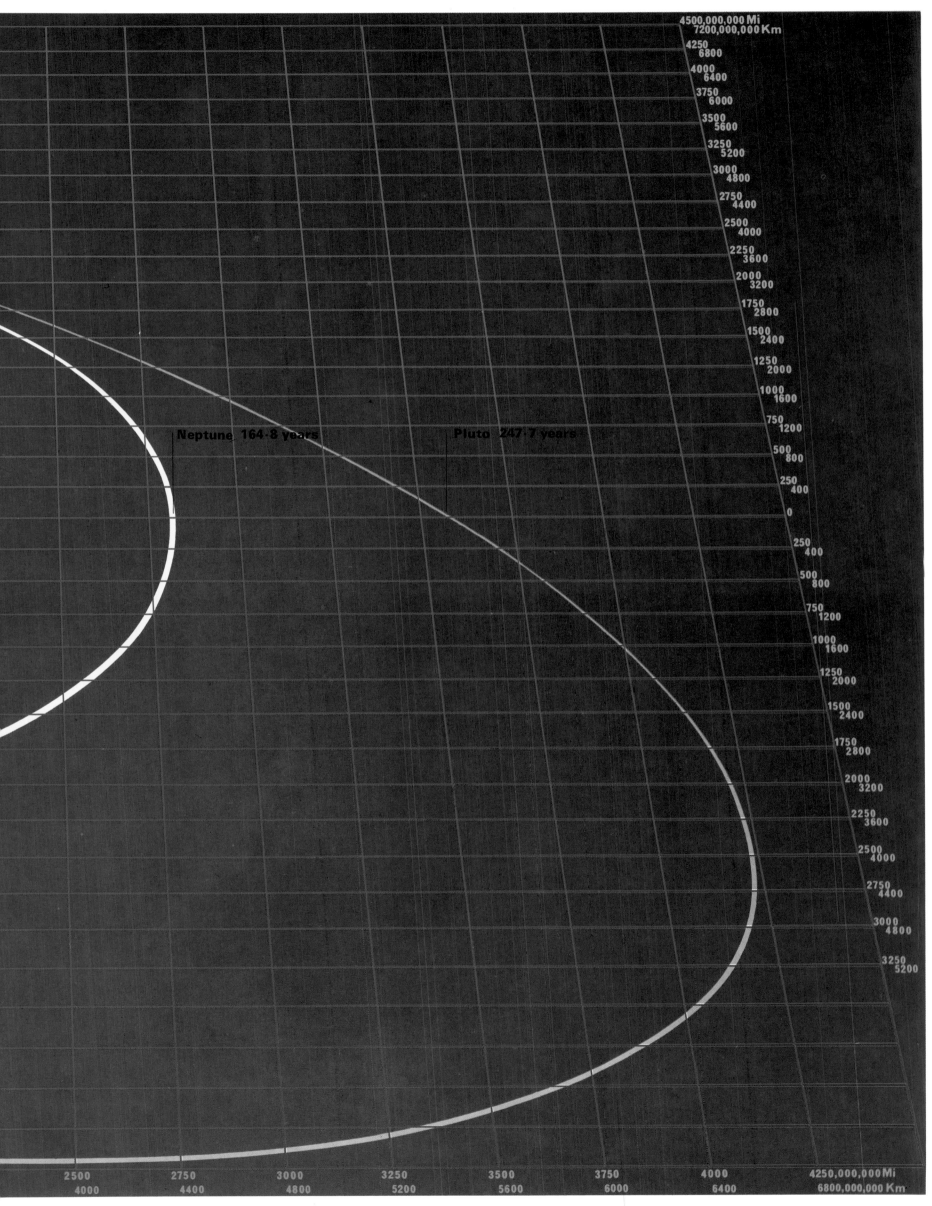

Neptune 164·8 years

Pluto 247·7 years

The Sun: Introduction

The Sun as a Star

The Sun is a star. It is undistinguished in the Galaxy, but to us it is all-important; without it, no life on Earth could survive. It is easy to understand why ancient peoples worshipped it as a god.

The idea that the Sun might be the centre of the planetary system goes back to Greek times, but there was no proof, largely because of the difficulty of measuring the Sun's size or distance. About 450 B.C. Anaxagoras, one of the greatest of the Greek philosophers, was expelled from Athens for daring to suggest that the Sun must be a red-hot stone larger than the Peloponnesus; earlier, it had been thought that the diameter of the Sun could hardly exceed 2 feet! Hipparchus, in the 2nd century A.D., estimated the Sun's distance as 1 million miles, but it was only in the 17th century A.D. that the first proper measurements could be made. We now know that the length of the astronomical unit, or Earth–Sun distance, is 93 million miles. Light takes 8·3 minutes to reach us from the Sun; therefore, we see the Sun not as it is "now", but as it used to be 8·3 minutes ago.

Compared with the Earth, the Sun is extremely large. Its diameter is 865,000 miles, and its volume is more than a million times greater than that of the Earth. The density is, of course, much lower; the specific gravity (that is to say, the density on a scale where water=1) is only 1·4. This is the mean value, but near the solar core the pressures and temperatures are very high, and the density also is high.

Sunspots and Faculæ

The Sun's surface is not smooth and featureless. There are darker patches known as sunspots, which are temporary, but which may become immensely large. The temperature of the bright surface, or photosphere, is 6000°C; that of an average sunspot is 2000° lower, which is why the spot seems dark. At times, many spot-groups may be seen simultaneously, whereas there are other periods when the disk is free of them. The Sun has a well-marked cycle of activity, spot-maxima occurring once in about 11 years.

Sunspot-watching is a favourite amateur occupation, but it must be stressed that there are dangers involved. To turn a telescope (or even a pair of binoculars) toward the Sun and then look direct is certain to result in permanent blindness, as has, unfortunately, happened on many occasions; Galileo, who was one of the first to observe sunspots (in 1610), damaged his sight in this way. Dark filters, to be attached to the telescope eyepiece, are unreliable. They cannot afford full protection, and are always liable to splinter abruptly, and so are best avoided. The only safe way to observe the Sun is to use the telescope as a projector, and view the solar image on a screen held or fixed behind the eyepiece.

Simple visual work can tell us how the sunspots behave, and there are other features to be seen, notably the bright faculæ, which lie in the Sun's upper atmosphere, and the granulations or mottlings on the disk. When the Moon passes in front of the Sun, at a total solar eclipse, the Sun's atmosphere flashes into view, and for a few moments we see the red chromosphere or "colour sphere" (the inner atmosphere) and the glorious pearly corona. But for regular solar work, the most important instruments are based on the principle of the spectroscope.

Solar Physics

In 1666 Isaac Newton split up sunlight by passing it through a glass prism. Much later, in 1814, J. Fraunhofer made the first scientific studies of the Sun's spectrum, and by now we have learned how the Sun shines; it is producing its energy not by burning, but by nuclear transformations near its core. It is losing mass at a rate of 4 million tons a second, but will not change much for at least 6000 million years.

Solar physics makes up a vitally important branch of modern astronomy. The Sun is an average star – and it is the only star which is close enough to be studied in detail with special instruments. In learning more about the Sun, we are also learning more about the stars. But for solar physics, our knowledge of the universe would still be painfully slender.

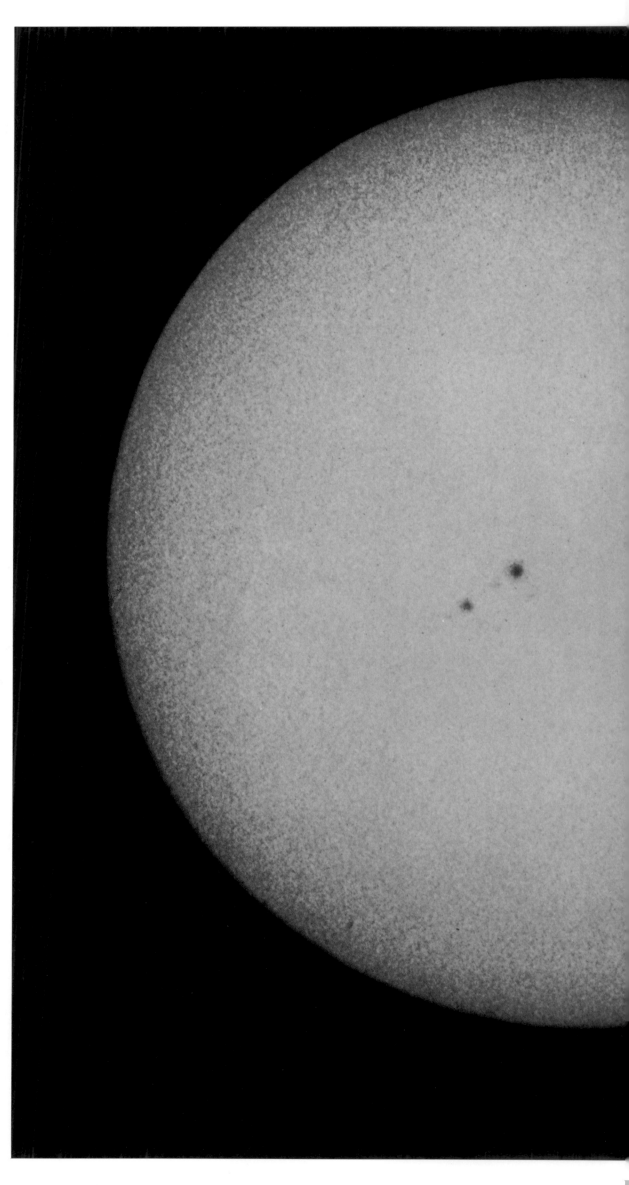

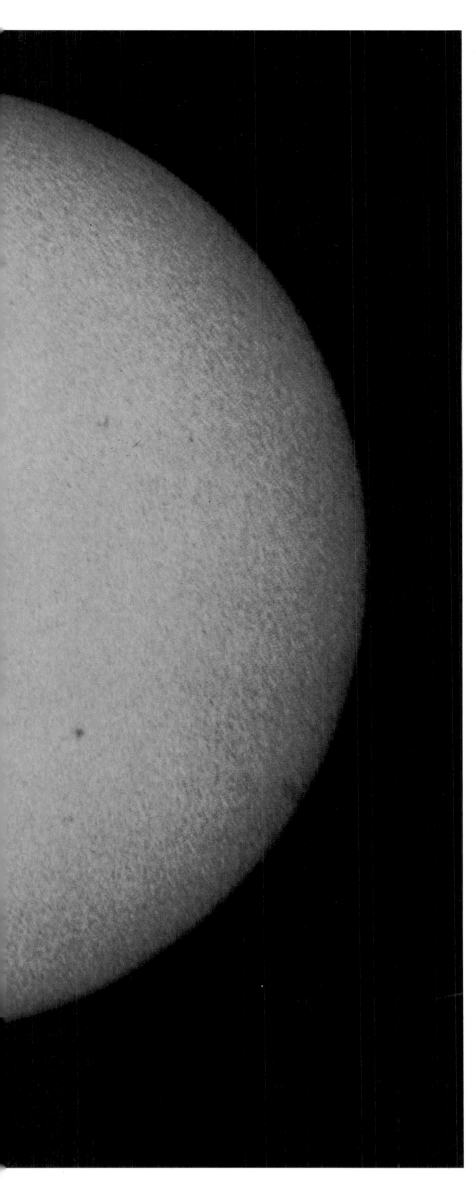

The Sun *left and above*
left Colour photograph of the Sun, showing the bright
photosphere, together with some spots; this is the view
that will be obtained through the telescope.
above The Sun's surroundings, as seen at the total eclipse
of 15 February 1961; photograph by W. Zünti,
Switzerland. The corona is well shown, extending to a
considerable distance; it is visible to the naked eye
only during totality.

Observing the Sun *right*
Warning! The Sun is so brilliant, and so hot, that to
look straight at it through any telescope is certain to result
in permanent blindness. Dark filters, for direct vision, are
unsafe, and should be avoided. The only safe method is to
project the Sun's image on to a screen, as shown in
the photograph (taken from Armagh during the partial
solar eclipse of 20 May 1966). In this photograph, the
screen was hand-held, but it is not difficult to fit up
a proper projection attachment. For solar work a refractor
is obviously more convenient than a reflector.

Size of the Sun *below*
The Sun is a large body; its diameter is 109 times that
of the Earth, as is indicated by the line-up of 109 Earths
across the bottom of the page. On the other hand, it is small
compared with some giant stars. Betelgeux in Orion,
for example, has a diameter of 250 million miles, as
against less than 1 million for the Sun; in the diagram,
immediately below, the Sun is shown together
with a segment of Betelgeux.

Sunspots

Discovery and Nature of Sunspots

In 1610, soon after the invention of the telescope, observations of sunspots were made by three men: Galileo and Fabricius in Italy, and Scheiner in Germany. Scheiner regarded them as small dark bodies moving round the Sun, but Galileo realized that they must be on the actual surface, the Sun not being so smooth or perfect as had been believed. Very large sunspots can be seen with the naked eye, and records of them go back to the time of ancient China. Since the Sun's surface is gaseous, a spot cannot be permanent; it may last for a few hours, days, or even weeks or months, but will then disappear.

A sunspot is not so dark as it looks. Its temperature is about 4000°C, roughly 2000° cooler than that of the bright surface or photosphere, so that it appears black by contrast; if it could be seen shining by itself, its surface brilliancy would be greater than that of an arc-lamp. Neither is it a true spot. There is considerable evidence that the average spot is depressed below the general level of the photosphere (Wilson effect).

A large sunspot is made up of a central area or umbra, surrounded by a lighter penumbra. Some spots are circular, but more often the umbra is irregular, the penumbra even more so; there may be many umbræ within an intricate area of penumbra.

Sunspot Groups

Single spots are not uncommon, but the general tendency is for spots to appear in groups. In a typical group there are two main spots: a leader and a follower, with smaller ones nearby. Though every spot-group has its own characteristics, an average sequence of events is as follows: The group begins as a pair of small "pores", which grow rapidly, and separate in longitude though remaining in much the same latitude. After 9 or 10 days, the group has reached a length of over 100,000 miles; the leader has become approximately circular, while the follower is irregular. Various other spots have appeared in the group. Then the decline starts, though it is less rapid than the increase. The follower is the first to go, together with the smaller members of the group; then it is the turn of the leader to disappear. Faculæ, which are bright patches on the Sun's surface, are often seen in areas where a spot-group is about to break out, and persist for some time after a group has disappeared.

No group can be followed continuously for more than about two weeks, because the Sun is spinning on its axis. The rotation period is just over 25 days at the solar equator, though it is appreciably longer at the poles. Therefore, a spot-group is carried across the disk from one limb to the other, taking $13\frac{1}{2}$ days to complete the crossing. It is then out of view, on the far side of the Sun, for a further $13\frac{1}{2}$ days before reappearing at the opposite limb. (A discrepancy is obvious here: twice $13\frac{1}{2}=27$, not 25, but the Earth is moving round the Sun at $18\frac{1}{2}$ miles per second, and so as seen from Earth the solar rotation appears to have a period of 27 days instead of 25.)

Sunspots and the Solar Cycle

No two spot-groups are alike, and there are rapid short-term changes. Also, a group seen near the limb will appear very foreshortened. Watching them from day to day, and taking photographs of them, is a fascinating pursuit. Yet there may be times when the disk is entirely free of them. There is a well-marked solar cycle of 11 years, which affects the numbers of sunspots. At solar maxima (as, for instance, in 1957 and 1968) there may be many groups visible at the same time; at minima (as in 1954 and 1965) there may well be none at all.

The cause of the solar cycle is not definitely known, and it must be admitted that the origin of sunspots is also uncertain. In 1926 Bjerknes, in Norway, suggested that below each hemisphere of the Sun there is a "primary vortex" under the globe, circling the solar globe as a parallel of latitude; when this vortex is bent upward and penetrates the photosphere, a spot appears. More recently H. Alfvén and C. Walén, in Sweden, have proposed a more complex picture, in which various zones inside the Sun lead to doughnut-shaped whirls which rise to the surface and open up as they reach the photosphere, producing sunspot pairs. In any event, sunspots are not simple holes in the bright surface. They are associated with strong magnetic fields.

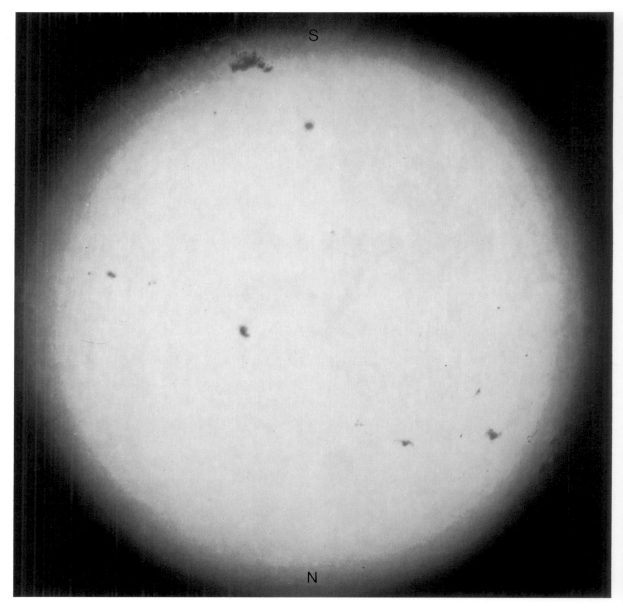

Size of Sunspots *below*
The illustration below shows the group of April 1947 ; this was the largest sunspot group ever recorded, and covered an area of 7000 million square miles. The Earth's comparative size is represented by the circle to the lower left.

A Solar Maximum *above*
Disk photograph of the Sun : 1 April 1958, taken at the Royal Greenwich Observatory, Herstmonceux. The solar maximum of 1958 was the most energetic ever recorded. The photograph shows a heavily spotted disk.

Sunspot Group *left*
The structure of a sunspot group can be very involved. The dark part of a spot is called the umbra ; the lighter surrounding area is the penumbra, but as seen on the illustration there may be many spots inside a large, irregular penumbral mass.

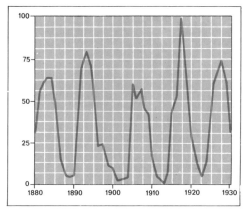

The Solar Cycle *above*
Spot and prominence frequencies show peaks every 11 years, though the cycle is not perfectly regular, and not all maxima are of equal intensity.

At the start of a new cycle, spots appear at high latitudes ; as the cycle progresses, the spots break out at lower and lower latitudes. Before the low-latitude spots of the new cycle have died out, high-latitude spots of the new cycle have already appeared. This is known as Spörer's Law, and is well shown in the "Butterfly Diagram" *above right*, in which latitudes of spots are plotted over more than a full cycle.

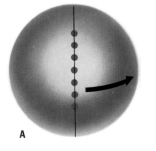 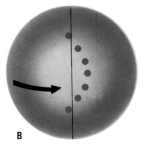

Apparent Drift *above*

The Sun does not rotate in the manner of a solid body. The rotation period at the equator is appreciably less than near the poles. Assume that originally the spots at various latitudes are lined up (Diagram A). After a spot on the equator has completed one full rotation, a spot at a higher latitude will not have had time to do so ; because of the slower rate of spin, it will lag behind (as in Diagram B). There is a difference of 1° per day in the Sun's angular velocity between the equator and latitude 35°.

Apparent Paths of Sunspots *below*

The Sun's equator is inclined to the ecliptic, and so the apparent paths of sunspots across the disk, due to the solar rotation, are not straight lines (B, C, D and F, G, H). Relative to a north–south line across the disk, the Sun's axis may lie at an angle of up to more than 26°E or W. The diagrams show the inclination of the solar axis, relative to the Earth, for different dates in the year. The paths of the spots across the disk appear to be straight lines only in early December (A) and early June (E).

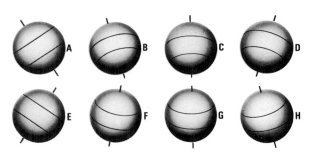

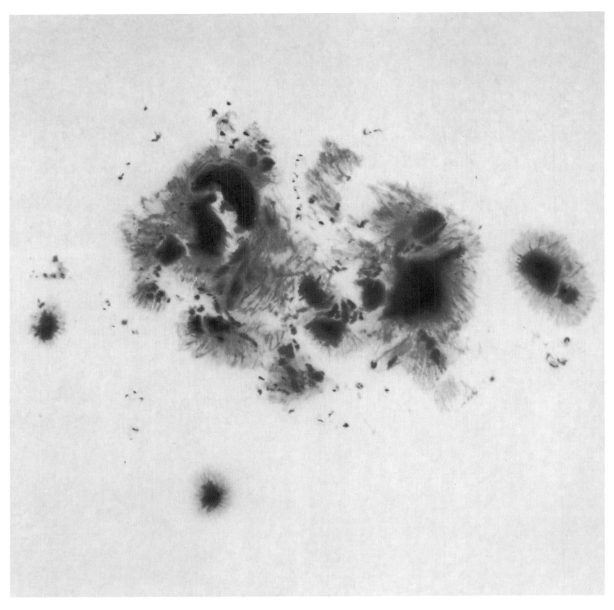

Magnetic Fields of Sunspots

The general magnetic field of the Sun is weak, but the magnetic fields of sunspots are strong. It has been found that with a two-spot group, the leader and the follower are of opposite polarity; in the opposite hemisphere, the situation is reversed. Thus if in the Sun's northern hemisphere the leading spot is a "north-seeker", the leading spot of a group in the southern hemisphere will be a "south-seeker".

At the end of one solar cycle, the situation will be reversed; the leader of a group in the northern hemisphere will be a "south-seeker", and so on. It seems almost as though the proper solar cycle is not 11 years, but 22. Single spots also have magnetic fields, and it has been found that the polarity is the same as would have been the case for the leader of a two-spot group.

There can be no doubt that the magnetic polarities of sunspots are of fundamental importance, but as yet we cannot claim to be in a position where we are able to interpret them fully.

Sunspot Group *above*

A sunspot group photographed on 17 May 1951 (Mount Wilson). At this time the Sun was well past the peak of its cycle of activity, which had occurred in 1947, but the group was extremely large and complex.

Granulations *right*

Each granule in the photograph has a diameter of about 500 miles ; they seem to be in constant motion, so that they are thought to be the tops of rising columns of hot, bright gas. Under good conditions, a modest telescope will show the mottled aspect of the Sun's disk.

The Wilson Effect *below*

In 1769 A. Wilson, of Scotland, found that with a spot which is near the limb (A) the penumbra always seems narrowest in the direction of the centre of the Sun's disk. Wilson concluded that the spots must be saucer-shaped, with the umbra at the bottom and the penumbra representing sloping sides. Measurements indicate that the average regular spot may be from 500 to 6000 miles below the level of the photosphere, but there is no general agreement among solar workers, and not all regular spots show the Wilson effect.

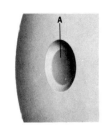  

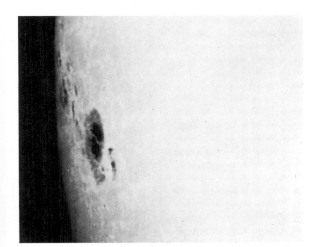

19 August
above A sunspot group near the limb, photographed on 19 August 1959 by W. M. Baxter (4-in. refractor). Note the Wilson effect, and the bright faculæ.

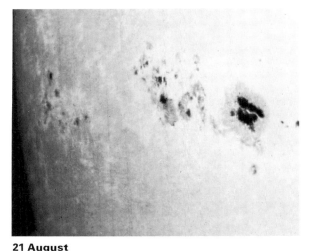

21 August
above The same group, 21 August ; it is now farther from the limb, and is well displayed. On this scale, the diameter of the whole Sun would be 36 inches.

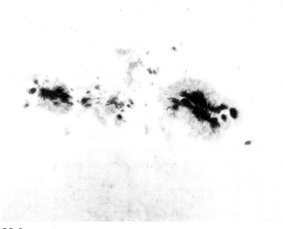

23 August
above The same group on 23 August. It is a typical two-spot group, and now that it is well away from the Sun's limb it can be seen to advantage.

The Source of the Sun's Energy

Solar Spectroscopy

The beginning of solar spectroscopy came in 1666, with some preliminary experiments carried out by Isaac Newton. Newton never followed up his work in this field, but in 1802 an English doctor, W. H. Wollaston, took matters further; he split up the Sun's light by using a glass prism, and saw that the resulting rainbow band, or spectrum, was crossed by dark lines. Wollaston failed to interpret these lines, and it was left to a young German optician, Josef Fraunhofer, to draw up the first reliable map of the solar spectrum. Fraunhofer's studies, which began in 1814, paved the way for modern solar physics.

A spectroscope is an instrument for analysing light. An incandescent solid, liquid, or gas at high pressure will yield a rainbow or continuous spectrum, while an incandescent gas at low pressure will produce isolated bright lines, forming what is termed an emission spectrum. Each line can be attributed to one definite element or group of elements. Thus two bright yellow lines in a special position (the D lines) are due to the element sodium; no other element can produce them. Each element and each compound has its particular set of trademarks which can be identified.

The Solar Spectrum

Spectroscopy can be applied to the Sun, but there is a difference; the continuous spectrum is crossed by lines which are dark, not bright. In 1859 the problem was solved by G. Kirchhoff, of Heidelberg. He found that under certain conditions, atoms are capable of absorbing radiation of the wavelength that they would normally emit. With the Sun, the bright surface (photosphere) produces a continuous rainbow. If they were shining on their own, the surrounding gases would produce bright emission lines; but since they lie between the photosphere and ourselves, they absorb the appropriate wavelengths and appear dark.

Consider, for instance, the double D line of sodium. In the yellow part of the Sun's spectrum there are two dark lines, corresponding in position and intensity to the D lines. There is no doubt that the association is genuine, and therefore there must be sodium in the Sun. By now, more than 70 elements have been identified in the solar spectrum.

The spectrum of the Sun is very complicated, and is not easy to interpret. Each element or compound may produce many lines – iron alone is responsible for thousands – and spectra of high dispersion are needed (that is to say, spectra which are widely spread out). Fortunately, the Sun is so bright that there is plenty of light, and special instruments have been devised to analyse the light when dispersed. In some cases the sunlight is collected by a mirror and directed into an underground laboratory, where the actual observations and analyses are made. Nothing of the kind is possible when studying the spectra of the stars, where the light amounts available for investigation are so much smaller.

Instruments based upon the principle of the spectroscope can do more than find out what materials are present in the Sun. In 1889 the American astronomer G. E. Hale invented what is known as the spectroheliograph, an instrument which enables the Sun to be studied in the light of one element only, usually hydrogen or calcium. Nowadays special filters, invented by the late Bernard Lyot of Paris, perform the same function. Moreover, by using spectroscopic equipment, the solar prominences can be studied at any time instead of only during the brief moments of a total eclipse. This has been possible ever since 1868, when the principle was discovered by Lockyer in England and, independently, by Janssen in France.

The Source of Solar Energy

Hydrogen is the most plentiful element in the Universe: the Sun contains a high proportion of it. Deep inside the solar globe, where the temperature rises to 14,000,000°C and the pressures are immense, the hydrogen nuclei are combining to form nuclei of helium; energy is released in the process, and it is this energy which keeps the Sun shining. As it radiates, the Sun is losing mass at the rate of 4 million tons a second. The Sun is so large and massive that it will not change much for several thousands of millions of years, but by stellar standards it is still no more than middle-aged.

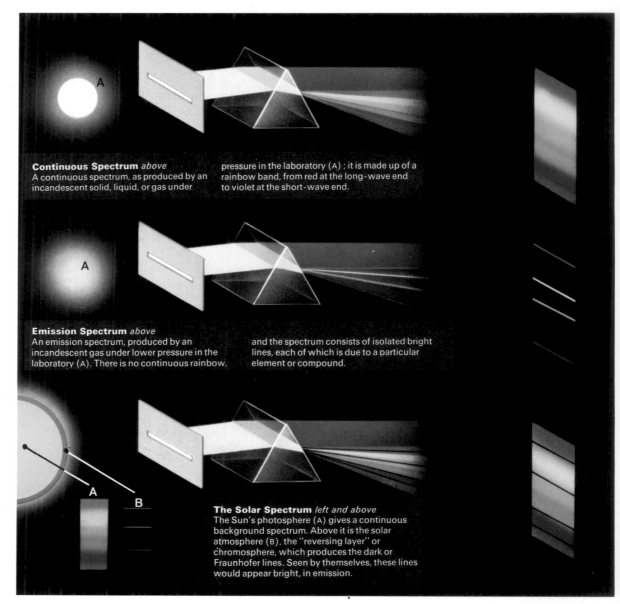

Continuous Spectrum *above*
A continuous spectrum, as produced by an incandescent solid, liquid, or gas under pressure in the laboratory (A); it is made up of a rainbow band, from red at the long-wave end to violet at the short-wave end.

Emission Spectrum *above*
An emission spectrum, produced by an incandescent gas under lower pressure in the laboratory (A). There is no continuous rainbow, and the spectrum consists of isolated bright lines, each of which is due to a particular element or compound.

The Solar Spectrum *left and above*
The Sun's photosphere (A) gives a continuous background spectrum. Above it is the solar atmosphere (B), the "reversing layer" or chromosphere, which produces the dark or Fraunhofer lines. Seen by themselves, these lines would appear bright, in emission.

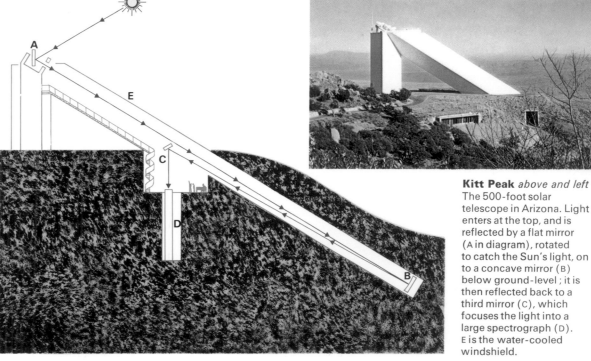

Kitt Peak *above and left*
The 500-foot solar telescope in Arizona. Light enters at the top, and is reflected by a flat mirror (A in diagram), rotated to catch the Sun's light, on to a concave mirror (B) below ground-level; it is then reflected back to a third mirror (C), which focuses the light into a large spectrograph (D). E is the water-cooled windshield.

Production of the Sun's Spectrum *below*
Solar spectrum, taken with the 13-foot spectrograph at Mount Wilson and Palomar Observatories. The range is from 3900 to 6900 Ångströms – i.e. from violet through to red. Many lines are shown, all of which have been identified; for example, the double D line at about 5900 Å is due to sodium. The lines are constant in position and intensity, and have been carefully mapped.

Over 70 elements have so far been identified in the Sun, some of which are more abundant than others. The most common substance is hydrogen. Interpretation of the complex spectrum of the Sun remains difficult, but modern instruments can make a highly detailed analysis.

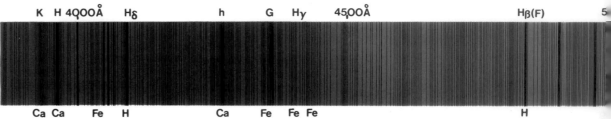

K H 4000Å   Hδ   h   G   Hγ   4500Å   Hβ(F)

Ca Ca   Fe   H   Ca   Fe Fe Fe   H

The Sun in Light of Different Wavelengths *below*

Toward the end of the 19th century, G. E. Hale invented the spectroheliograph, which made it possible to take photographs of the Sun in the light of one element only (usually hydrogen or calcium); the visual counterpart of the spectroheliograph is the spectrohelioscope, and nowadays Lyot or monochromatic filters are used to give similar results. It is possible to study the distribution of the various elements over the disk, and to observe certain definite features such as the flocculi, patches on the Sun's surface which are invisible in ordinary light. Dark flocculi are usually made up of hydrogen, bright flocculi of calcium.

left Photograph of the Sun taken in ordinary or integrated light; 15 September 1949, Mount Wilson and Palomar Observatories. The indicators show the position of the Sun's poles. Several spots are seen.

left A photograph taken at the same time, but in hydrogen light. The overall aspect is different, as no other gases apart from hydrogen are recorded photographically.

left Another photograph taken at the same time, this time in calcium light with the spectroheliograph. The distribution of the calcium is seen to be different from that of the hydrogen.

The Energy of the Sun

Early men thought that the Sun must be burning. However, a Sun made up of coal, and burning fiercely enough to emit as much energy as the real Sun, could not continue shining for more than a few millions of years at most (even if the process were possible), whereas we have irrefutable evidence that the age of the Sun must be as great as that of the Earth (4700 million years) and is probably more.

Lord Kelvin, the great British physicist of the late 19th century, believed that the Sun's energy was drawn from gravitational contraction. This would lead to a possible age of 50 million years, but even this is inadequate. Much more plausible was the theory of the annihilation of matter. It was formerly thought that a proton (unit positive charge) and an electron (unit negative charge) might annihilate each other if they collided, releasing energy. This would provide enough reserve for the Sun to continue shining for millions of millions of years, but such a time-scale was as clearly too long as earlier ones had been too short. Moreover, it is now known that straightforward annihilation in this way is not possible.

Just before World War II, H. Bethe in America and C. von Weizsäcker in Germany suggested a different solution. The most abundant element in the Sun, as it is in the universe generally, is hydrogen. According to the new theory, nuclei of hydrogen were combining to make nuclei of helium. It takes 4 hydrogen nuclei to make one nucleus of helium, and the difference in mass is transformed into the energy which maintains the Sun as a radiating object. The mass-loss during the course of this process amounts to 4 million tons a second.

The process of changing hydrogen into helium is not simple; there are various "steps", but the end result is that helium is being built up in the Sun's core. Eventually the supply of available hydrogen will become exhausted, but this is not likely to happen for another 6000 million years.

The Structure of the Solar Corona *left*

The outer part of the Sun's atmosphere is the corona (A), visible to the naked eye only during a total eclipse, though the inner part of it may be studied at any time with a coronagraph. The corona is extensive, and has no sharp boundary. It simply thins out into space, and there have even been suggestions that traces of it reach the Earth.

The corona has a very high temperature, at least 1,000,000 °C, and it is the source of radio emissions. Yet it would be wrong to suppose that the corona contains more "heat" than the photosphere, whose temperature is a mere 6000 °C. The scientific definition of temperature is not the same as our everyday meaning of "heat". Temperature is measured by the velocities at which the atoms and molecules move about. In the corona, the velocities are very great, and so the temperature is high, but there are so few atoms and molecules that the heat is slight. There is an analogy here with a firework sparkler. Each spark is white-hot, but contains so little mass that there is no danger in holding it in the hand.

Chromosphere

The Sun's chromosphere (B) rises to 6000 miles above the photosphere. Its density is between 1/1000 and 1/10,000 that of the photosphere, and it is where the Fraunhofer lines are produced.

The upper part is not uniform. It is made up of spicules, high-temperature gases shot up at great speeds which penetrate the corona. There is continuous agitation, so that particles escape into space and produce the solar wind. At any given moment there must be about 100,000 spicules in the chromosphere.

Interior of the Sun

Heat is transferred to the surface of the Sun through the outer layers (C) by convection. Lower down, however (D), heat probably travels as pure radiation.

Core

In the Sun's core (E), the temperature rises to 14,000,000 °C. It is here that the nuclear transformations take place.

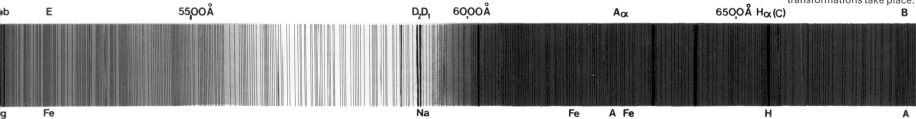

	Mi	Km
	50,000	80,000
	100,000	160,000
	150,000	240,000
	200,000	320,000
	250,000	400,000
	300,000	480,000
	350,000	560,000
	400,000	640,000
	432,475	695,800

Outbursts on the Sun

Prominences

In 1733 there was a total eclipse of the Sun, visible from Sweden. It was watched by an observer named Wassenius, who described "red flames" apparently rising from the limb of the Moon. This was the first report of the solar prominences; it was only in 1851 that astronomers became certain that they were associated with the Sun and not with the Moon. In 1868, Lockyer and Janssen independently discovered that by spectroscopic methods it was possible to study the prominences at any time, without having to wait for an eclipse to occur.

Prominences are made up of incandescent gas, and are phenomena of the Sun's chromosphere, which lies above the brilliant photosphere. They may exceed 100,000 miles in length, and are of two basic types: eruptive and quiescent. Eruptive prominences are in violent motion, and have been followed out to more than half a million miles above the Sun's surface; quiescent prominences are much more stable, and may hang in the chromosphere for days or even weeks before breaking up. There are also surges, which may be described as jets of luminous material which rise to heights of many tens of thousands of miles and then fall back, often along their original paths. Prominences are best seen when jutting out from the Sun's limb. They can also be seen when full on the disk, when the Sun is examined in hydrogen light; they then look dark, and are called filaments.

Prominences, like sunspots, are affected by the solar cycle, and are commonest near maximum. They are often, though not always, associated with spot-groups. On occasions the material may move at over 400 miles per second.

Solar Flares

Also associated with prominences are solar flares, which are brilliant outbreaks. They are seldom seen in integrated light – that is to say, with ordinary telescopes; a few such observations have been made (the first was by two British amateurs, Carrington and Hodgson, in 1859), but in general flares are shown only in hydrogen light. A flare is as violent as its name suggests. It appears without warning, and may last for no more than a few minutes. Each flare is made up of patches of various size from a million square miles upward; the total area covered is very large, and the emission is correspondingly intense.

Flares emit ultra-violet radiations as well as charged particles. When the short-wave radiation reaches the neighbourhood of the Earth, it causes magnetic storms and interference with radio transmission, as well as displays of auroræ. It has been suggested that particle emission from solar flares may present a real hazard to astronauts who are in space or on the unprotected surface of the Moon. Flares also emit cosmic rays, which are atomic nuclei.

Other Solar Emissions

Quite apart from these violent outbreaks, the Sun also emits what is rather misleadingly called "solar wind", made up of constant streams of low-energy particles sent out in all directions. It is this emission which has so strong an effect upon the tails of comets, forcing them to point away from the Sun.

The Sun emits long wave-length radiations as well as visible light. Radio emission from the quiet Sun was detected before World War II by Southworth in the United States. Flare outbursts at radio wavelengths were first found by British research workers in 1942 (for a time, they were thought to be due to enemy action!). Since then the radio waves from the Sun have provided invaluable information. Whenever there is violent activity, as with a flare, the radio emission increases to hundreds or thousands of times its "quiet" value. Moreover, the "radio Sun" is larger than the "optical Sun".

The corona, which represents the Sun's outer atmosphere, is a strong radio source – and it also seems to be at a very high temperature of several millions of degrees. In spite of this, however, the atoms in the corona are so sparse that there is very little "heat". The radio emission from the corona has a wave-length much longer than that coming from the bright surface of the Sun.

Radio emission, too, is affected by the solar cycle, but it is always present, even when the Sun is at the minimum of its cycle of activity.

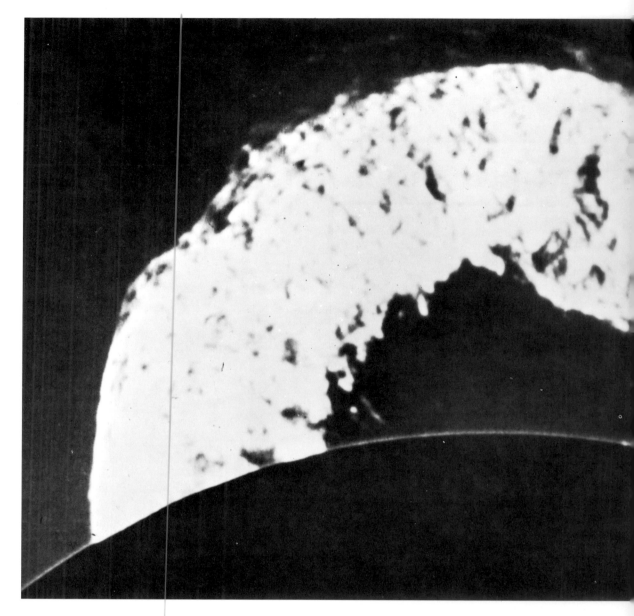

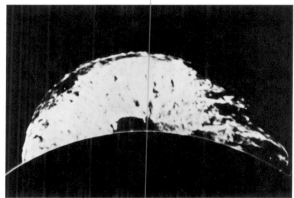

Eruptive Prominence, 16.03 and 16.36 Hours *above*
The great eruptive prominence of 4 June 1946, as photographed by Roberts. The first photograph (*above*) shows it at 16.03 hours, when it took the form of a huge arch. The second and larger photograph (*top*), taken at 16.36 hours, shows it at its maximum development.

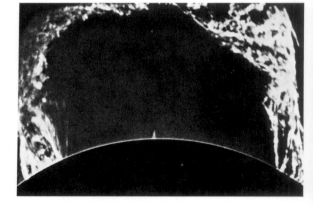

Prominence at 17.03 Hours *above*
The same prominence, photographed at 17.03 hours. By now the arch has blown upward, and much of the material of the prominence has been sent out to over 200,000 miles above the Sun's surface. The remains of the arch are still very striking.

"Anteater" Prominence
left
The so-called "Anteater" Prominence, seen at the total solar eclipse of 20 May 1919 from Sobral, Brazil. This was a really impressive structure, comparable with the explosive arch of 1946. The colour was strikingly red.

Quiescent Prominence
right
Solar prominence, photographed by W. M. Baxter on 2 May 1968, using a Lyot filter on the 4-in. refractor at his observatory in Acton, London. This was a quiescent prominence.

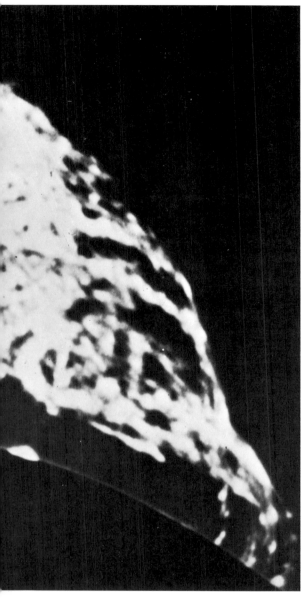

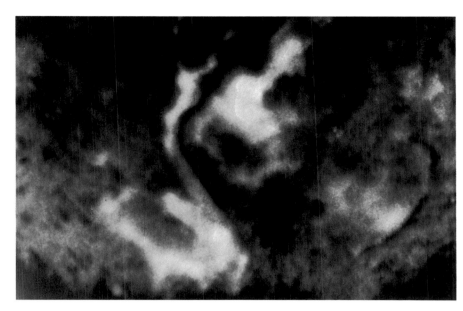

Solar Flares *right*
A solar flare, photographed in colour. The average flare does not last for long, and cannot be photographed in integrated light ; usually a flare is visible only in hydrogen light, so that a monochromatic filter or some equivalent instrument has to be used. Active spot-groups may produce flares, but so far it has not proved possible to predict exactly when or where a flare will break out.

The particle emission produced by solar flares could form a potential hazard to astronauts, once beyond the protection of the Earth's atmosphere.

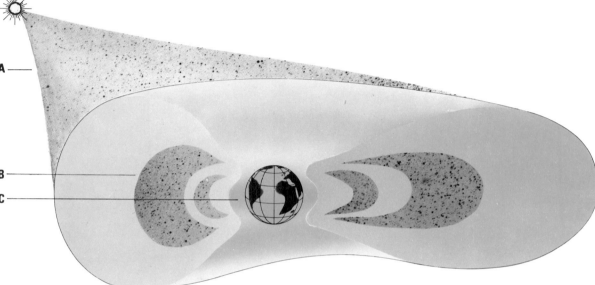

Prominence at 17.23 Hours *above*
A photograph of the same prominence, taken at 17.23 hours. Little now remains of the great arch, and the prominence has dispersed ; the main activity lasted less than two hours. This was one of the most spectacular prominences ever seen.

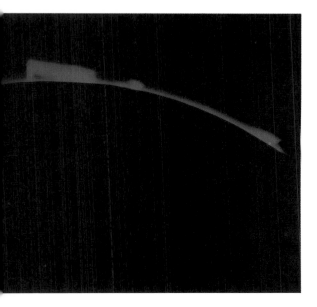

Polar Lights
above and right
Auroræ, or Polar Lights, are the beautiful glows seen in the sky during nights at high latitudes. They are a consequence of activity on the Sun, and are commonest at times of solar maxima. The "auroral curtain" in the four top photographs is typical. The lower photograph shows a brilliant display over Washington D.C.

Auroræ occur in the upper atmosphere, at heights ranging from 600 miles down to as little as 60. Sometimes the lights take the form of regular patterns ; at other displays there may be arcs, streamers, curtains, "draperies", and the aptly called "flaming surges".

It used to be thought that an aurora was due to charged particles from the Sun (particularly from flares) entering the upper air and causing a glow. This explanation is now known to be too straightforward. When a flare occurs, it seems (as shown in the diagram above) that the stream of emitted particles (A) from the Sun enters the outer Van Allen zone (B) of charged particles ; the particles are somehow accelerated downward, entering the upper atmosphere (C) and giving rise to auroræ.

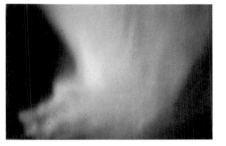

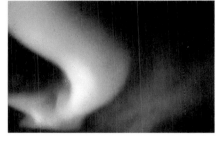

Solar Eclipses

Ancient Eclipses

Eclipses were observed by the ancient Chinese, who were alarmed by them. (It is said that two astronomers, Hsi and Ho, were executed by the Emperor for failing to predict an eclipse, though the story is probably apocryphal.) The Chinese did not know that the Moon is responsible; they thought that a dragon was trying to devour the Sun. In 585 B.C. the first great philosopher of Ancient Greece, Thales, is said to have predicted an eclipse, which duly took place and put a halt to a battle between the armies of the Lydians and the Medes. By the end of Greek times, the causes of solar eclipses were well known to contemporary Greek scientists.

How an Eclipse Happens

A total eclipse of the Sun is probably the most magnificent sight in all nature. For a brief period, as the Moon hides the brilliant solar disk, the Sun's atmosphere flashes into view; the red prominences and the pearly corona dominate the scene, and the sky darkens, so that stars may be visible.

The Moon revolves round the Earth (or, to be more precise, around the barycentre), while the Earth revolves round the Sun. There must be times when the three bodies move into a direct line, with the Earth in the mid-position. By coincidence – so far as we know, it is nothing more – the Moon and the Sun appear almost the same size in the sky, so that the lunar disk is just big enough to cover up the bright surface of the Sun, producing an eclipse. If the alignment is exact, the eclipse is total. If only a portion of the Sun is covered by the Moon the result is a partial eclipse.

When the Earth is at perihelion, the apparent diameter of the Sun is $32' 35''$; at aphelion, the value is $31' 31''$. The apparent diameter of the Moon ranges between $33' 31''$ and $29' 22''$. The Moon can therefore hide the Sun completely when it is near perigee, but it is moving steadily in its orbit, and totality can never last for more than 8 minutes or so, while at most eclipses totality is much shorter. Moreover, the track of totality on Earth can never be more than 169 miles wide, so that as seen from any one place a total eclipse is a rare event. To give a typical example: from England, the last total eclipse occurred in 1927, while the next will not take place until 1999. Some eclipses are not total anywhere on the Earth, and at annular eclipses, when the alignment is exact but the Moon is near apogee, the Sun's disk is too big for it to be completely covered; a ring of sunlight is then left showing round the dark body of the Moon.

A solar eclipse can happen only at new moon. If the orbit of the Moon lay in the same plane as that of the Earth there would be an eclipse every month. However, the lunar orbit is inclined by just over 5°. The points at which the Moon's orbit crosses the ecliptic are called the nodes. In order for an eclipse to occur, new moon must occur when the Moon is very near a node, which happens about every six months.

The Saros Cycle

Because of the gravitational pull of the Sun, the nodes of the lunar orbit shift round slowly and regularly. After a period of 18 years 11·3 days, the Earth, Moon and Sun return to almost the same relative positions, so that a solar eclipse may be followed by another eclipse 18 years 11·3 days later, though there will be others in the intervening period. This is the Saros Cycle. Known in ancient times, it provided a rough means of predicting eclipses. The period is not exact; for instance, the 1927 eclipse was total over part of England, but the "return" eclipse of 1945 was only partial. All that can be said is that the Saros gives a general indication of the time when an eclipse is due.

Importance of Total Eclipses

Total eclipses are important because there are scientific investigations which cannot be carried out at any other time. The outer part of the corona is very tenuous, for instance, and is hard to study except at totality. Astronomers are ready to go on long journeys to take the best possible advantage of the few fleeting moments when the Moon covers the Sun. Partial and annular eclipses are of little astronomical importance, but no one who has seen the full glory of a total eclipse will ever be likely to forget it.

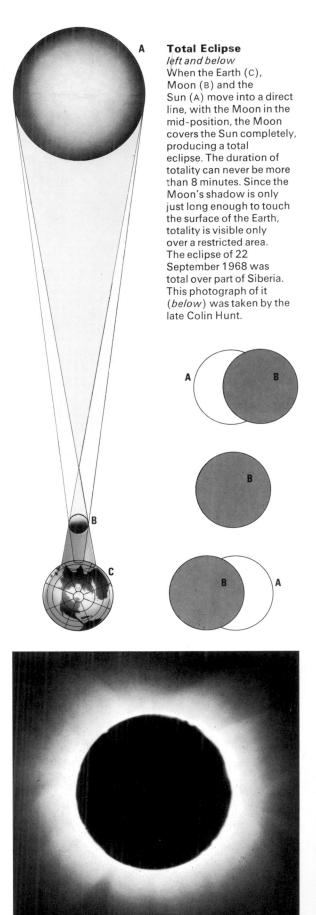

Total Eclipse
left and below
When the Earth (C), Moon (B) and the Sun (A) move into a direct line, with the Moon in the mid-position, the Moon covers the Sun completely, producing a total eclipse. The duration of totality can never be more than 8 minutes. Since the Moon's shadow is only just long enough to touch the surface of the Earth, totality is visible only over a restricted area. The eclipse of 22 September 1968 was total over part of Siberia. This photograph of it (*below*) was taken by the late Colin Hunt.

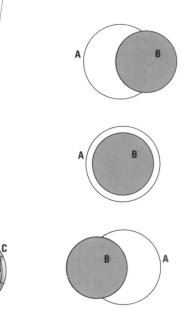

Annular Eclipse
left and below
When the Moon (B) is near apogee, its apparent diameter seen from the Earth (C) is smaller than that of the Sun (A), and even if the alignment becomes exact, the Moon is not big enough to hide the Sun completely; a ring of sunlight is left showing, producing an annular eclipse, as shown in the photograph below.

The glorious phenomena of totality cannot be seen, and astronomers do not regard annular eclipses as being of any real importance, though they are certainly always interesting to watch.

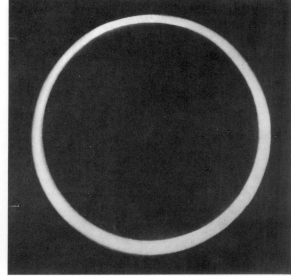

Total and Annular Eclipse *right*
The shadow of the Moon is only just long enough for it actually to reach the Earth. If the Moon were any smaller or more distant, no total eclipses could take place, and we might still be in a position where we know nothing at all about the existence of the Sun's corona.

An eclipse which is total over a restricted area (B) may be annular along the rest of the central track (A), as the diagram on the right makes clear; this was in fact the case with the Siberian eclipse of 22 September 1968.

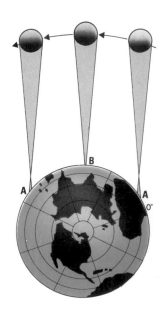

Partial Eclipse *below*
Partial eclipse of the Sun, 22 September 1968, as seen from Selsey, in Sussex.
While the eclipse was total over parts of Siberia, from Britain the phase never exceeded 40 per cent.

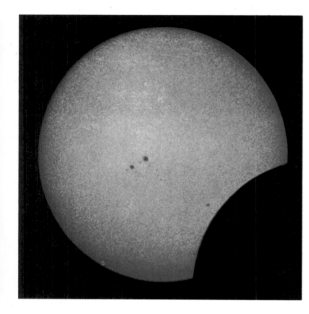

Partial Eclipse, 21 November 1966 *above*
Photographed by Henry Brinton (4-in. refractor, Selsey).
Several spots can be seen. Though a partial eclipse is
interesting to watch, it is not of any real importance to
astronomers, because the glorious phenomena of totality
cannot be seen ; for the corona to be observed, the Sun
must be completely hidden.

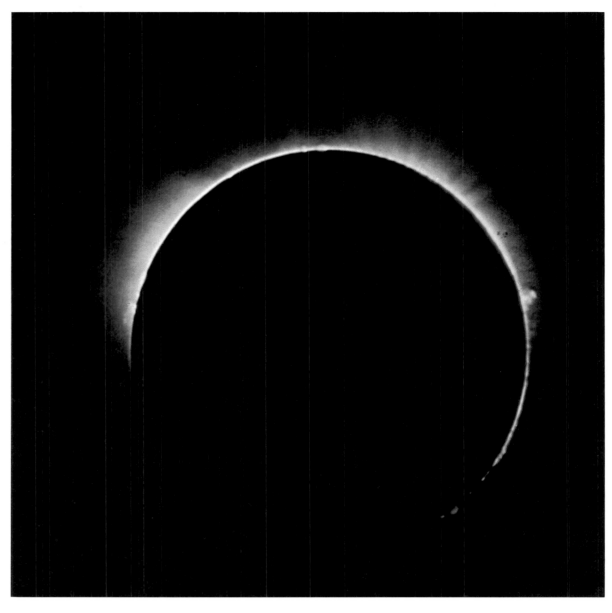

Total Solar Eclipse, 20 July 1963 *right*
Photographed by Father F. J. Heyden from Ellesworth,
Maine. Totality is just beginning. To the right, the redness
of the chromosphere is striking, and there is a
conspicuous prominence ; to the left and above, the
corona is coming into view, and some structure can be
seen in it. This was a short-exposure photograph. To show
the outer corona, a longer time is needed, and this of course
means that the bright inner parts are over exposed. The 1963
eclipse was total over parts of Canada, but was not visible
from anywhere in Europe.

Total Eclipse of 15 February 1961 *above*
Photographed by W. Bohenblust in Switzerland.
Exposure : 1/5 second on Ektachrome high-speed, 160
ASA. The inner corona is visible, and there are many
prominences round the Sun's limb.

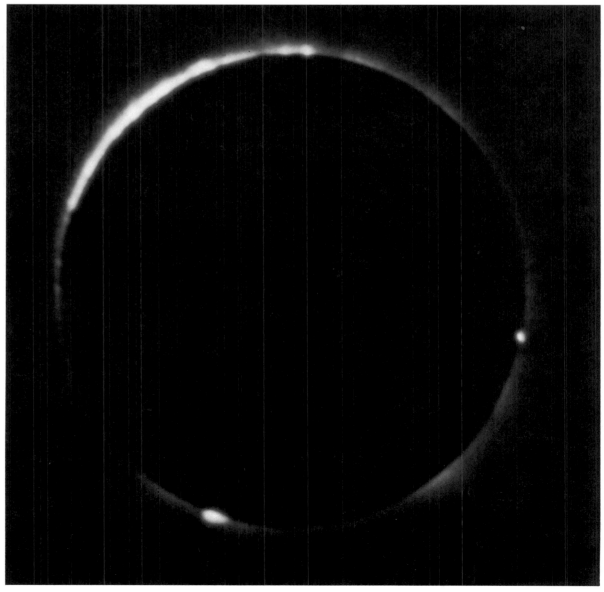

Total Eclipse Seen from the Air *right*
The eclipse of 21 November 1966, photographed from a
jet aircraft over the south Atlantic. The chromosphere, the
inner corona and some prominences are well shown.
Eclipse photography from any aircraft has advantages, for
there can be no interference from clouds and as the aircraft
can "follow" the Moon's shadow, and it is now even
possible to take up heavy and complicated scientific
equipment.

The Moving Planets

Early Ideas and Kepler's Laws

In ancient times it was thought that the orbits of all celestial bodies must be circular, because the circle was the perfect form, and nothing short of perfection could be allowed in the heavens. Yet it was clear that the planets could not move round the Earth in perfect circles at uniform velocities. They were not regular in motion; sometimes they would stop, turn, and move backward (east to west) in retrograde motion before resuming their eastward movement.

The first astronomer to break free from the idea of circular orbits was Johannes Kepler, who announced his first two Laws of Planetary Motion in 1609 and the third in 1618. After long and complex calculations, based on the observations of the Danish

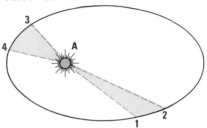

Diagram of Kepler's Laws *above*
If the planet moves from 1 to 2 in the same time that it takes to move from 3 to 4, A representing the Sun, then the area 1A2 must be equal to the area 3A4. The planet moves at its quickest when at perihelion (nearest point to the Sun) and slowest when at aphelion (farthest point from the Sun). It follows that the velocities of the planets decrease in order of distance from the Sun; Mercury has the greatest orbital velocity, Pluto the least.

astronomer Tycho Brahe, Kepler found that the orbits of the planets were ellipses. Kepler's Laws are so fundamental that it is worth giving them in full:

1. The planets move in elliptical orbits; the Sun is situated in one focus of the ellipse, while the other focus is empty.
2. The radius vector, or imaginary line joining the centre of the planet to the centre of the Sun, sweeps out equal areas in equal times. (In other words, a planet moves at its fastest when it is at its closest to the Sun.)
3. The squares of the sidereal periods of the planets are proportional to the cubes of their mean distances from the Sun. (Sidereal period is the time taken for a planet to complete one journey round the Sun – in the case of the Earth, 365¼ days.)

Though the planetary orbits are elliptical, most of them are not far from circular; the Earth, for instance, has an orbital eccentricity of only 0·017. Of the main planets, only Pluto has an orbit which is markedly elliptical, though those of both Mars and Mercury are more eccentric than that of the Earth.

It is also significant that (again with the exception of Pluto, which seems to be in a class of its own), the planetary orbits lie in approximately the same plane. Relative to the plane of the Earth's orbit, the inclination is 7° for Mercury, 3°24' for Venus, and less than 2° for all the rest. This means that the planets can be seen only in certain parts of the sky. They never move very far away from the ecliptic, which may be defined as the projection of the Earth's orbit on to the celestial sphere. (Since the Earth's axis is tilted by 23½° to the perpendicular, the angle between the ecliptic and the celestial equator is also 23½°.)

Planets in the Zodiac

The belt round the sky in which the main planets are always to be found is called the Zodiac. In the pseudo-science of astrology, the Zodiacal signs were given mystical significance, but astrology has no scientific or logical basis.

Mercury and Venus, closer to the Sun than we are, are always seen in the neighbourhood of the Sun, and so are never visible throughout a night. The superior planets, beyond the orbit of the Earth, are easier to follow as they move among the constellations. Mars, of course, moves quickest; Jupiter takes almost 12 years to go round the sky, so that on average it moves by one Zodiacal constellation a year; Saturn is even slower. Yet the movements of the planets can be tracked after only a night or two, which is how the ancients recognized them as being "wandering stars".

The Terrestrial Planets
left
The four planets closest to the Sun – Mercury (A), Venus (B), the Earth (C), and Mars (D) – are known collectively as the terrestrial planets, because they are relatively small, and are all solid bodies. Their mean distances from the Sun range between 36 million miles (Mercury) and 141·5 million miles (Mars); the orbits of Venus and the Earth are practically circular, while those of Mercury and Mars are more eccentric. The asteroid belt lies beyond the orbit of Mars, between the orbits of Mars and Jupiter.

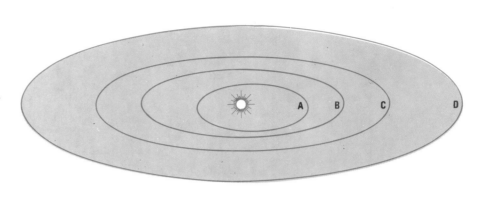

The Outer Planets *left*
Beyond the asteroids move the four giant planets, Jupiter (B), Saturn (C), Uranus (D), and Neptune (E), and beyond them Pluto (F). Mars (A) is shown for comparison. The scale of the Solar System may not be immediately apparent from the diagram; a spacecraft travelling from Earth to Neptune would pass the half way mark only when nearing the orbit of Uranus! Pluto, the outermost planet, has a much more eccentric and inclined orbit. At its closest to the Sun it comes within the orbit of Neptune.

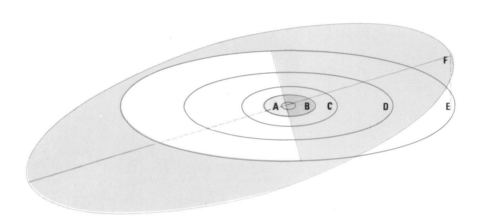

The Zodiac *below*
The ecliptic is the projection of the Earth's orbit on to the celestial sphere, and may also be defined as the apparent yearly path of the Sun among the stars. It passes through 12 Zodiacal constellations, together with a small part of a 13th, Ophiuchus. In the diagram, the Sun moves along the ecliptic (A), and the Moon and planets keep within 7° of it; with the exception of Pluto. The figures marked along the ecliptic show the position of the Sun at various times in the year.

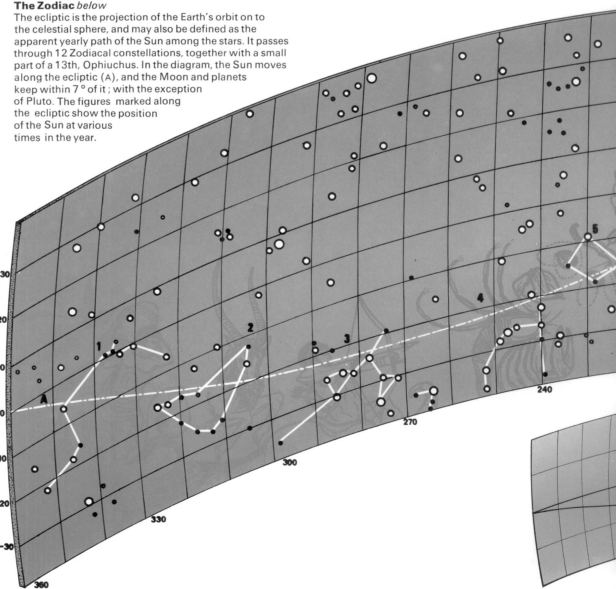

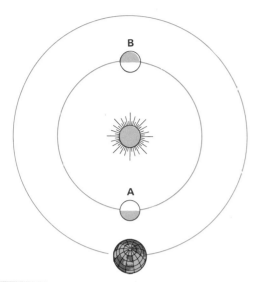

Inferior and Superior Conjunction *left*

The inferior planets, Mercury and Venus, come to inferior conjunction (A) when between the Sun and the Earth; if the alignment is exact, the planet is seen in transit against the Sun. The diagram reproduced here shows Venus in its orbit.

At superior conjunction (B), the planet is on the far side of the Sun, at full phase but unobservable. The synodical period (mean interval between successive inferior conjunctions) is 115·9 days for Mercury, 585·9 days for Venus.

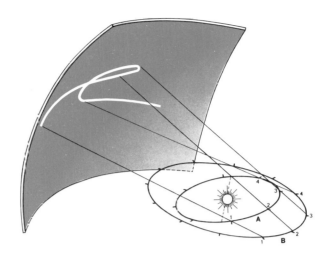

Apparent Motion of an Inferior Planet *left*

The motion of Venus, as seen from Earth. Because the orbit of Venus lies within the orbit of the Earth, it is always seen in the same area of the sky as the Sun. Mercury behaves in a similar manner.

The apparent movements are rather complicated, and appear as shown in the diagram:

A = orbit of Venus
B = orbit of Earth

The numbers indicate corresponding positions for the two planets.

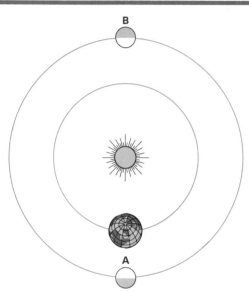

Superior Conjunction and Opposition *left*

With a superior planet, opposition (A) occurs when the planet is on the far side of the Sun as seen from Earth; it is then due south at midnight local time, and is well placed for observation. At superior conjunction (B) the planet is on the far side of the Sun, and is unobservable.

The mean synodical period (defined as the interval between successive oppositions) is 779·9 days for Mars, but much less for the outer planets (399 days for Jupiter, only 366·7 days in the case of Pluto).

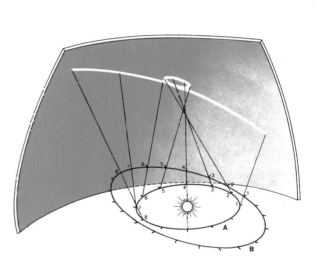

Apparent Motion of a Superior Planet *left*

Retrograding of a superior planet (B) — that is to say, a planet beyond the orbit of the Earth (A) in the Solar System. As the Earth "catches up" and passes the superior planet, the combined effect of the movement of the two bodies is to make the planet appear to an observer watching it from a position on the Earth to move temporarily in a retrograde or east–west direction before it resumes its normal direct movement.

As in the diagram above, the numbers indicate positions for the two planets.

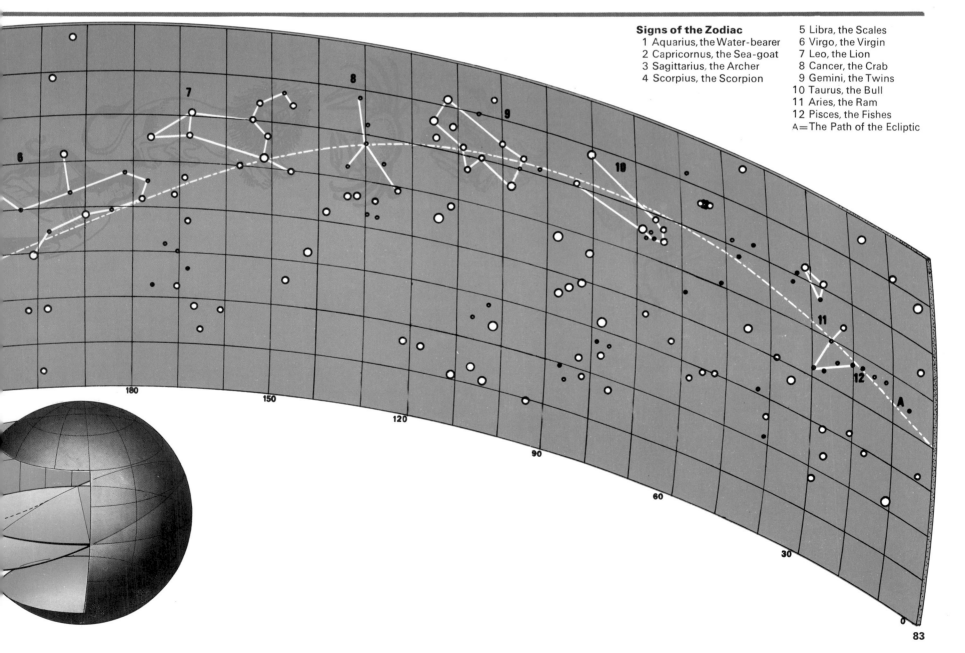

Signs of the Zodiac

1 Aquarius, the Water-bearer
2 Capricornus, the Sea-goat
3 Sagittarius, the Archer
4 Scorpius, the Scorpion
5 Libra, the Scales
6 Virgo, the Virgin
7 Leo, the Lion
8 Cancer, the Crab
9 Gemini, the Twins
10 Taurus, the Bull
11 Aries, the Ram
12 Pisces, the Fishes
A = The Path of the Ecliptic

Mercury 1

Appearance of Mercury

Five planets were known in very ancient times. Of these, four – Venus, Mars, Jupiter and Saturn – are very conspicuous with the naked eye, and appear as brilliant star-like objects. Mercury is far less prominent. It is much closer to the Sun than any of the other planets, and is never striking.

The maximum elongation of Mercury from the Sun – that is to say, its greatest angular distance – never exceeds 27°, so Mercury and the Sun are always in the same part of the sky. Consequently, Mercury is a naked-eye object only when it is low down in the west after sunset ("evening star") or low down in the east before sunrise ("morning star"). Not unnaturally, early peoples mistook these appearances for two planets, but the Greeks realized that there is in fact only one; they named it after the swift-footed Messenger of the Gods, Hermes (Mercury).

Yet Mercury is quite bright, and rivals Sirius, brightest of the stars. The difficulty of finding it is simply that it is always seen against a bright background, and is never visible for very long after sunset or before sunrise. Also, when it is a naked-eye object it is bound to be low down, and its light will be coming to us by way of a thick layer of the Earth's atmosphere. Mercury twinkles obviously, and it is not strongly coloured, though many observers claim that it has a somewhat pinkish hue.

Observation of Mercury

Mercury is a difficult planet to observe scientifically. There is little point in trying to study it with a telescope when it is visible with the naked eye, because seeing conditions are certain to be bad owing to the low altitude of the planet. The only solution is to study it during broad daylight, when Mercury is at least high above the horizon. This was the principle followed by Schiaparelli, the famous Italian planetary observer of the 19th century who drew up the first real map of Mercury; all later astronomers have adopted the method. Small or even moderate telescopes are hopelessly inadequate. Mercury is a small planet, and never comes closer than about 50 million miles, so that large instruments are needed to show much upon its surface.

Another difficulty is that Mercury, as an inferior planet (that is to say, closer to the Sun than we are), shows lunar-type phases. When Mercury is at its closest to the Earth, its dark side is turned toward us; Mercury is then new, and cannot be seen. As the phase increases, the apparent diameter decreases, until by the time that Mercury is full it is on the far side of the Sun. All these adverse conditions make mapping the planet particularly difficult, and it is hardly surprising that even the best modern charts do not agree well.

Surface Features

There is one compensation: the surface details, faint and elusive though they may be, are real. They represent markings on the solid surface of Mercury, and are not mere clouds. Mercury is probably devoid of atmosphere. Its escape velocity is only 2.6 miles per second, and this is too low for an appreciable atmosphere to be retained. Some observers (Dollfus in France and Moroz in the U.S.S.R.) have suspected traces of an atmosphere made up of carbon dioxide, but these results have not been confirmed, and the present tendency is to regard the planet as being in effect "airless".

The existence of any form of life on Mercury is improbable in the highest degree. In addition to the lack of atmosphere, the temperatures are extreme, and conditions are in every way unsuitable. Whether there are any mountains, valleys or craters remains to be seen. It seems very likely that there are craters on Mercury, similar in type to those on the Moon and Mars, but they are not visible from the Earth, and confirmation – or otherwise – must await the launching of a photographic probe. Plans are already being made for a Mariner probe to Mercury. The vehicle is scheduled to by-pass the planet in the spring of 1974.

Like Venus, Mercury has no satellite. Telescopic searches have been made from time to time, and any satellite of Mercury more than a mile or two in diameter would certainly have been found.

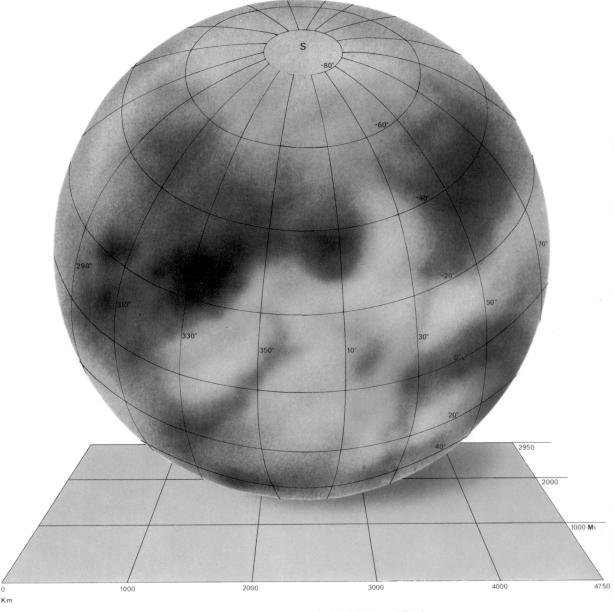

Dimensions of Mercury *above*
Mercury is the smallest of the principal planets. Its diameter is uncertain within narrow limits. Older measurements give 3100 miles, while Camichel, the French planetary astronomer, prefers 2900 miles. In any case, 3000 miles is certainly not far from the truth. There are suggestions that Mercury may once have been a satellite of Venus, though this is no more than a highly speculative hypothesis.

The mass of Mercury is 0.05 that of the Earth, and the volume 0.06. The escape velocity, 2.6 miles per second, is intermediate between those of the Moon and Mars.

Mercury and the Earth Compared
right
The discrepancy in size shows up clearly. The diameter of Mercury is not much greater than the width of the Atlantic Ocean at its widest point.

Distance from Sun and Earth *below*
The orbit of Mercury is more eccentric than that of the Earth. The distance from the Sun ranges between 29 million miles (perihelion) and 43 million miles (aphelion), and this changing distance has a powerful effect upon Mercury's surface conditions.

The orbital velocity of Mercury also varies, reaching 36 miles per second at perihelion and dropping to only 24.4 miles per second at aphelion. At its greatest distance from the Earth, Mercury recedes to 136 million miles; it is then at superior conjunction, on the other side of the Sun.

In the diagram, the mean distances of the four inner planets are shown to scale. On the same scale, the first of the giant planets, Jupiter, would be 4½ feet away from the Sun.

Orbit / Sidereal Period
Mercury takes 87.9 days to complete one revolution round the Sun. Very little can be said about the seasons which it may experience, as the angle of inclination of the planet's axis is not known with any certainty. In any case, they must be strongly influenced by the planet's changing distance from the Sun as a result of its orbital eccentricity.

Until comparatively recently it was thought that the axial rotation period must also be about 88 days. This would indicate a captured or synchronous rotation. One area of Mercury would always be facing the Sun, and would have eternal day; another area would always be turned away from the Sun, and would have permanent night.

Because of effects analogous to those of the librations of the Moon, there would be a relatively narrow "twilight zone" between these two extremes over which the Sun would seem to rise and set, always keeping fairly close to the horizon. However, it has now been found, by radar measurements, that the rotation period is not captured. It amounts to only 58½ days, so that every part of the planet is in sunlight at some time or other.

Density of Mercury *right*
Mercury is relatively dense. According to the latest measurements, its specific gravity is 5.4 (that is to say, unit mass of Mercury would be 5.4 times as great as that of an equal volume of water). Unfortunately the value for Mercury cannot be regarded as very reliable, as measurements are difficult to make. All we can really say is that the density of Mercury is very similar to that of the Earth.

Mercury

36,000,000 mi
57,920,000 km

Mapping Mercury

The first serious telescopic observations of Mercury were made early in the 19th century by J. H. Schröter, at Lilienthal (near Bremen). From these, F. W. Bessel deduced a rotation period for the planet of 24h.0m.53s. It is now known that this is completely wrong – and it is unlikely that any reliance can be placed upon Schröter's drawings; his telescopes were not adequate for this particular task.

In 1877, G. V. Schiaparelli, at Milan, produced a map showing dark features and bright regions. His observations were made in daylight, and were far better than any previously made. He believed the rotation period to be captured (88 days), and though this is not correct, it does not necessarily invalidate his observations or his map, since every time that Mercury is best placed for observation it has the same hemisphere turned in our direction.

A more detailed chart was produced by E. M. Antoniadi, a Greek astronomer who spent most of his life in France. Antoniadi's main observations were carried out between 1924 and 1927 at the Observatory of Meudon, near Paris, with the 33-in. refractor – one of the largest and best refracting telescopes in the world. Antoniadi's nomenclature is still in use, and his skill as an observer and draughtsman is beyond question.

The dark markings are certainly permanent, and it is not impossible that they are analogous to the lunar maria. Rather surprisingly, Antoniadi believed that he could trace local obscurations on Mercury which, in his words, were "more frequent and obliterating than those of Mars"; this would indicate dust-storm activity in a thin Mercurian atmosphere. However, later work has not confirmed these obscurations, and it is now thought that Mercury has no atmosphere capable of suspending dust particles.

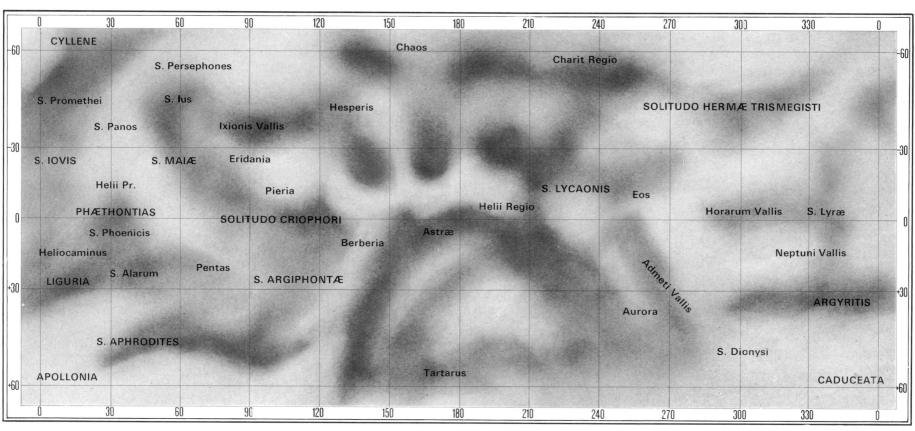

Names of Mercurian Features

The nomenclature now in common use for the features on Mercury is basically that of Antoniadi, slightly modified by G. N. Katterfeld and his colleagues in the U.S.S.R. When probe results provide more accurate maps, it will be necessary to revise the nomenclature.

Photography of Mercury is difficult, and the best mapping results are obtained by visual observation with large telescopes. Even so, the agreement between charts drawn up by various observers is not good. One of the most reliable charts – if not the most reliable – has been compiled by H. Camichel and A. Dollfus at the Pic du Midi Observatory, in France, from both photographic and visual data. This map is shown above, and may be compared with Antoniadi's; main features like the Solitudo Criophori, are recognizable on both charts, but there is considerable difference with minor features.

Most of the work carried out by Camichel and Dollfus was undertaken with the 24-in. refractor at the Pic du Midi, 10,000 feet up in the French Pyrenees, where observing conditions are excellent.

Schiaparelli's Map of Mercury *right*
Schiaparelli's map of Mercury was the first attempt at a chart. Though the agreement with the later maps is imperfect, some of the main features can be identified, notably the dark Solitudo Criophori. Schiaparelli shows some linear features; he had a tendency to draw faint details as harder and more regular than they really are.

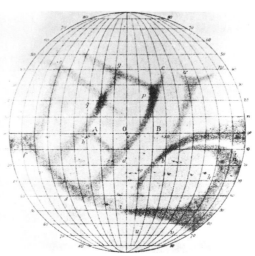

Map by Camichel and Dollfus *above*
The map drawn by Camichel and Dollfus shows all the features whose existence may be regarded as reasonably established. In 1953 Dollfus reported that he had detected signs of a very tenuous atmosphere, but this has not been confirmed. The bright areas may be of the same basic type as the reddish-ochre tracts of Mars, though they are much less highly coloured. Lunar-type craters may be expected.

Map by Chapman *left*
Compiled by C. Chapman, of the Massachusetts Institute of Technology, from a selection of 130 drawings and photographs made by many observers. There is some measure of agreement with the Camichel–Dollfus map, but the charts differ in many respects. It must be admitted that our knowledge of the topography of Mercury is still rudimentary. Even giant telescopes can never show Mercury as clearly as the Moon can be seen with the naked eye.

Earth

Mars

,000 mi
00,000 km

92,957,200 mi
149,600,000 km

141,6,000,000 mi
228,000,000 km

Mercury 2

There is a legend – probably apocryphal – that the great astronomer Copernicus, whose book published in 1546 began the great controversy about the status of the Earth in the universe, never saw Mercury in his life, because of mists rising from the River Vistula near his home. Whether this is true or untrue, it is a fact that few people living in densely populated countries will see Mercury unless they go out and deliberately search for it at a suitable moment. Binoculars are helpful in picking it up in the dawn or evening sky, but great care should be taken until the Sun has dropped below the horizon. If a search is being made, and the Sun enters the field of view of the binoculars, permanent damage to the observer's eyes is certain to result.

Yet when Mercury has been found, it seems so bright that one wonders how it could have been overlooked. It is at its best when at the crescent stage; by the time the phase has become half (dichotomy), the distance from Earth is greater. As with Venus, Mercury is waning when it is an evening star, and waxing when it is a morning star.

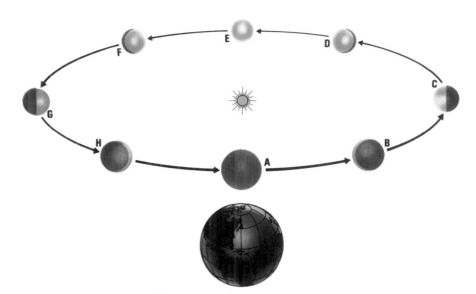

Phases *left*
Diagram to illustrate the phases of Mercury:

A = inferior conjunction (new)
E = superior conjunction (full)
C and G = elongation (half)
D and F = gibbous
B and H = crescent

Mercury is always most brilliant when in the crescent stage.

Charts of Mercury *below*
Charts of Mercury by A. Dollfus (Pic du Midi), drawn with the 24-in. refractor.

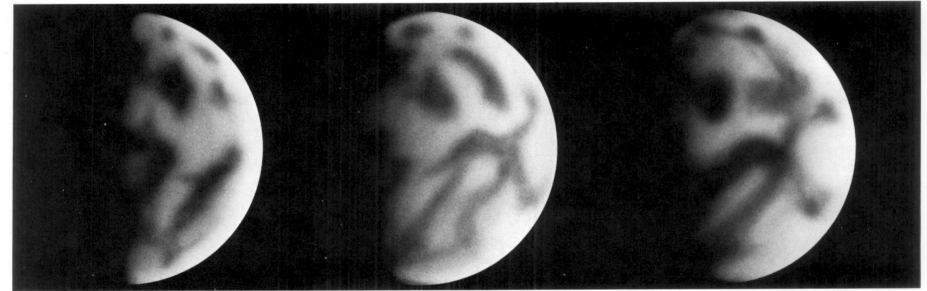

Period

The average synodic period of Mercury – that is to say, the period after which it returns to the same position in the sky relative to the Sun – averages just under 116 days. The maximum elongation for any one apparition may be as much as $27°5'$ or as little as $17°50'$, so that Mercury is never seen very far away from the Sun.

In the diagram at the top of the page, the Earth is shown (not to scale), and Mercury is drawn at various positions in its orbit. At A it is new, and invisible; at B it is crescent.

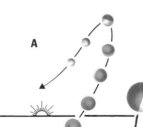

Elongations of Mercury *above*
Diagram to show the morning (A) and evening (B) elongations of Mercury. As the planet's phase increases, its distance from us also increases, and the apparent diameter becomes smaller.

At E, Mercury is at superior conjunction, and is full, but as it is then almost behind the Sun in the sky it is to all intents and purposes unobservable.

The terminator of Mercury appears smooth, almost certainly because of the fact that we cannot see it clearly enough to show up any irregularities. There is every reason to believe that Mercury has a mountainous, probably cratered surface, and in this case the terminator will be as uneven and rough as that of the Moon; but from Earth, these irregularities cannot be detected.

Because Mercury is so difficult to observe, it cannot yet be said that our knowledge of the surface features is at all satisfactory. Neither can we tell whether or not the surface is essentially similar to that of the Moon.

Rotation of Mercury *below*
For many years it was thought that Mercury must have a captured (88-day) rotation period, so that the same hemisphere was always illuminated by the Sun, the other hemisphere being constantly turned toward the cold and dark. Then, in 1962, W. E. Howard and his colleagues at Michigan measured radio emission from Mercury, and showed that the dark hemisphere was much warmer than could possibly have been the case if it were always turned from the Sun.

Radar measures were then used, initially by Dyce and

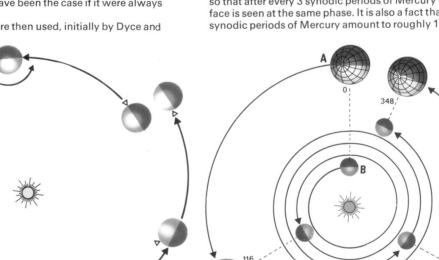

Pettingill in the United States, and it was found that the real rotation period of Mercury, relative to the stars, is 58.6 days. This means that every part of the surface will be exposed to sunlight at some time or other. The solar "day" on Mercury (that is to say, the average time between successive sunrises as seen from any point on the planet) is 176 terrestrial days, or 2 Mercurian years. In the diagram reproduced here a fixed marker on Mercury is seen experiencing one daylight period of 88 days.

Appearance of Mercury from Earth *below*
If the rotation period of Mercury is not a captured one of 88 days, as was previously thought, we must explain why the maps of older observers showed apparently fixed markings on the disk.

The synodic period of Mercury is 116 Earth-days. The Mercurian "day" is 176 Earth-days, or 2 Mercurian years. This is approximately equal to 1½ synodic periods, so that after every 3 synodic periods of Mercury the same face is seen at the same phase. It is also a fact that 3 synodic periods of Mercury amount to roughly 1

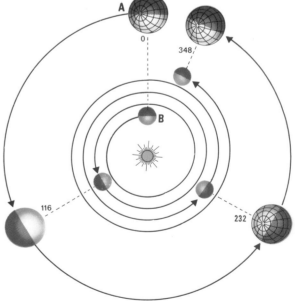

Earth-year.

It is for this reason that the best times for studying Mercury recur every 3 synodic periods – and at these times, the same surface features are seen at the same positions on the disk! The agreement is not in fact absolutely exact, but it was near enough to mislead all astronomers up to the development of radar techniques during the 1960s. The diagram shows the relative positions of Earth (A) and Mercury (B) over 3 synodic periods (348 days).

The Sun from Mercury

Because of the eccentricity of the orbit of Mercury, the apparent diameter of the Sun as seen from the planet varies considerably: 1°·6 at perihelion, only 1°·1 at aphelion (see diagrams below).

From a point on Mercury's equator, at a longitude where Mercury is at perihelion when the Sun is overhead, the Sun will rise when it seems smallest and is moving most rapidly. As it nears the overhead point it will increase in size; after passing the zenith it will retrograde briefly, and it will set 88 Earth-days (one Mercurian year) after it had risen. From a longitude 90° away, the Sun will be at its largest when rising or setting; when rising it will retrograde for a short period and almost disappear below the horizon before it begins to move toward the zenith. When near the overhead point, it will be smallest and fastest-moving. On setting, it will pause, rise, and then vanish below the horizon, again after an interval of 88 Earth-days since sunrise. Stars will move across the sky at three times the Sun's speed. Clearly, the calendar of Mercury is very unfamiliar by terrestrial standards.

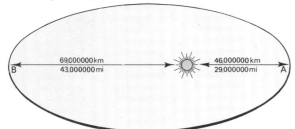

Apparent Size of the Sun from Mercury *left*
The two circles represent the Sun as seen from Mercury at perihelion (A) and aphelion (B) indicated in the diagram *above*.

Transits of Mercury

The orbit of Mercury is inclined to the ecliptic by 7° (more than that of any other planet, apart from Pluto). This means that at most inferior conjunctions it passes either above or below the Sun as seen from Earth. Occasionally, however, the three bodies move into a direct line. Mercury then appears in transit, and is seen in the form of a black disk against the brilliant face of the Sun.

The only planets which can appear in transit are Mercury and Venus. This was realized by Johannes Kepler: in 1627, in what was destined to be his last work, he published a set of tables in which it was forecast that both planets would transit the Sun in 1631—Mercury on 7 November, Venus on 6 December. By then Kepler was dead, but Mercury's transit was successfully observed by the French astronomer Pierre Gassendi.

Transits of Mercury are less uncommon than those of Venus. The last were on 7 November 1960, 9 May 1970, and 10 November 1973. The next will be on 12 November 1986, 5 November 1993, 7 May 2003, 8 November 2006, and 9 May 2016.

When in transit Mercury appears as a black spot, much darker than any sunspot, but too small to be seen with the naked eye. Since the planet has virtually no atmosphere, there is no "Black Drop" effect as is seen at transits of Venus (page 91), and the limb of Mercury appears sharp and clear-cut, with no hint of fuzziness. Occasional bright points on the planet's disk reported during transits are certainly due to observational or instrumental error.

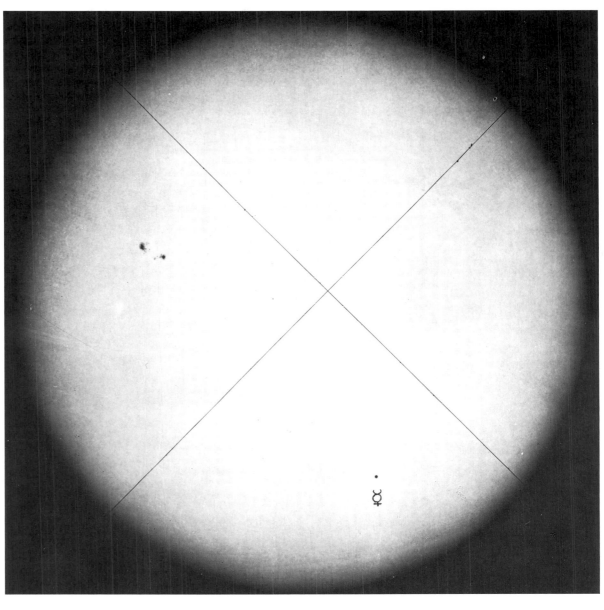

Transit of Mercury *above*
Transit of Mercury, observed on 7 November 1914. This photograph was taken with a 4-in. photoheliograph at the Royal Greenwich Observatory. Mercury is indicated by the symbol ☿. At this time there was little sunspot activity, and only one group was visible (upper left). When Mercury in transit can be compared with an adjacent sunspot, the more intense darkness of the planet's disk is evident.

Transits of Mercury, 1960 to 2016 *left and lower left*
It is seen that not all transits are of equal duration; for instance, that of 2016 will last for much longer than that of 2003. As is shown in the bottom diagram, transits of Mercury can occur only during two months in the year, May (X) and November (Y).

The planet's orbit is inclined at an angle of 7° to the ecliptic, and only during relatively short periods is there any chance that the Sun, Mercury (A) and the Earth (B) will become aligned when Mercury is at its inferior conjunction.

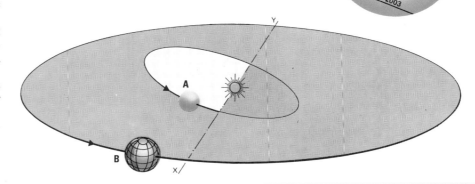

Vulcan
During the 19th century it was believed that a small planet moved round the Sun at a distance less than that of Mercury. It was even given a name—Vulcan, in honour of the blacksmith of the gods. Urbain Le Verrier, the great French astronomer, believed in its existence, and thought that it might be glimpsed either during a transit across the face of the Sun, or at a total solar eclipse.

Le Verrier's belief was based on the supposed influence of Vulcan upon the motion of Mercury. A French amateur, Lescarbault, claimed to have seen Vulcan in transit; and in 1878 two Americans believed they had seen it during an eclipse. However, it is now certain that the movements of Mercury can be explained without the need to introduce an extra planet, and that Vulcan does not exist.

Notes on Data
The diameter of Mercury is uncertain within narrow limits; it has been given as 3100 miles, but the recent determinations, mainly by French astronomers, give only 2900 miles. In any case, 3000 miles is certainly not far from the truth. Indications of an atmosphere, reported by Dollfus from the Pic du Midi Observatory and by V. Moroz in the U.S.S.R., have not been confirmed. There seems to be no doubt that Mercury is without any satellites.

Data
Distance from Sun max. 43,000,000 miles mean 36,000,000 miles min. 29,000,000 miles
Sidereal period 87·9 days
Mean orbital velocity 29·8 miles per second
Axial rotation period 58·5 days
Synodic period 115·9 days
Orbital eccentricity 0·206
Orbital inclination 7°0'
Equatorial diameter 3000 miles (?)

Mass (Earth=1) 0·05
Volume (Earth=1) 0·06
Density (specific gravity) 5·4
Surface gravity (Earth=1) 0·37
Escape velocity 2·6 miles per second
Albedo 7 per cent
Oblateness Inappreciable
Apparent diameter max. 12″·9, min. 4″·5
Maximum magnitude −1·9
Maximum surface temperature about 770 °F

Venus 1

The Mysterious Planet

"Venus, the Planet of Mystery". This was a phrase often heard before the launching of the first space-probes. Before 1962 Venus was indeed mysterious; almost nothing was known about the surface conditions there, and even the length of the axial rotation period was unknown. Today Venus is still very much of a mystery, but for different reasons. We know more *what* it is like – but not *why*.

Apart from the Moon, Venus is the nearest natural body in the sky. At its closest it can approach us to 24 million miles, which is 10 million miles nearer than Mars can ever come. Moreover, Venus is almost the twin of the Earth in size and mass, whereas Mars is much smaller. It may seem strange, then, that our knowledge in the pre-Space Age was so slight. The answer is that Venus has a dense, cloudy atmosphere which hides the surface completely, and which never clears away. When we observe Venus, we see nothing but the upper part of a cloud-layer, which accounts in part for its great brilliance.

Venus must have been known since the dawn of human history. It is brighter than anything else in the sky except for the Sun and the Moon, and at times it may cast a perceptible shadow; under good conditions it is visible with the naked eye in the middle of the day. Like Mercury, it is at its brightest in the west after sunset and in the east before dawn, but it is much farther from the Sun, and so it can often be seen against a dark background. Even so, the best telescopic views of it are obtained during daylight, partly because Venus is then higher in the sky, and partly because there is less glare from the brilliant, cloud-covered disk. It is not surprising that the ancients named the planet in honour of the Goddess of Beauty. It looks beautiful; and, until recently, we had no idea of its exceptionally hostile nature.

Surface

Originally it was thought that the clouds might protect the surface of Venus from the intense solar heat, and that the temperature there might be no more than pleasantly warm. The existence of oceans was regarded as probable, and early in our own century the Swedish physicist and chemist Svante Arrhenius believed that Venus might be a world in what might be called a "Carboniferous" state, similar to the condition of the Earth 200 million years ago. According to Arrhenius, there could have been swamps, seas, and abundant vegetation of primitive type, together with amphibians and reptiles of the kind which became extinct on Earth many millennia ago. This was quite a reasonable suggestion at the time when it was made, but doubts were cast upon it in 1932, when Adams and Dunham, two American astronomers, proved spectroscopically that the main constituent of Venus' atmosphere is the heavy gas carbon dioxide.

Very little was added for several decades, and even the length of the planet's rotation period remained unknown. In the early 1960s, two theories were current. According to F. L. Whipple and D. H. Menzel, of the United States, the upper clouds of Venus were composed of H_2O, and the surface of the planet was covered with water; this water would have been penetrated by the carbon dioxide in the atmosphere, producing what might be called seas of soda-water. In such a case, primitive life might exist (just as life began in the warm seas of the Earth in pre-Cambrian times, when the terrestrial atmosphere contained much more carbon dioxide than it does now). Venus, then, could be a world where life was beginning to evolve. Other astronomers regarded Venus as a wild dust-desert, scorching hot, and with no trace of water on the surface.

Probes to Venus

In 1962 the first successful planetary probe, the U.S. vehicle Mariner 2, by-passed Venus at 21,600 miles and sent back information which showed the surface to be fiercely hot. In 1969 two Soviet vehicles sent back signals during parachute descent through the atmosphere of Venus, and it now seems overwhelmingly probable that the conditions on the surface are intolerable for any advanced life-forms.

These conclusions have been supported by Venera 7, the first probe to make a successful soft landing on the planet. Venus is not so welcoming as it looks!

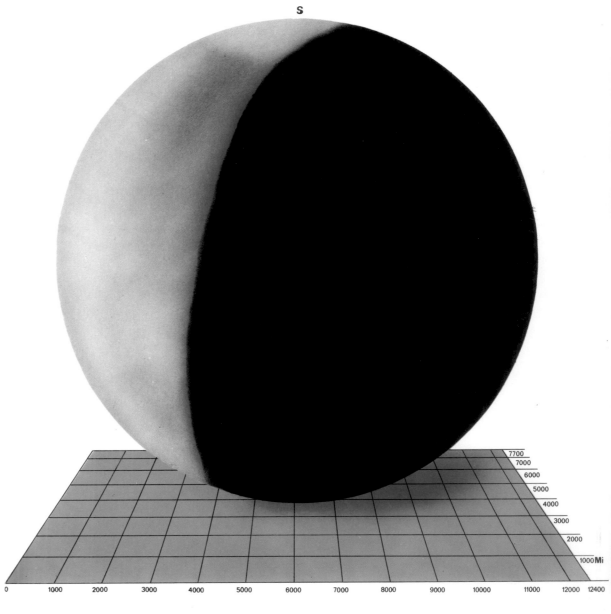

Size of Venus *above*

In size, mass and density Venus and the Earth are very alike. If the Earth were represented by a marble, Venus would be another marble, so like the first that careful weighing would be needed to tell which was which.

Though the diameter of Venus is not known with precision, because we can never see the actual surface and the depth of the atmospheric layer is uncertain, it is approximate to 7700 miles, as against 7926 for the Earth. The polar compression of Venus is inappreciable. The escape velocity of Venus is 6·3 miles per second.

Venus and the Earth Compared

right

On this scale, the difference between the two planets is too slight to be noticeable without careful measuring.

Distance from Earth and Sun *below*

Venus is the second planet in the Solar System; its average distance from the Sun is 67,200,000 miles. This varies only a small amount as the orbit is nearly circular. At perihelion the distance is 66,700,000 miles, while at aphelion it is 67,600,000 miles.

Venus can approach us much more closely than Mars can ever do. The minimum distance from Earth is 24 million miles, while Mars' closest approach is 34·6 million miles. Unfortunately, Venus is "new" when at its closest, and cannot be seen except on the rare occasions when it transits the face of the Sun.

Any satellite of Venus would have been discovered a long time ago if it were of appreciable size. Various reports of a satellite were made in the 19th century, but these have not been confirmed, and it is now certain that Venus, like Mercury, is moonless.

Orbit / Sidereal Period

The sidereal period or "year" of Venus is 224·7 days. If the axis is almost at right angles to the plane of the orbit, as is possible (the most recent measures give a value of 84°), there will be no marked seasons. As the orbit is nearly circular, the apparent size of the Sun in the sky would not vary much – though in fact the clouds would make it impossible to see the sky at all from the surface of the planet. The synodic period is on average 584 days.

Until the development of modern space-probe and radar techniques, the axial rotation period of Venus was unknown. Estimates ranged between 22 hours and 224 7 days – the latter being equal to Venus' sidereal period, so that the rotation would be captured or synchronous. Mariner 2, in 1962, indicated that the period must be long. More recent work, carried out both by probes and by radar measures from Earth, indicate that Venus spins on its axis in a period of 243 days, and that the direction of rotation is retrograde. In that event the "day" on Venus would be longer than the "year"; but to an observer on the planet the interval between one sunrise and the next would be only 118 days. The Sun would rise in the west and set in the east.

Density of Venus *right*

The mass of Venus is 0·81 of that of the Earth, its volume 0·92 that of the Earth and its density 4·99 (water=1).

If Venus and the Earth are so alike in size, mass and density, why are they so different in character? Venus is closer to the Sun, its higher temperature causing the planet's whole development to be different, which must be why the Earth has produced life – while Venus has not.

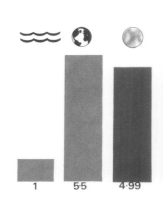

Mercury

36,000,000 mi
57,920,000 km

Mapping

The surface of Venus is never visible from Earth; it is permanently hidden by the dense layers of atmosphere. Therefore, the only way to study it is by means of radar, and very interesting results have been obtained, mainly by American researchers. Some of the first features detected by radar measurements were thought to be mountain areas, or perhaps plateaux, but the most sensational advances were announced in August 1973. From the Goldstone tracking station, with its powerful radar equipment, Venus was found to have craters – at least in one large area, and so presumably in others also. The craters are large (some of them over 100 miles in diameter) but relatively shallow. It is quite probable, though as yet unproved, that the whole surface of Venus may be as crater-scarred as the Moon and Mars. The shallowness of the structures is presumably due to erosion; the craters can hardly be of impact origin, in view of the dense atmosphere, so that they are presumably volcanic.

Map by Boyer *below*
Composite chart of Venus; C. Boyer (France) on the basis of 386 photographs (1953–62), assuming a rotation period of only 4 days.

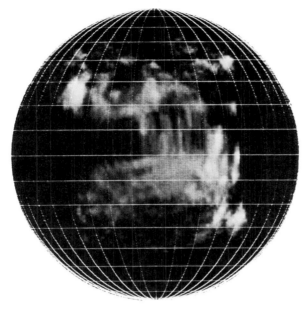

Atmosphere of Venus

The atmosphere of Venus is obviously extensive, and since 1932 it has been known to consist largely of carbon dioxide. However, measures from Earth could investigate only the upper part of the cloud-layer, and our best information has come from rocket probes. Mariners 2 and 5 have by-passed Venus at relatively close range, while several Russian vehicles have landed there, and have continued transmitting for some time after arriving on the surface of the planet.

The results seem to be quite conclusive. The temperature at the surface is approximately 990°F, and the atmospheric pressure is about 90 times that of the Earth's air at sea-level. The atmosphere is almost pure carbon dioxide; there is practically no free oxygen, though it is true that water-vapour exists at high altitude above the surface. The reason why Venus' atmosphere is so very different from that of the Earth remains unknown. The reason must be associated with the lesser distance of Venus from the Sun, but as yet there is no really satisfactory explanation.

The Structure of Venus' Atmosphere
right

Our knowledge of the structure of Venus' atmosphere is incomplete, since all we have to guide us are the results of the probes. Originally there was a discrepancy between the Mariner 5 and Venera 4 results, but this was resolved when it became clear that Venera 4 had stopped transmitting when still well above the planet's surface – probably because its instrumentation had been put out of action by the high temperature and atmospheric pressure. The next Russian probes, Venera 5 and 6, were designed to withstand much more extreme conditions, but even so they may have stopped transmitting before impact on the surface. If the available information is analysed, it seems that according to Venera 5, the atmospheric pressure at the surface is 60 times as great as on Earth, but Venera 6 gives the much higher value of 140 Earth-atmospheres. It has been suggested that Venera 5 landed over a mountainous area, while Venera 6 came down over a depression; this might go some way toward clearing up the discrepancy. Later Veneras had given full confirmation of these results. The pressure is in the region of 90 times that of our own air; this may indicate the conditions at the mean height of the landscape.

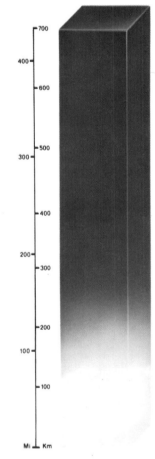

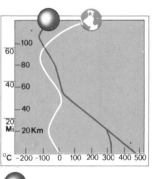

lower right
Two graphs showing the temperature and pressure conditions in Venus' atmosphere as compared with the Earth.

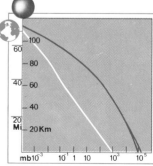

Phases of Venus *left and right*
Photographs of Venus by H. R. Hatfield (12-in. reflector). Taken in ordinary light, they show no surface detail; the curve of the terminator is smooth. (*Top left*) 18 January 1969. (*Bottom left*) 8 February 1969; the apparent diameter of the planet has increased, but the phase has decreased. (*Right*) 5 March 1969. Venus is now at the crescent stage, and the apparent diameter is much greater. Inferior conjunction was reached on 8 April, after which Venus became a morning star.

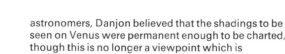

Drawings by Danjon *below*
Composite drawings of Venus, made in 1943 by the late A. Danjon, former Director of the Meudon Observatory. In common with other French planetary astronomers, Danjon believed that the shadings to be seen on Venus were permanent enough to be charted, though this is no longer a viewpoint which is widely held.

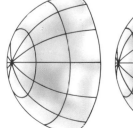

Earth

Mars

,000 mi
0,000 km

92,957,200 mi
149,600,000 km

141,6,000,000 mi
228,000,000 km

Venus 2

Telescopic Appearance

The first definite statement that Venus showed "spots" was made in 1645 by an Italian amateur astronomer, F. Fontana. Other observations followed, and in 1727 F. Bianchini, also in Italy, went so far as to compile a map showing what he regarded as oceans and continents! Yet since he was using a very feeble telescope (a long-focus 2½-in. refractor), it seems certain that he could have recorded no genuine detail on Venus.

Even good modern telescopes generally show the disk blank. Elusive shadings are visible at times, but the outlines of these are always indefinite, and most attempts to derive a rotation period from observations of them have been completely unsuccessful. More definite are the whitish caps seen at the planet's horns, or cusps. It is tempting to refer to them as "polar caps", and that is what they may well be; but they are certainly not a result of any surface deposit – and ice or frost could not exist in the torrid climate of Venus. The cusp-caps are more likely to be due

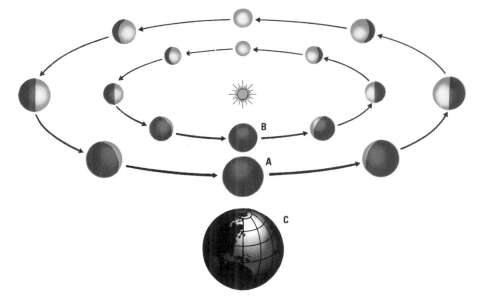

Phases of Venus *left and below*
Venus (A) shows movements of the same basic type as those of Mercury (B), but from the Earth (C) the maximum elongation from the Sun is 47° instead of only 27°. Dichotomy (half-phase) can be predicted, but when Venus is waning, dichotomy is always early; when Venus is waxing, dichotomy is late. Known as Schröter's effect it was first mentioned by J. H. Schröter in 1797. Its cause is associated with the planet's atmosphere.
below Venus at various phases, as drawn by A. Dollfus (Pic du Midi).

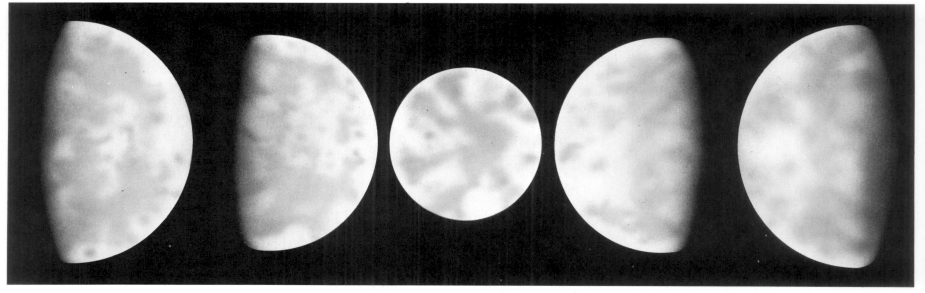

to some peculiarity of the atmospheric circulation. If they are indeed polar, this is less difficult to understand.

Visual observations of the shadings, and of the appearance and disappearance of the cusp-caps, led G. P. Kuiper, in the United States, to the conclusion that Venus must have a rotation period of about 30 Earth-days; this was the generally accepted value until the flight of Mariner 2 in 1962. Even now it would be premature to claim that the problem has

Ashen Light *left*
When Venus is at the crescent stage, the night side may be seen shining faintly. This is known as the Ashen Light. It is not certain whether it is a real phenomenon, or whether it is due to contrast effects.

been settled. From their visual and photographic studies, French observers believe that far from being very long, the rotation period is in reality only 4 days, retrograde.

If Venus is photographed in ordinary light, or in red light, no detail is seen; but in blue or violet light some darker patches show up. It is logical to suppose that these are phenomena of the upper part of the planet's atmosphere. Shifts in the position of these patches, plus a generally somewhat banded appearance, have been followed by the French astronomers. If they are right, and the radar measurements from the United States are wrong, our ideas about Venus are going to have to be revised yet again; however, at the present time a majority of astronomers believe that the rotation period really is 243 Earth-days, and that the shifts in the dark features must be explained in some other way. Venus must be regarded as a disappointing object when seen through a telescope; usually, almost nothing will be seen apart from the characteristic phase.

Rotation of Venus

Assuming that the radar measurements are correct, and that the rotation period of Venus is 243 Earth-days (retrograde: that is to say, in the opposite sense to the rotation of the Earth), the calendar would be very unfamiliar to an observer used to the terrestrial rotation.

As seen from the surface of the planet, a solar "day" will be as long as 118 Earth-days – provided that the axial inclination is not far from perpendicular with respect to the orbit.

If the tilt is 6° to the perpendicular to the orbital plane,

the pole star of Venus will be Phi Draconis, which is of the 4th magnitude; of course, it would never be possible for either it or any other star to be seen in practice from the planet's surface. owing to the dense, cloudy atmosphere.

To an observer on Venus, there would be less than two complete solar days in every "year". With a 243-day rotation period in the direct sense, he would find that the interval between successive sunrises on Venus would have taken place over the equivalent of 12 Earth-years.

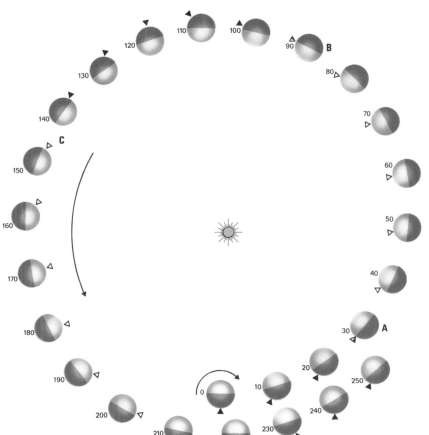

Rotation of Venus *left*
The diagram shows the consequences of the slow, retrograde rotation of Venus. The position of the planet in its orbit is shown at intervals of 10 Earth-days.

At the bottom of the drawing (day 0) the point on Venus which is directly turned away from the Sun and where it is so to speak "midnight" is indicated by a mark △

From the drawing, it will be seen that this point has a period of sunlight equal to approximately 59 Earth-days (from A to B), followed by an equal period of darkness (B to C); there are in fact less than two "days" in the "year".

The orbit of Venus is so nearly circular that the planet's velocity in its path is close to being constant, and so there are no complicated libration-type effects, as happen in the case of Mercury.

Transits of Venus

Like Mercury, Venus can pass in transit across the face of the Sun. The inclination of the orbit of Venus to the ecliptic is 3°24', much less than that of Mercury, and this makes transits less frequent. There were transits in 1631 and 1639, 1761 and 1769, and 1874 and 1882; the next pair will be in 2004 and 2012, followed by another in 2117 and 2125. At other inferior conjunctions, Venus passes above or below the Sun as seen from Earth, and there is no transit.

During a transit, Venus is prominently visible with the naked eye. All the transits since 1639 have been observed, and they were once regarded as being of great importance; observations of them gave a means of working out the value of the astronomical unit, or Earth–Sun distance. The method is now obsolete, and the transits of 2004 and 2012 will not be regarded as of more than academic interest.

Transits of Venus *right*
Transits of Venus, from 1761 to 2012. The transit of 2004 will be visible from Central European latitudes, but only the end of the 2012 transit will be seen there, as the Sun will rise while Venus is still silhouetted against the disk.

History of Transit Observation

The 1639 transit was seen by only two observers, J. Horrocks, a Lancashire curate, and his friend W. Crabtree; but in 1761, and again in 1769, major international programmes were organized. What had to be done was to time the moment when Venus passed fully on to the Sun's face. Unfortunately, the accuracy of the method was ruined by the phenomenon known as the "Black Drop", which made it impossible to give an accurate timing for the start of

"Black Drop" *right*
When Venus passes on to the Sun, it draws a strip of blackness after it, an effect which hampered early timings of transits. When it disappears, the transit is already in progress. The "Black Drop" is due to the atmosphere of Venus.

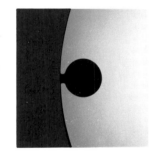

the transit. In 1761 M. Lomonosov, in Russia, was observing the transit when he noticed that the limb of Venus was hazy, and correctly deduced that this was due to the presence of an atmosphere; in 1769 Captain Cook was dispatched to the South Seas to observe the transit – and later discovered Australia.

Similar programmes were organized in 1874 and 1882, but again the accuracy of the method was ruined by the Black Drop, and nothing comparable is likely to be attempted at the next transit.

Occultation of a Star by Venus *right*
When Venus occults a star, the star flickers and fades before disappearing because its light is coming to us through Venus' atmosphere. This effect can help in estimating the height of Venus' atmosphere.

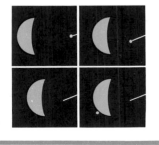

Mariner, Zond and Venera Probes

The first attempt at a planetary probe was made by the Russians on 12 February 1961, when they sent a vehicle out toward Venus. The launching was successful, but radio contact was lost at an early stage, and all that can be said is that the probe may have passed within 65,000 miles of Venus around May of the same year.

The first American attempt, with Mariner 1 (22 July 1962), was a failure, but Mariner 2, launched on 17 August, was a triumph; on 14 December it went past Venus at 21,600 miles, and sent back the first really reliable information about the surface temperature, as well as providing other invaluable information. There followed another Russian failure, Zond 1 (April 1964), and then two more Soviet probes. Venera 2 (launched 12 November 1965) went within 15,000 miles of Venus in February 1965, but without sending back useful information, while Venera 3 may have

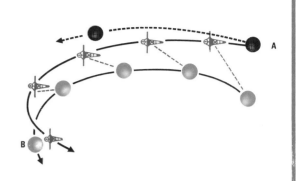

Venera 4 in Orbit *above*
Transfer orbit of Venera 4. In the diagram A = the launch from Earth on 12 February 1967, B = the encounter point in mid-May. In order to travel from the Earth to Venus, the probe makes a journey which carries it almost half-way round the Sun.

Mariner 2 Passes Venus *right*
Mariner 2's encounter with Venus, showing the approach to Venus (A), the scan zone (B) and direction of the Sun (C). The probe was never intended to land on Venus (and it was not sterilized). It had been meant to by-pass the planet within 9000 miles. Slight errors in launching and guidance meant that the minimum distance from Venus was 21,600 miles, and it was a great triumph that such good results were obtained.

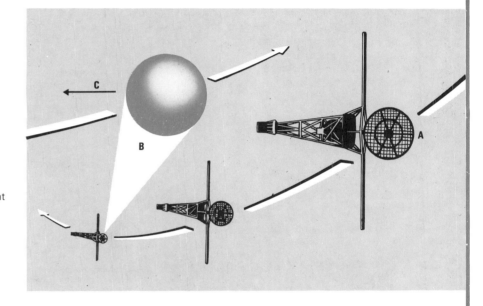

crash-landed on the planet on 1 March 1966. In 1967 the Russians brought down Venera 4 gently on to the surface, using parachute braking; since then there has been the American fly-by, Mariner 5, as well as three more Russian soft-landers. Venera 7, the most successful of these, sent back transmissions after arrival on Venus.

Transfer Orbits

In sending a probe to Venus, the vehicle must be slowed down relative to the Earth, so that the probe will start to swing in toward the orbit of Venus – meeting the planet at a pre-computed point. (This is known as a transfer orbit.) This kind of path is the most economical of fuel, but it is by no means the shortest route. Until the development of better fuels there is, unfortunately, no alternative; most of the journey to the target planet, whether it be Venus or Mars, has to be done in unpowered free fall.

A Super-Refractive Atmosphere *right*
At the surface of Venus the atmosphere may be super-refractive, so that an observer would seem as though standing at the bottom of an immense bowl.

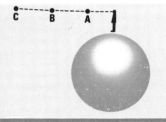

Future Exploration

Before any positive knowledge about the surface conditions had been gained, Venus was rated a more promising astronautical target than Mars. This has proved not to be the case. The intense heat on Venus must make manned landings in the foreseeable future extremely improbable. Following the revelations from the space-probes, Venus has been tacitly abandoned as a potential landing-site, and the attention of astronautical planners has been turned back to Mars.

This does not mean that exploration of Venus will be given up, and no doubt further probes will be sent there; but these can hardly be expected to add a great deal to our knowledge of the planet unless they can either transmit radio signals from the surface, or else undertake detailed analysis from above the clouds.

All probes intended to land on Venus are carefully sterilized; but it is not likely that any life-forms exist there.

Data

Mean distance from Sun	67,200,000 miles
(0·7233 astronomical units)	
Maximum distance from Sun	67,600,000 miles
Minimum distance from Sun	66,700,000 miles
Sidereal period	224·701 days
Synodic period	583·92 days
Axial rotation period	243 days
Orbital eccentricity	0·007
Orbital inclination	3°24'
Apparent diameter	max. 66".0, min. 9".6
Maximum stellar magnitude	−4·4
Diameter in miles	7700
Mass (Earth=1)	0·81
Density (specific gravity)	4·99
Volume (Earth=1)	0·92
Escape velocity	6·2 miles per sec (10 km. per sec.)
Maximum temperature	400 °C

Venus with the Naked Eye

Venus is so brilliant when seen with the naked eye that it is apt to be regarded as a telescopic disappointment. There has been considerable discussion as to whether its phases may be seen without optical aid. This would be possible only at the crescent stage, and there is no general agreement on the subject, but there are various well-authenticated cases in which the phase has been reported. The phases are easily visible in binoculars.

Mars 1

Mars is the most distinctive of all the planets. Its strong red colour led the ancients to name it after Ares or Mars, the mythological God of War; at its most brilliant, it can outshine every other star and planet apart from Venus. Yet Mars is not a large world. Its diameter (4219 miles) is little over half that of the Earth, and its brightness is due to the fact that it is relatively close. It may approach us to within 35 million miles. Of the planets, only Venus can come nearer.

Early Drawings

The first drawings to show detail on Mars were made by the Dutch astronomer C. Huygens in 1659. One of Huygens' sketches, made on 28 November of that year, shows recognizable detail, and there has never been any serious doubt that the dark patches on Mars are true surface features.

19th-Century Maps

More accurate observations were made during the 19th century, and maps of the Martian surface were drawn up. Three kinds of features were outstanding. There were the dark areas, which seemed to be permanent; there were the bright reddish-ochre regions, often called "deserts"; and there were the whitish caps covering the poles, thought by most astronomers of the time to be made up of some icy or frosty deposit. It was quite reasonable to assume that the bright regions were land, while the dark areas were seas. Mars appeared to be a reasonably friendly world, and it was thought quite probable that life might exist there.

Early Theories

Yet even before modern techniques were developed, it had become evident that conditions on Mars were very different from those on Earth. The temperature was certain to be lower, because Mars is much farther away from the Sun; but at noon on the Martian equator in midsummer, a thermometer would register 70°F, so that the planet is not permanently frozen. On the other hand, it was found that the polar caps could not be analogous to the deep snow and ice-caps of the Earth's poles. They shrank rapidly during the Martian spring and early summer, and yet released very little water vapour into the atmosphere, so that at best they could hardly be more than a few inches thick. Mars had to be regarded as being extremely short of water, and by 1877 it had become clear that whatever the dark patches were, they were not seas. The French astronomer Liais put forward the alternative theory that the dark regions were tracts of vegetation, which was accepted until recently.

Satellites and Canals

1877 was a notable year in the history of Martian observation. In America, Asaph Hall discovered the two dwarf satellites, Phobos and Deimos; and in Italy, G. V. Schiaparelli drew attention to what he described as a network of straight, artificial-looking lines crossing the "deserts". He called his lines "canali", and ever since then they have been known as the Martian canals.

What were the canals? Inevitably, the suggestion was made that they were artificial. The American astronomer Percival Lowell believed the network to be the work of highly intelligent Martians, striving to save every scrap of water on their arid planet.

The canals are not pure illusions; they have a basis of reality, but they are certainly neither so straight nor so regular as Lowell supposed.

New Knowledge from Mariner Probes

Before 1965, Mars still seemed reasonably welcoming. It was thought that the atmosphere must be made up chiefly of nitrogen, with polar caps composed of frosty material, and organic matter in the depressed dark regions; the bright deserts were assumed to be coated with some coloured mineral, possibly limonite or felsite. However, the Mariner probes of 1965, 1969 and 1971 have caused a change of our outlook. The Martian surface is covered with lunar-type craters; the atmosphere is unexpectedly thin, and is made up chiefly of carbon dioxide; and there is a good chance that the polar caps are made up of carbon dioxide in solid form. Moreover, the dark areas are not depressions; some of them are high plateaux or mountainous regions. We can no longer be at all confident that there is any life on Mars.

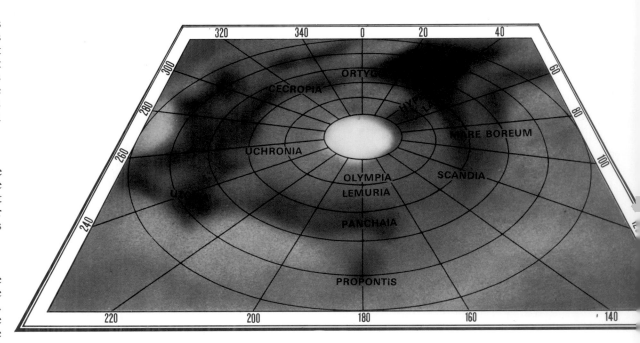

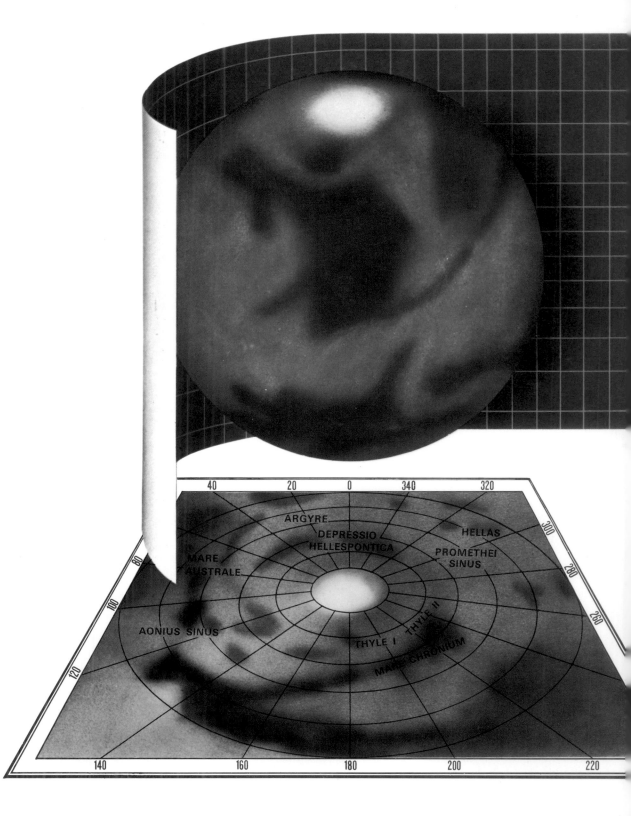

Mercury

36,000,000 mi
57,920,000 km

The North Pole *left*
The winters in the northern hemisphere of the planet are neither so long nor so severe as those in the southern hemisphere. This is why the northern polar cap never vanishes, though at midsummer it does become very small.

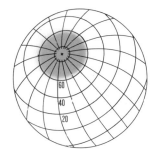

I.A.U. Map of Mars *below*
Many maps of Mars have been drawn up. It cannot be said that the agreement between them is uniform, but at least they are in accord with regard to the main dark features. This chart is a coloured interpretation of the official map approved by the International Astronomical Union. A Mercator Projection maps the temperate and equatorial regions of Mars. It is complemented by two perspective zenithal projections which map the polar regions. For clarity the globe is drawn at a relatively enlarged scale. North is at the top.

The most prominent of the dark areas are the Syrtis Major in the southern hemisphere, and the Mare Acidalium in the north. Both these may be seen with a small telescope when Mars is favourably placed, as happens every alternate year for a few weeks to either side of opposition. Other conspicuous dark regions are the Mare Erythræum, Mare Sirenum and Sinus Sabæus all in the southern hemisphere. Note also the bright, almost white area Hellas, south of the Syrtis Major, which is often very brilliant, and which unwary observers have been known to mistake for the polar cap.

On the I.A.U. map, various streaky canals are shown; these are delicate features, and many observers of Mars doubt their existence in such a form. There are also many minor spots, such as the Nix Olympica, which is shown on the Mariner photographs of 1969 as a huge crater.

The dark areas are essentially permanent. Local variations occur in both outline and extent, but these are usually minor and temporary.

19th-Century Map of Mars *right*
Map of Mars, drawn by N. E. Green in the 1870s. His nomenclature was different from that of today; for instance our Syrtis Major was his "Kaïser Sea", while our Hellas was his "Lockyer Land" The modern nomenclature was introduced by Schiaparelli in 1877.

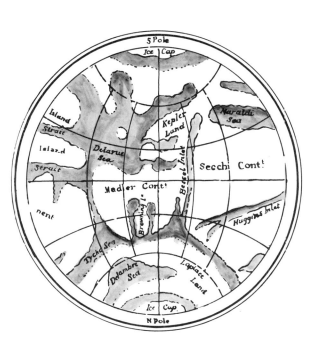

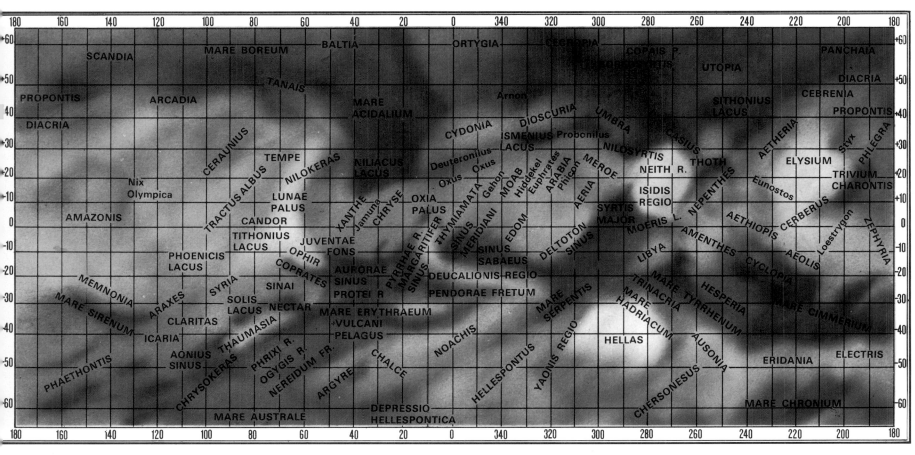

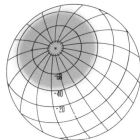

The South Pole *left*
The southern cap of Mars, covering the pole, can become larger than its northern counterpart because of the longer and colder winter; on the other hand, it has been known to vanish completely during the shorter, warmer summer.

Size *right*
Mars is much smaller than the Earth; its diameter is only a little over half that of our world. The surface area is 0.28 that of Earth, but it must be remembered that this is made up completely of "land", since there are no seas on Mars.

With Earth, the shortest season (northern winter, southern summer) amounts to 89 Earth-days, and the longest (northern summer, southern winter) to 93½ days; but it must be remembered that the orbit of the Earth is much more circular than that of Mars. Moreover, the greater amount of water in the southern hemisphere of the Earth has an influence upon the climate which is much greater than that of the changing distance from the Sun.

The Position of Mars *below*
Mars lies well beyond the Earth in the Solar System. Its mean distance from the Sun is 141,600,000 miles, and so it receives much less solar energy than does the Earth. The orbit is much more eccentric, so that the distance from the Sun ranges between 129,500,000 miles (perihelion) and 154,400,000 miles (aphelion). The tilt of the axis is only slightly greater than that of the Earth (25°), so that the Martian seasons are of the same general type, though they are of course much longer.
Mars is situated at the outer boundary of the Sun's "ecosphere". The Martian climate therefore is forbidding by terrestrial standards.

Sidereal Period
The Martian "year", or sidereal period, is 687 Earth-days. The length of the axial rotation period, or "day", is 24h. 37m., only about half an hour longer than ours. The seasons will, of course, be affected by the orbital eccentricity, and they will be of unequal length in the two hemispheres. In Martian days, the lengths will be as follows:

Northern spring, southern autumn: 194.2 Martian days
Northern summer, southern winter: 176.8 Martian days
Northern autumn, southern spring: 141.8 Martian days
Northern winter, southern summer: 155.8 Martian days

Density *(left)*
Mars is appreciably less dense than the Earth; its specific gravity (density on a scale in which water = 1) is 4.12
This leads to a fairly low escape velocity (3.1 miles per second), adequate to retain only a thin atmosphere. The surface gravity on Mars is 0.38 of that of the Earth, so that here also Mars is intermediate between Earth and Moon.

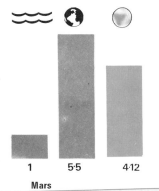

1	5·5	4·12

Mars

Earth

,000 mi	92,957,200 mi	141.6,000,000 mi
0,000 km	149,600,000 km	228,000,000 km

Mars 2

In 1971 the Americans sent out Mariner 9, a vehicle designed to orbit Mars and remain active for several months. (A twin vehicle, Mariner 8, had failed on launch.) Mariner 9 proved to be triumphantly successful. Until the autumn of 1972 it continued to send back detailed pictures, as well as miscellaneous information of all kinds.

For the first time, the Martian structures were seen in their true forms. Huge volcanoes were shown, associated with extensive drainage systems; there were valleys which looked as though they had been watercut, and many craters and rills. Mars was not the sort of world we had believed it to be, and there have been some recent suggestions that it may go through periods of greater atmospheric density, so that precipitation can occur. Certainly there can be no liquid water on the surface of Mars now, because the pressure of the atmosphere is too low. However, the valleys must have been cut by some liquid, presumably water, and the relative lack of erosion indicates that they cannot be very old on the "geological" time-scale.

Mariner 9 was not the only probe of 1971-72. The Russians sent up Mars 2 and Mars 3, both of which went into circum-Martian orbit; Mars 3 soft-landed a capsule on the planet's surface, though unfortunately signals from it ceased after only 20 seconds. In July and August 1973 the Russians dispatched four more probes to Mars two of which were "landers". The rest The next American Mars probe will be the Viking of 1975-6.

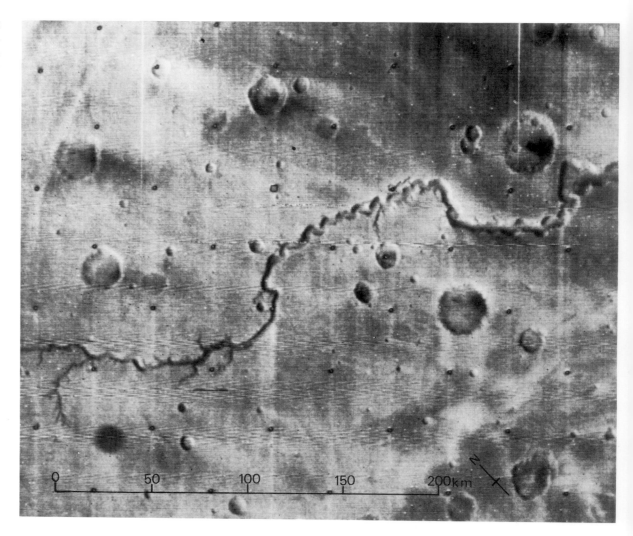

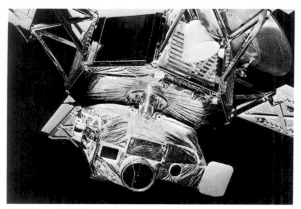

Mariner 9. This was the first probe to be put into an orbit round Mars. The cameras were programmed from Earth.

Tithonius Lacus Mosaic *below.*
The total length of the system is 3,500 miles, and the width is up to 250 miles. Not all of the system is shown in this mosaic; it runs from the lofty volcanoes right through to the region of Chryse (where the first American soft-landing of an automatic probe is expected in 1976). There is evidence of tributaries running from the main valley, though there is no chance that liquid water can exist on Mars today.

Sinuous Valley North of Argyre *above*
This remarkable valley is 190 miles (300 km) long and some 3 to 4 miles broad in the section shown here. It has been compared with an arroyo, a watercut gully found in parts of America. It lies 200 miles north of the boundary of the depressed basin of Argyre, one of the most clearly defined features on Mars. The photograph was taken from Mariner 9 on 2 February 1972.

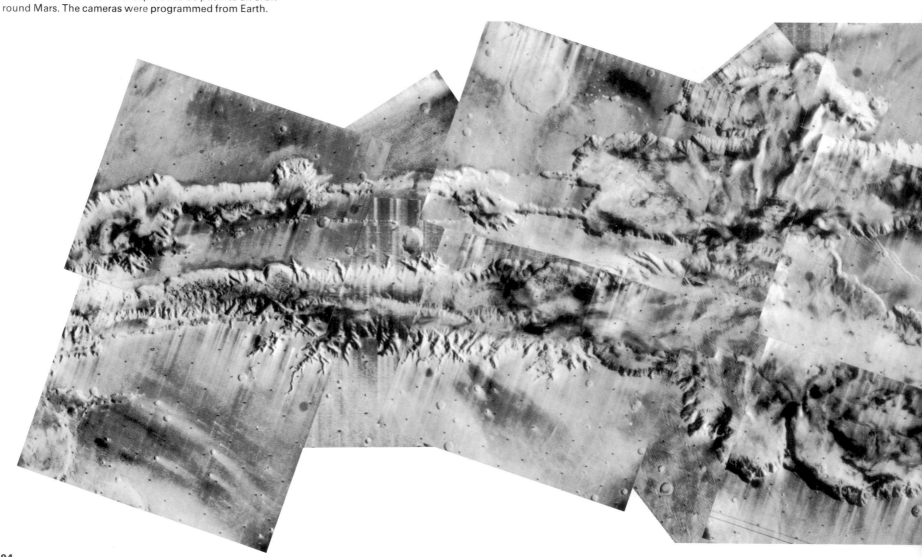

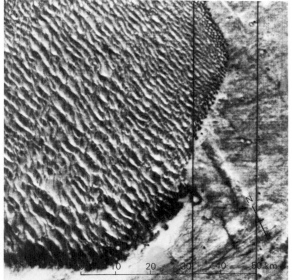

"Sand-Dune" Crater in Hellespontus *left*
This is a remarkable crater, on the cratered region of Hellespontus, which slopes down to the deep basin of Hellas. In the crater we see what look very like sand-dunes. The exact nature of the material is uncertain. The prevailing wind direction is from the top right of the photograph. The slope of the ground is, no doubt, an important factor here since Hellas is known to be the deepest point on the entire surface of Mars —instead of being a raised plateau, as used to be thought in pre-Mariner days.

The "Chandelier" *right*
This was the nickname given early on to this great system of canyons in the Noctis Lacus. Each canyon has an average width of almost 20 km, fully comparable with the Grand Canyon of the Colorado. The Chandelier marks the head of the great system of rift valleys lying to the east. Photograph taken from Mariner 9 on 9 January 1972, from an altitude of 8150 km. Despite the great size of the canyons, there is no chance of their being visible from Earth, even with powerful telescopes, so they are not identical with Lowell's canals.

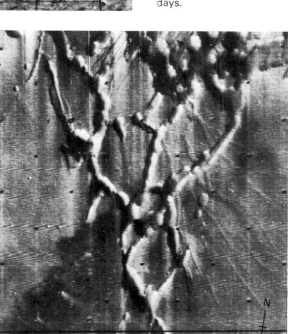

The Satellites of Mars
right Phobos and Deimos, the two attendants of Mars, were photographed from Mariner 9 at reasonably close range. Both have proved to be cratered, and are irregular in shape. Phobos is roughly elliptical, and has considerable craters on it which may be due either to blow-hole activity or to meteoritic impact; Deimos is similar, though smaller. They are probably not genuine satellites but captured asteroids.

Orbits of Phobos and Deimos *right*. Both satellites are too faint to be seen in small telescopes. Phobos moves in an almost circular orbit (A); its distance from the centre of Mars ranges between 5,924 miles and 5,726 miles, or less than 3,700 miles from the planet's surface. Phobos takes $7\frac{1}{2}$ hours to go once round Mars: from the planet itself, it would seem to rise in the west, cross the sky, and set in the east $4\frac{1}{2}$ hours later. Deimos, at 12,400 miles above the Martian surface, has a period of 30 hours 18 minutes, and would rise in the east and set in the west, staying above the horizon for $2\frac{1}{2}$ Martian days consecutively – though from high latitudes neither satellite would ever be visible. Phobos is also the only known natural satellite in the Solar System to have a sidereal period shorter than the rotation period of its primary.

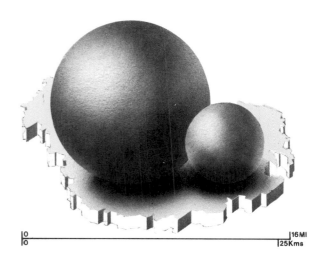

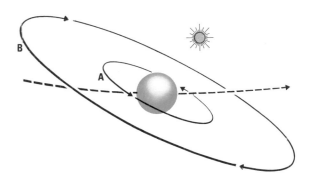

Satellite Data
Phobos and Deimos are difficult telescopic objects owing to their closeness to Mars. Phobos is only 5800 miles, and Deimos 14,600, miles from the centre of the planet. Their sizes are not accurately known, but have been estimated at 10 and 5 miles diameter respectively. Both satellites have orbits inclined at 2°, with eccentricities of 0·017 (Phobos) and 0·003 (Deimos) The sidereal period of Phobos is 7h. 39m.; of Deimos 1d. 6h. 18m.

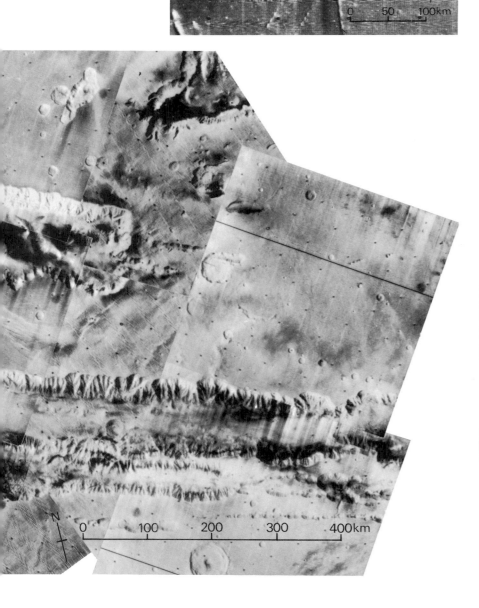

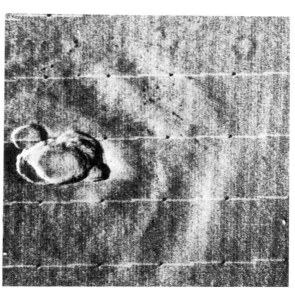

Ascraeus Lacus. *left*
This crater complex was photographed from Mariner 9 in late 1971, when the dust-storm was still in progress. The main crater, 13 miles wide, is visible above the top of the dust-storm, and the light arcs to the right are due to dust in the Martian atmosphere. Major dust storms of this kind are capable of obscurring much of the planet. Another occurred in the latter part of 1973, when the veiling was just as great as it had been at the time of Mariner 9's arrival in late 1971.

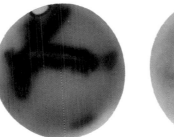

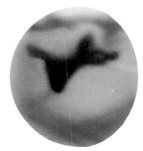

Dust Storm of 1971
20 September 1971: before the storm, the dark markings show up.

The date is 12 October 1971, and the planet is shrouded by a dense dust cloud.

Four months later, on 8 February 1972, the dust had cleared, making surface features visible.

Mars 3

1 *below*
Mars from Johannesburg, 1954.
Photographs by W. S. Finsen at the Republic
Observatory, Johannesburg. The first picture was
taken on 1 July 1954, at 21.05 U.T. The south polar
cap is well shown, and the dark mass of the Mare
Acidalium is prominent. North is at the top.

2 *below*
5 July 1954, 20.56 U.T. The Sinus Meridiani
(Martian longitude 0°) is to the right of the photograph.
The dark area below the centre of the picture,
extending toward the polar cap, is the Mare Erythræum,
and the Mare Acidalium is very prominent in the
northern hemisphere.

3 *below*
11 July 1954, 20.15 U.T. The Sinus Meridiani
is now to the left, and the famous triangular Syrtis
Major is to the right-hand side. The Syrtis Major is the
most conspicuous of all the dark areas on Mars ;
it used to be nicknamed the "Hour-Glass Sea", and
was drawn by Huygens in 1659.

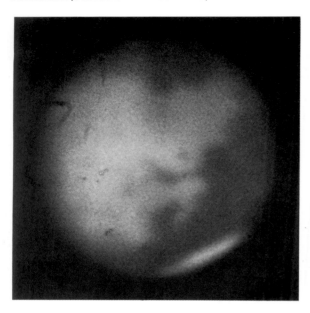
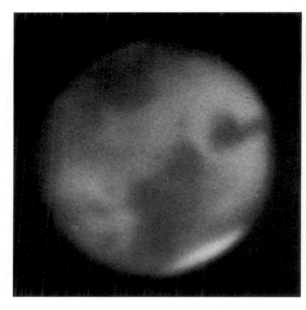
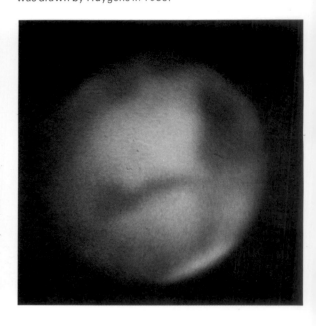

4 *below*
30 August 1956, 21.56 U.T. The opposition of 1956
was particularly favourable, since Mars was at perihelion.
This photograph shows the Sinus Sabæus
(right of centre) and some of the dark areas in the
far north.

5 *below*
31 August 1956, 22.04 U.T. The southern polar cap
is much smaller than on illustrations 1 and 2,
since the southern hemisphere of Mars was having
its summer. The dark markings near the equator do not
appear so intense as on some of the earlier pictures.

6 *below*
26 September 1956, 21.52 U.T. Much of
Mars is obscured – not by clouds in the atmosphere of
the Earth, but by obscuration above Mars. Only
the polar cap and part of the Mare Cimmerium
can be seen clearly.

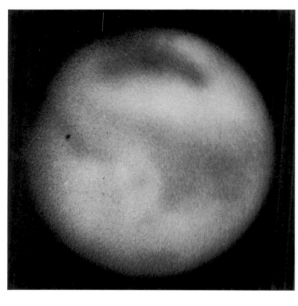
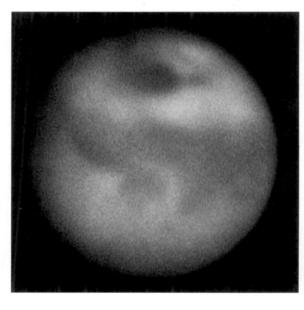
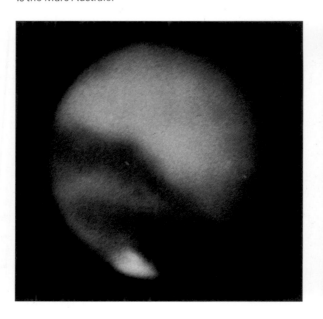

7 *below*
14 July 1954, 18.49 U.T. The curved streak
running upward from the Syrtis Major, left of centre, is
the Nepenthes, one of the Martian "canals". It joins the
Syrtis Major to Thoth. It is nothing like a sharp, narrow
canal, and gives no impression of artificiality !

8 *below*
17 August 1956, 23.52 U.T. Taken in 1956
before the development of the widespread obscuration.
In the south (lower centre) the dark features are
prominent. The dark region bordering the polar cap
is the Mare Australe.

9 *below*
29 August 1956, 23.24 U.T. A good general view of Mars.
The features can be identified on the Mercator map on
page 93. Fine detail on Mars can be seen only with
powerful telescopes ; small instruments will not show nearly
so much as Finsen's series of photographs.

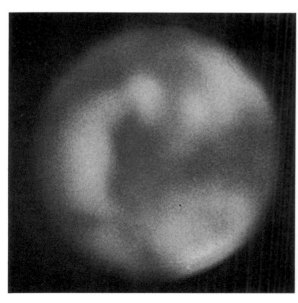
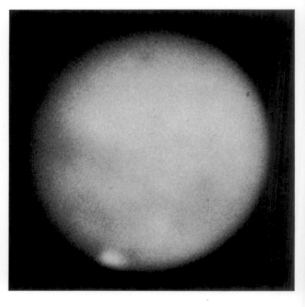
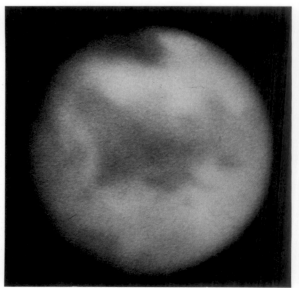

Mars from Mount Wilson, 1956 *above*
The phase is evident, with the terminator to the right
The dark markings near the centre are in the Mare
Cimmerium—Mare Sirenum area. The border of the

polar cap is well-defined. There is no trace of a
dark peripheral band, often seen during times when
the cap is shrinking, and formerly attributed to moisture
temporarily dampening and so darkening the ground.

The Martian Terrain
There seem to be three types of terrain on Mars. There are
heavily-cratered regions; "chaotic" areas marked by
jumbled ridges and valleys but almost without craters; and
occasional regions which appear to be devoid of features.
The best example of a featureless area is the bright
circular plain Hellas, south of the Syrtis Major (see map on
page 93).

Radar Profile *below*
This cross-section of the planet shows radar measurements
of height. The area features lies between Niliacus Lacus
and Syrtis Major, at latitude 21°N., and was obtained at a
wavelength of 3.8 cm. The measurements show no
correlation between height and colour. The bright area of
Elysium (E) is high, but so is the dark area of Syrtis Major (G).

**Radar Profile
Features**
A Niliacus Lacus
B Lunæ Palus
C Ceraunius
D Trivium Charontis
E Elysium
F Thoth
G Syrtis Major

Atmosphere of Mars
right The existence of an
atmosphere round Mars
has been known for a long
time. Before the age of
space-probes, it was
generally thought that the
surface pressure must be
in the region of 85
millibars, and that the
main constituent was
likely to be nitrogen.
However, experiments
carried out with the Mars
Mariners since 1965 have
shown that the pressure is
below 10 millibars, and
that the Martian
atmosphere is almost pure
carbon dioxide. This is no
more than the pressure at
over 100,000 feet above
sea-level on the Earth, and
indicates that there is no
effective screen against
the various radiations
coming from space.
Obviously this alters our
whole idea of Mars as a
world, and makes the
existence of any form of
life there much less
probable, particularly as it
also seems that the white
polar caps are mainly
composed of solid carbon
dioxide. The Martian
clouds are not in the least
like our own, but we can
observe major dust-storms
which are capable of
hiding the surface
features for weeks at a
time. One such storm
occurred in 1971, and
another in late 1973.
The rarefied atmosphere
means that liquid water
cannot exist on the surface
of Mars; there is not
enough pressure. In
theory, it would just be
possible for water to exist
in liquid form at the very
bottom of Hellas, where
the pressure is greater than
at any other point on the
Martian surface—but no
scientist will seriously
suggest that it really does
exist there.

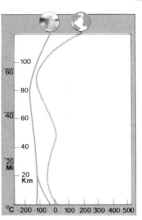

Earth/Mars Charts *left*
The diagrams compare
the temperature and
pressure gradients for the
atmospheres of the Earth
and Mars.
left Temperature in the
atmospheres of Earth and
Mars: in both cases the
temperature is high at
great altitudes, but
'temperature' here
depends upon the rate at
which the atoms and
molecules are moving
about, and is no indication
of what we normally call
'heat'.
Left Pressure. At the
Martian surface the
atmosphere pressure is
below 10 mb, and in some
areas is below 3 mb. With
increasing height above
the surface the pressure
decreases still further, just
as with the Earth; but the
Martian atmosphere is less
concentrated toward the
ground, because of the
weaker gravitational pull,
and the pressure falls
away less quickly. It is
impossible to estimate the
height above Mars at
which the density has
become inappreciable.

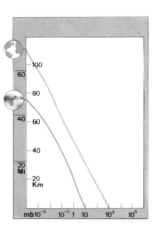

Data
Mean distance from Sun 141,600,000 miles (1·524 astronomical units)	**Apparent diameter** max. 25".7, min. 3".5
Maximum distance from Sun 154,400,000 miles	**Maximum stellar magnitude** —2·8
Minimum distance from Sun 129,500,000 miles	**Diameter** 4219 miles
Sidereal period 686·98 days	**Mass (Earth=1)** 0·11
Synodic period 779·94 days	**Density (specific gravity)** 4·12
Axial rotation period 24h. 37m. 23s.	**Volume (Earth=1)** 0·15
Orbital eccentricity 0·093	**Number of satellites** 2 (Phobos, Deimos)
Orbital inclination 1°51'	**Escape velocity** 3·1 miles per sec. (5·0 km. per sec.)
	Maximum temperature 22°C
	Minimum temperature —70°C

The Asteroids

Discovering the Minor Planets

Even a casual glance at a scale map of the Solar System will show that the planets are divided into two groups. Between the orbits of Mars and Jupiter, there is a wide gap of over 300 million miles. A mathematical relationship discovered in the 18th century by J. D. Titius, and popularized by J. Bode, led to the suggestion that there might be a planet moving in this part of the Solar System, and at the instigation of a Hungarian amateur astronomer, the Baron von Zach, a group of observers formed themselves into an association of "planet-hunters". They worked out a method of systematic searching, but in the event they were forestalled. On 1 January 1801– the first day of the new century – G. Piazzi, at Palermo, was making routine observations for a new star catalogue when he discovered a telescopic object which proved to be the expected planet. It was named Ceres. It moves round the Sun at a mean distance of 257 million miles, so conforming well to Bode's Law.

However, the planet-hunters were not satisfied. Ceres seemed too small to be a proper planet (we now know that its diameter is a mere 427 miles); there might be others. Three more – Pallas, Juno and Vesta – were found between 1802 and 1808, but no more seemed to be forthcoming, and the planet-hunters disbanded. Then, in 1845, a German amateur named Hencke discovered the fifth minor planet, Astræa. Since 1848 no year has passed without there being several new discoveries; the number of known asteroids now amounts to thousands – and according to one estimate, there may be well over 40,000. In modern times, the discoveries are made photographically, since the asteroids move perceptibly over relatively short periods, and show up as trails on star-plates.

The Largest Minor Planets

Ceres is the largest of the asteroid swarm, and of the rest only Pallas and Vesta have diameters of more than 250 miles. No asteroid can be of sufficient mass to retain any vestige of atmosphere, and most are mere lumps of material a few miles or even only a few hundred yards across; probably they are not even approximately spherical.

Vesta, with a probable diameter of 380 miles, is the brightest of the minor planets. It has a mean distance from the Sun of only 219,300,000 miles, so that it is much closer to us than Ceres; it also has a higher albedo or reflecting power. It can reach magnitude 6, so that it is then just visible without optical aid.

Clusters and Gaps

Minor planets tend to "cluster", and there are certain regions which are free of them. At distances in which a minor planet would have a revolution period of a simple fraction of the period of Jupiter (particularly one third, two fifths, and one half of Jupiter's period), cumulative perturbations by Jupiter's pull of gravity will force any asteroid into a different path. These gaps in the belt are known as Kirkwood gaps. Some asteroids have highly inclined orbits (over 34° in the case of Pallas) and some orbits are eccentric, but no asteroid with retrograde motion is known.

Exceptional Orbits

The most interesting of the asteroids are those which depart from the main swarm. Some, such as Eros, may come close to the Earth; Eros, an irregularly shaped body with a longest diameter of approximately 18 miles, by-passed us at 15 million miles in 1931, and will do so again in 1975. In 1937 a 1-mile body, Hermes, brushed past at a mere 485,000 miles from the Earth; Icarus, which can approach us to within 4 million miles, has an orbit which carries it closer to the Sun than Mercury; Hidalgo has a path which is so eccentric that its aphelion is not far from the orbit of Saturn. The Trojans move in the same orbit as Jupiter, but at distances of 60° or so from it, so that there is no fear of collision.

The origin of the asteroids is uncertain. Either they represent the debris remaining after the formation of the main planets – that is to say, material which never collected into a large body, because of the disruptive influence of Jupiter – or else they are the remains of a former planet (or planets) which met with some disaster in the remote past, and suffered disruption.

Vesta

left Vesta, brightest of the asteroids (photograph by F. C. Acfield). Vesta is between the two arrows; the cross marks its position 24 hours later.
above and below Two sketches of Vesta made by Patrick Moore at a 24-hour interval in 1969.

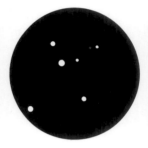

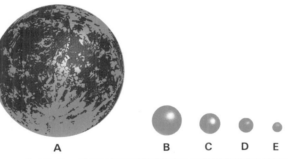

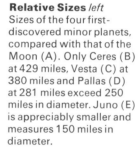

Relative Sizes *left*
Sizes of the four first-discovered minor planets, compared with that of the Moon (A). Only Ceres (B) at 429 miles, Vesta (C) at 380 miles and Pallas (D) at 281 miles exceed 250 miles in diameter. Juno (E) is appreciably smaller and measures 150 miles in diameter.

Even if the asteroids were combined into one body, the mass would still be smaller than the Moon's.

Asteroid Trails *below*
in this time-exposure by Max Wolf, the motion of the asteroids relative to the background of fixed stars creates small trails on the photographic plate.

Mars	Asteroids	Jupiter	Saturn
141,600,000 mi 227,800,000 km		484,300,000 mi 779,100,000 km	886,100,000 mi 1,426,000,000 km

Asteroid Belt *right*

Orbits of minor planets in the main swarm. The tendency toward grouping is clearly shown, and there are definite regions of avoidance — the Kirkwood gaps, due to the perturbing influence of Jupiter.

Orbit of Icarus *below*

Orbit (A) of the exceptional asteroid Icarus, discovered by the late W. Baade on a photograph taken on 26 June 1949 at Palomar. Icarus is only about 1 mile in diameter, with an orbital inclination of 23°. The revolution period is 409 days ; the orbital eccentricity is 0·83, and at perihelion the asteroid is only about 19 million miles from the Sun — within the orbit of Mercury (B). The temperature must then be over 500°C, enough to make the surface material glow.

Icarus is not an ''Earth-grazer'', but it can approach us to a distance of 4 million miles, as happened in 1969. During this approach Icarus became bright enough to be visible in a moderate telescope (it attained the 12th magnitude) and radar echoes were obtained.

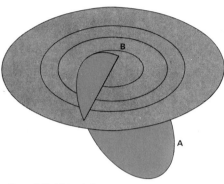

Exceptional Orbits *right*

The diagram shows some asteroid orbits of relatively high eccentricity. Icarus is the only asteroid to pass within the orbit of Mercury, but Hidalgo (1) recedes beyond the orbit of Jupiter, and its orbit — highly inclined as well as eccentric — is so like that of a comet that unsuccessful searches have been made for any trace of a cometary-type nucleus around the asteroid. Orbits of this kind are exceptional in the minor planet swarm, and it is significant that so far no minor planets with retrograde motion have been discovered.

Minor Planet Data

	Year of Discovery	Diameter (miles)	Sidereal Period (years)	Orbital Inclination (degrees)
Ceres	1801	429	4·60	10·6
Pallas	1802	281	4·61	34·8
Juno	1804	150	4·36	13·0
Vesta	1807	384	3·63	7·1
Astræa	1845	112	4·14	5·3
Hebe	1847	106	3·78	14·8
Iris	1847	94	3·68	5·5
Flora	1847	77	3·27	5·9
Metis	1848	133	3·69	5·6
Hygiea	1849	222	5·60	3·8
Parthenope	1850	75	3·84	4·6
Victoria	1850	94	3·56	8·4
Egeria	1850	123	4·14	16·5
Irene	1851	98	4·16	9·1
Eunomia	1851	146	4·30	11·8
Psyche	1852	201	4·99	3·1
Fortuna	1852	100	3·82	1·5
Massalia	1852	112	3·74	0·7
Calliope	1852	156	4·96	13·7
Euterpe	1853	94	3·60	1·6
Amphitrite	1854	114	4·08	6·1
Urania	1854	56	3·64	2·1
Lactitia	1856	160	4·60	10·4
Harmonia	1856	56	3·41	4·3
Nemausa	1858	94	3·64	10·0
Ausonia	1861	92	3·70	5·8
Nausicaa	1879	120	3·72	6·9
Bamberga	1892	122	4·40	11·3
Dembovska	1892	160	5·0	8·3
Eros	1898	11	1·76	10·8
Hidalgo	1920	27	3·96	42·5
Amor	1932	2	2·67	11·9
Apollo	1932	2	1·81	6·4
Adonis	1936	1	2·76	1·5
Hermes	1937	1	1·47	4·7
Icarus	1949	1	1·12	23·0

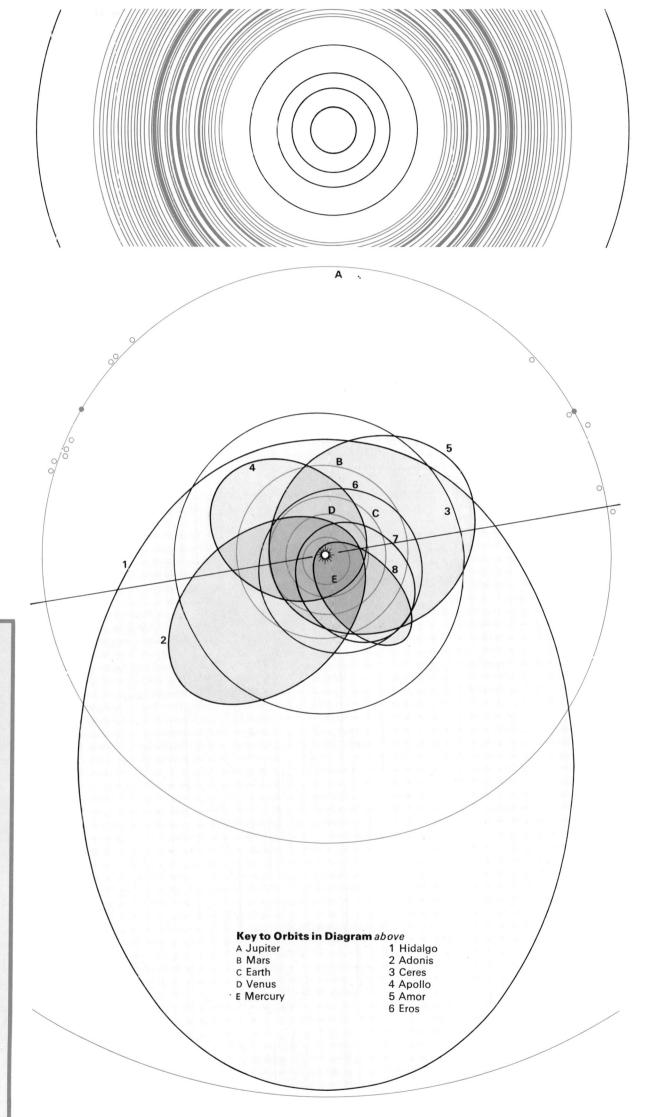

Key to Orbits in Diagram *above*

A	Jupiter	1	Hidalgo
B	Mars	2	Adonis
C	Earth	3	Ceres
D	Venus	4	Apollo
E	Mercury	5	Amor
		6	Eros

Uranus

1,783,000,000 mi
2,870,000,000 km

Neptune

2,793,000,000 mi
4,493,000,000 km

Pluto

3,666,000,000 mi
5,898,000,000 km

Jupiter 1

The Largest Planet

Jupiter is the largest member of the Sun's family. Its mass is greater than that of all the other planets combined, and it was appropriately named after the ruler of the mythological Olympus. Despite its great distance from us – always at least 370 million miles – it shines more brilliantly than any other planet apart from Venus, except on very rare occasions when it may be temporarily outshone by Mars. Its maximum magnitude on the stellar scale is —2·5.

The difficulties of observation are much less than with Venus, which is "new" when at its closest to us, or Mars, which can be well studied only for a brief spell to either side of opposition. Jupiter is well placed for several months in every year, and it reaches opposition every 13 months (more precisely, every 399 days).

Telescopically, Jupiter is seen to be a very different world from the Earth. It is, of course, very much larger (its volume is 1312 times that of our world), and its globe is obviously flattened, due to the comparatively rapid rotation; a "day" on Jupiter amounts to less than 10 hours, so that the equatorial regions tend to bulge out. The difference between equatorial and polar diameter is more than 5000 miles. Jupiter is also less dense than Earth. The outer layers, at least, are composed of gas.

There is a fundamental difference between the clouds of Venus and those of Jupiter. With Venus, we see a relatively shallow atmosphere lying over a solid globe. With Jupiter, it is impossible to tell how deep the atmosphere is. There may be no well-defined solid surface, though deep inside the planet the pressures must be great and the temperatures high. On its surface, Jupiter is cold, with a temperature of —130°C (—200°F).

Telescopic Appearance

Jupiter is a fascinating object when seen through a telescope. The flattened, rather yellowish globe is seen to be crossed by streaky "cloud belts", and there is a tremendous amount of detail to be seen, including wisps, festoons, and spots. Of the latter, the most famous is the Great Red Spot, which appears to be a semi-permanent, if not permanent, feature – unlike most of the other spots, whose lifetimes are only a few weeks or months. Several belts are always to be seen, and are given special names (see diagram on right).

Because Jupiter's surface is gaseous, the rotation period is not the same for the whole planet. The equatorial zone has a period about 5 minutes shorter than that of the rest of Jupiter, but various features, such as the Red Spot, have individual periods, and drift about in longitude. Amateur observers have done invaluable work in measuring the rotation periods of the different belts and zones. The method is to wait until a feature is carried by rotation on to the central meridian of Jupiter. The central meridian can be found, because of the cloudbelts and the polar flattening. If the meridian transit is timed, reference to tables will give the longitude on Jupiter of the feature concerned, and successive transit timings will provide an accurate rotation period.

Nature of Jupiter

Up to half a century ago it was thought probable that Jupiter must be hot, giving off light on its own account, and behaving in the manner of a feeble star. This is now known to be wrong. There is a fundamental difference between a stellar and a planetary body; large and massive though Jupiter may be, it is not massive enough spontaneously to generate or maintain internal nuclear reactive processes and so qualify as a star. No nuclear energy is being produced at its core. The outer gas is chiefly hydrogen – not surprisingly, since hydrogen is much the most abundant element in the universe. Hydrogen compounds, like ammonia and methane, with a considerable quantity of helium and smaller amounts of other elements, are also found in its atmosphere.

Jupiter's hydrogen-rich atmosphere is due to the fact that it is so massive; it has been able to retain all its original material. However, there are many problems about it which remain to be solved. In May 1971 a probe, Pioneer 10, was sent out on a mission to it; it is scheduled to fly past the planet, at within 100,000 miles, in late 1973. Pioneer 11, launched in 1972, will arrive in early 1975. After by-passing Jupiter, they will leave the Solar System altogether.

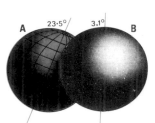

Size *right*
The equatorial diameter of Jupiter is 88,700 miles, and Jupiter is by far the largest and most massive of the planets. The Red Spot (A in the diagram) is 25,000 miles long and 7000 miles broad when at its greatest extent, so that its surface area is equal to that of the Earth (B).

Axial Inclination *left*
The axial inclination of Jupiter (B) is only 3°·1, much less than that of any other planet; it is strikingly less than that of the Earth (23½°) (A). Jupiter would therefore have no "seasons" of terrestrial type.

Density *right*
The mass of Jupiter is greater than that of all the other planets combined, and is 318 times that of the Earth. It is composed very largely of the lightest elements, hydrogen and helium, and in consequence its density is low; the specific gravity is only 1·33, so that the density is comparable with that of water. This alone is enough to show that the constitution of Jupiter must be very different from that of the Earth, but the globe is not uniform, and near the centre the temperatures, pressures and densities must be very high.

1	5·5	1·3

Mars	Asteroids	Jupiter	Saturn
141,600,000 mi		484,300,000 mi	886,100,000 mi
227,800,000 km		779,100,000 km	1,426,000,000 km

Nomenclature *left*

Each of the belts and zones of Jupiter numbered in the diagram to the left has its own rotation period. The periods given in this list are means only, and are subject to marked fluctuations.

Belts and Zones of Jupiter

1 South Polar Region (SPR) 9h. 55m. 30s. Lat. −90° to −45°. Includes the South South South Temperate Belt (SSSTB)
2 South South Temperate Zone (SSTZ) 9h. 55m. 7s. } In the SS Temperate Current
3 South South Temperate Belt (SSTB) 9h. 55m. 7s. } In the SS Temperate Current
4 South Temperate Zone (STZ) 9h. 55m. 20s. } In the S Temperate Current
5 South Temperate Belt (STB) 9h. 55m. 20s. } In the S Temperate Current
6 Great Red Spot 9h. 55m. 35s. Lat. approx. −22°
7 South Tropical Zone (STrZ) 9h. 55m. 36s. Lat. −26° to −21°
8 South Equatorial Belt. S. component (SEBs) 9h. 55m. 39s. S edge, lat. −19°. Interrupted by the Red Spot Hollow.
9 South Equatorial Belt, N. component (SEBn) 9h. 55m. 39s.
10 Equatorial Zone (EqZ) 9h. 50m. 24s. Great Equatorial Current
11 Equatorial Band (EqB) 9h. 50m. 24s. An elusive feature
12 North Equatorial Belt (NEB) 9h. 54m. 9s. Lat. +13°
13 North Tropical Zone (NTrZ) 9h. 55m. 29s. Lat +14° to +22°
14 North Temperate Belt (NTB) 9h. 53m. 17s. Lat. +27°. In N Tropical Current
15 North Temperate Zone (NTZ) 9h. 56m. 5s.
16 North North Temperate Belt (NNTB) 9h. 55m. 42s. Lat. +36° to +40°
17 North North Temperate Zone (NNTZ) 9h. 55m. 42s.
18 North North North Temperate Belt (NNNTB) 9h. 55m. 20s. Lat. +43°
19 North Polar Region (NPR) 9h. 55m. 42s. Lat. +47° to +90°

The region between the north edge of the South Equatorial Belt and the south edge of the North Equatorial Belt is called System I; the rest of the planet is System II. When working out observations, two separate sets of tables are needed. In this diagram, and in subsequent drawings and photographs of the outer planets, south is at the top and north at the bottom, corresponding to the telescopic view.

Hypothetical Cross-Section *left*

Structure of Jupiter according to a model proposed by R. Gallet in 1964. This cross-section is still very tentative, and assumes that the temperature at the base of the atmosphere is very high. At the moment, we must admit that there is no generally accepted theory of the constitution of Jupiter and the other giant planets.

On Gallet's hypothesis, the temperature of the outer atmosphere (A on the main diagram) increases toward the interior. The substances present may hence occur in their gaseous forms lower in the

atmosphere, and as cirrus-like clouds of crystals in the outer zones.

In the expanded diagram of the atmosphere above (layer A on the main diagram), 1 = ammonia crystals, 2 = ammonia droplets, 3 = ammonia vapour, 4 = ice crystals, 5 = water droplets, 6 = water vapour.

Under the atmosphere the surface lies at ·94 of the radius of the planet. It marks the upper limit of a layer of liquid and/or solid hydrogen (layer B on the main diagram) extending down to 4/5 of the planet's radius. Below this gaseous layer, the pressure would reach about a million atmospheres. A transition zone (C) would mark the onset of a solid "metallic" state (D) unknown on Earth, and extending down to perhaps a quarter of the radius. By analogy with their cosmic abundances, a core (E) of dense metallic elements and rocky silicates is assumed to lie beneath layer C. This core alone would have a total mass perhaps 10 times that of the Earth.

The Changing Aspects of Jupiter *left*

top left 25 January 1968, 1.25 GMT (G. P. Kuiper, 61-in. reflector). Many belts are visible, and there is a great deal of fine structure. This series of photographs shows how markedly the surface aspect can change from one year to another, though usually the North Equatorial Belt is dominant.

centre left 31 October 1964, 05.41 GMT (P. Glaser, 8-in. reflector). The most striking feature is the yellow hue of System 1 between the Equatorial Belts; the Red Spot is prominent, on the central meridian.

centre right 20 January 1967 (G. P. Kuiper, 61-in. reflector). The yellow hue of the equatorial region is no longer evident; at the central meridian, south of the South Temperate Belt, is one of the characteristic white spots of the South Temperate Zone. The ragged outline of the main belts is strongly evident.

bottom left Jupiter from Pioneer 10. On 3-4 December 1973 the probe Pioneer 10 by-passed Jupiter at only 81,000 miles, and sent back close-range photographs. This picture, shows the planet with a definite phase – something never seen from Earth.

below 28 September 1963, 23.10 GMT (W. Rippengale, 10-in. reflector). Shadow of Ganymede to upper left.

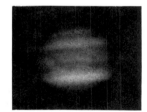

below 17 October 1963, 23.20 GMT (Rippengale).

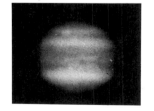

below 1 November 1964, 01.34 GMT (H. E. Dall, 10-in. reflector). Europa's shadow is visible against Jupiter.

Uranus

Neptune

Pluto

1,783,000,000 mi
2,870,000,000 km

2,793,000,000 mi
4,493,000,000 km

3,666,000,000 mi
5,898,000,000 km

Jupiter 2

The Red Spot

The Great Red Spot, which can be traced on drawings made as long ago as 1631, is a prominent feature on Jupiter under normal conditions. It became very conspicuous in 1878, and was described as brick-red, measuring 30,000 miles long by 7000 wide, so that its surface area was greater than that of the Earth. Since then it has fluctuated in visibility; the colour ranges from brick-red to pale orange. It is not fixed in longitude, but drifts about slowly.

The precise nature of the Red Spot is still unknown. On the "raft" theory, it is some sort of solid body, or group of solid or semi-solid bodies, floating in Jupiter's outer gas; when it disappears, as sometimes happens, it is assumed to sink in level, so that it is covered up. Only the associated "Hollow" then remains visible. Alternatively, it may be the top of a Taylor Column – a column of stagnant gas due to the atmospheric winds flowing round some feature on a solid surface well below the visible cloud-layer.

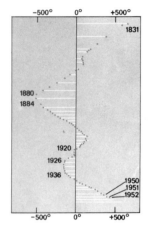

Motion of the Red Spot *left*
Changes in longitude of the Great Red Spot, 1831–1955. Up to 1880 the longitude decreased, and increased, with fluctuations, thereafter. In 1970 the longitude on System II was 025°.

Two Theories of the Red Spot

(1) The first theory is the "raft" theory that the Spot is floating in the Jovian atmosphere. Changes in visibility can be explained by differences in level.

(2) The second is the "Taylor Column" theory, that the Spot is the top of a column of stagnant gas. This assumes a solid surface below the cloud layer.

The Red Spot is in any case exceptional inasmuch as it seems to be semi-permanent, and it has a marked influence on Jupiter's southern hemisphere. Between 1901 and 1940 the South Tropical Disturbance (STrD) in the same latitude was very prominent. As it had a revolution period shorter than that of the Spot there were periodical interactions; but the STrD faded away, and after 1940 was lost, probably permanently. The two theories are illustrated below

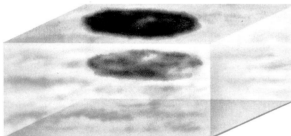

Raft Theory *above*
The Spot is assumed to be a solid or semi-solid body floating in Jupiter's outer atmosphere. When the Spot sinks, it becomes invisible, as it is covered by a thicker layer of gas.

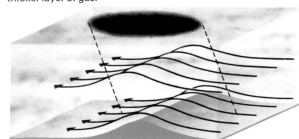

Taylor Column Theory *above*
The Red Spot is assumed to be the top of a column of stagnant gas produced by the interruption of the atmospheric circulation, at low level, by some topographical feature of major proportions on the actual surface of Jupiter.

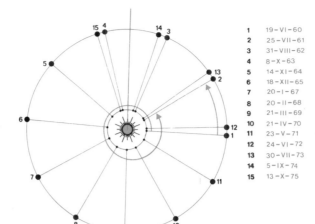

1	19 – VI – 60
2	25 – VII – 61
3	31 – VIII – 62
4	8 – X – 63
5	14 – XI – 64
6	18 – XII – 65
7	20 – I – 67
8	20 – II – 68
9	21 – III – 69
10	21 – IV – 70
11	23 – V – 71
12	24 – VI – 72
13	30 – VII – 73
14	5 – IX – 74
15	13 – X – 75

Orbit and Oppositions *left*
Jupiter is so far from the Earth and from the Sun that the orbital eccentricity makes no appreciable difference so far as observation is concerned; the apparent diameter at opposition is always over 40" of arc, and even when Jupiter is at its most distant from the Earth the apparent diameter still exceeds 30" of arc.

The opposition magnitude varies between −2·0 and −2·5. The opposition distance from Earth ranges between 367 million miles when Jupiter is at perihelion to 413 million miles when Jupiter is at aphelion. In 1975 Jupiter will approach as closely as it can ever do. The diagram shows the opposition positions between 1960 and 1975.

Jupiter in Six Different Lights *above and below*
Photographs of Jupiter, taken on 19 September 1927 by W. H. Wright at the Lick Observatory in the United States. These photographs are taken in the light of different colours: ultra-violet, violet, green, yellow, red, and infra-red. Note that the different wavelengths show different surface aspects, but in all cases the North Equatorial Belt (below the centre of the disk) is the most prominent feature.

Photographs taken in the light of different wavelengths bring out differences in the belt structure. When Jupiter is observed through a small telescope, the belts appear straight and regular; with increased magnification, their outlines are seen to be ragged, with much fine detail. Generally the outlines of the spots, including the Great Red Spot, are much more regular. When these photographs were taken, the South Equatorial Belt was relatively obscure.

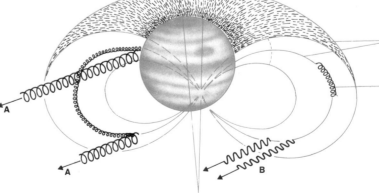

Radio Emissions *left*
A speculative diagram, drawn up to account for some of the characteristics of the radio emissions from Jupiter:

A Circular- polarized radiation
B Plane-polarized radiation
C Jupiter's rotation axis
D Jupiter's magnetic axis
E Magnetic field lines
F Paths of trapped particles

The Jovian Magnetosphere

It seems that Jupiter has a powerful magnetic field, and it is also a very strong radio source. Intense bursts of decametric radio radiation were discovered by Burke and Franklin in 1954, but have no certain explanation. The theories proposed to account for the radiation include lightning discharges in the Jovian atmosphere, plasma oscillations, shock waves caused by disturbances of a volcanic nature occurring below the cloud layer, and chemical explosions in the Jovian atmosphere. The radio emission seems to be associated with the positions in orbit of Io, Jupiter's inner large satellite.

Satellites

Jupiter has a total of 12 satellites. Of these, four are bright enough to be seen with any telescope; good binoculars will show them. They would be easy naked-eye objects if they were not overpowered by the brilliant light of Jupiter itself. They were discovered in the winter of 1609–10 by Galileo and independently by Simon Marius. They have been named Io, Europa, Ganymede and Callisto. Io and Europa are about the size of our Moon; Ganymede

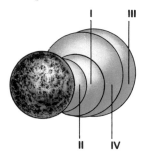

Jupiter's Four Largest Satellites *left*
Four of Jupiter's satellites are bright enough to be seen with any telescope; Io (I) and Europa (II) are comparable in size with our Moon, while Ganymede (III) and Callisto (IV) are larger, though less dense.

and Callisto are about the same size as Mercury, but are much less dense.

The remaining satellites are of different type. Rather surprisingly, they have not been officially named, but some years ago names were proposed for them and may come into general use. Satellite V, Amalthea, was discovered by W. H. Pickering in 1892, and is the closest to Jupiter of the entire satellite family. Beyond Callisto come three more satellites, VI, VII and X (Hestia, Hera and Demeter). At almost twice this distance are four more satellites, XII, XI, VIII and IX (Adrastea, Pan, Poseidon and Hades) which move in a retrograde direction. They are so far from Jupiter, and are so strongly influenced by the gravitational pull of the Sun, that their orbits are not even approximately elliptical. Poseidon was in fact "lost" for many years after its original discovery and was only found again in 1940.

Satellite Orbits *above*
The first group comprise the four large satellites and the innermost faint satellite, Amalthea. Then come two groups, of which the outermost is made up of satellites with retrograde motion. All the outer 7 satellites may be captured asteroids.

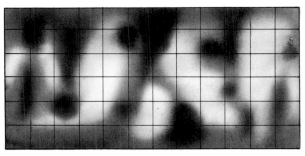

Io *above*
Surface detail on Io; a Mercator chart, drawn by A. Dollfus and his colleagues at the Pic du Midi. Various dark patches and lighter areas are shown.

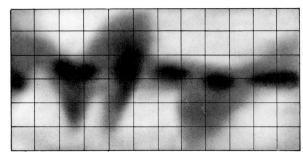

Europa *above*
Mercator chart of Europa (A. Dollfus, Pic du Midi). The main feature on this chart is a dark region near the satellite's equator.

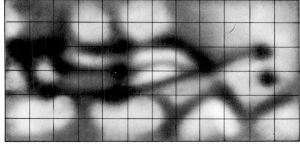

Ganymede *above*
Mercator chart of Ganymede (Dollfus). The details here are more certain. The general colour is distinctly yellowish.

Callisto *above*
A Mercator map of Callisto's surface. Though Callisto is about the same size as Ganymede, it is much less dense, and has a lower escape velocity.

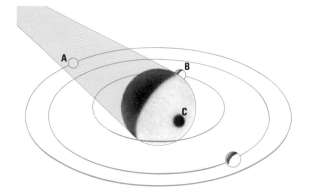

Satellite Eclipses and Occultations *above*
Eclipses (A) and occultations (B) of the four Galilean satellites are easy to observe with a small telescope. Shadow transits (C) are also visible.

Shadow Transit of Ganymede *right*
This photograph, taken with the 200-in. reflector at Palomar, shows Ganymede together with its shadow on Jupiter's disk. The Great Red Spot is also visible.

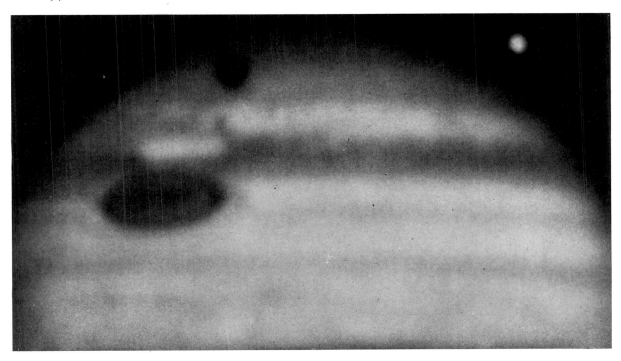

Jupiter and Satellite Data

Distance from Sun max. 506,800,000 miles
mean 484,300,000 miles   min. 459,800,000 miles
Sidereal period 11·86 years
Synodical period 398·9 days
Orbital eccentricity 0·048   **Inclination** 1°18'
Mean orbital velocity 8·1 miles per second
Axial rotation period equatorial 9h. 51m.
Axial inclination 3°.1
Equatorial diameter 88,700 miles
Mass (Earth=1) 318   **Volume (Earth=1)** 1312
Density (specific gravity) 1·3
Surface gravity (Earth=1) 2·54
Escape velocity 37·1 miles per second
Albedo 0·44   **Oblateness** 0·062
Maximum surface temperature −130°C (−200°F)

		Mean distance from centre of Jupiter	Period d.	h.	m.	Orb. eccentricity	Orb. inclination		Diameter (mil)	Magnitude
V	Amalthea	113,000		11	57	0·003	0°24'		150	13
I	Io	262,000	1	18	18	0·0	0		2310	5·5
II	Europa	417,000	3	13	14	0·0	0		1950	5·7
III	Ganymede	666,000	7	3	43	0·0	0		3200	5·1
IV	Callisto	1,170,000	16	16	32	0·0	0		3220	5·8
VI	Hestia	7,120,000	250	26		0·158	27°38'		100	13·7
VII	Hera	7,290,000	259	26		0·207	24	6	35	17
X	Demeter	7,300,000	260	23		0·130	29	1	15	18·8
XII	Adrastea	13,000,000	625			0·169	147		14	18·9
XI	Pan	14,000,000	700			0·205	164		19	18·4
VIII	Poseidon	14,600,000	739			0·378	145		35	16
IX	Hades	14,700,000	758			0·275	153		17	18·6

Saturn 1

Early Views of Saturn

Saturn, the outermost of the planets known in ancient times, is a giant world of the same basic type as Jupiter. Its equatorial diameter is 75,100 miles, so that although smaller than Jupiter it is very much larger than the Earth. Yet it is not so massive as might be expected. Its density is surprisingly low – only 0·7 that of water – and it has been said that if Saturn could be dropped into a vast ocean, it would float! The mass is 95 times as great as that of the Earth.

Because Saturn is much farther away than Jupiter, and is smaller, it is not so brilliant. Even so, it still appears as a bright, slightly yellowish star. The motion against the starry background is relatively slow, and the ancients regarded the planet as sluggish and baleful. It was only with the invention of the telescope that Saturn could be revealed as the most beautiful object in the entire sky.

The first telescopic observations were made by Galileo in 1609–10. His telescope gave a magnification of only 32 diameters, and by modern standards the definition was poor: it is therefore not surprising that his view of Saturn was puzzling. He thought that the planet must be triple, and it was only in 1655 that Huygens, with a much better telescope, was able to show that the curious aspect could be explained on the assumption that Saturn is surrounded by a system of rings. There are three rings altogether, two of which are bright, while the third, the Crêpe Ring (discovered in 1848), is much fainter and more transparent than the other two.

Saturn's Rings

The rings are not solid or liquid sheets. They lie within the Roche limit for Saturn, and so no continuous ring could exist; it would be disrupted by the powerful gravitational pull. During the 19th century J. E. Keeler, following up theoretical investigations by Clerk Maxwell, showed that the rings must be composed of small particles, moving round Saturn in the manner of dwarf satellites, and giving the misleading impression of a solid mass. Recent work carried out in America by G. P. Kuiper indicates that the ring-particles are made up of ammonia ice. It has been suggested that the rings represent the débris of a former satellite which approached within the Roche limit, and was broken up; this may be so, but in any case Saturn still has ten satellites left. The largest of them, Titan, is visible with a small telescope, and is of planetary size. Four more (Iapetus, Rhea, Dione and Tethys) were discovered before the end of the 17th century; the rest are much fainter.

The Globe of Saturn

The glory of the ring-system tends to divert attention from the globe of the planet, but here too there is much of interest. The belts are less prominent than those of Jupiter, and are less easy to observe; when the ring-system is well displayed it hides a considerable part of the disk. Spots on Saturn are relatively rare, and are usually short-lived; there is nothing comparable with the Great Red Spot on Jupiter. In general, the surface of Saturn seems to be comparatively inactive. This may well be due to the lower temperature: —180°C (—290°F), or 50°C (90°F) below that of Jupiter. The idea that Saturn might be faintly self-luminous is now known to be wrong. Neither is it a strong source of radio emission.

The outer surface is, of course, gaseous. There is more detectable methane and less ammonia than with Jupiter, because the lower temperature has frozen more of the ammonia out of the planet's atmosphere. Hydrogen in the molecular state is abundant, and there must also be a large quantity of helium. Not much is known about the internal composition; W. de Marcus has recently calculated that Saturn contains 63 per cent of hydrogen by mass, and that the pressure at the centre of the globe may be as much as 50 million atmospheres, but we still cannot tell whether or not it possesses what could be described as a sharply defined solid surface.

There seems no prospect of our finding life either on (or in) Saturn, or elsewhere in the Saturnian system. Neither is it practicable to consider manned flight there. However, plans are being made to send out an automatic probe, and within the next ten years it may well turn out to be possible to obtain close-range television pictures.

Ring A

The outermost ring, measuring 169,000 miles from one side to the other. It is bright, though not so reflective as Ring B. In it is Encke's Division, which seems to be a genuine gap even though it is much less prominent than Cassini's Division. In 1909 French astronomers reported an extra dusky ring outside Ring A. It is known as Ring D, and has been described by various observers from time to time. However, there is no proof of its existence, and its reality is doubtful.

Cassini's Division

In 1675 G. D. Cassini discovered that the "ring" which had been described by Huygens is split by a distinct gap, now known as Cassini's Division. It is 2500 miles wide, and separates Ring A from the brighter Ring B. It is kept swept clear of ring-particles by the gravitational effects of Saturn's inner satellites.

60375	56000	48000	40000	32000	24000	16000

Ring B

Inside Cassini's Division comes Ring B, which is 16,000 miles wide. It is the brightest of the rings, and has a reflective power or albedo greater than Saturn's. Where the rings cross the planet, they produce a strong shadow.

Ring C

More generally known as the Crêpe or Dusky Ring, Ring C was first reported in 1850 by Bond in America and almost simultaneously by the English amateur W. Lassell. It is 10,000 miles wide, and extends to within 9000 miles of

Magnitude

Saturn is not of equal brightness at every opposition. Much depends upon the angle at which the ring-system is presented, the rings having an albedo higher than that of the globe. When the rings are wide open, the magnitude attains −0·3, Saturn then being brighter than any star apart from Sirius and Canopus. When the rings are edge-on, the opposition magnitude is only +0·8.

Size of Saturn *right*

Though Saturn is not so large as Jupiter, with its diameter of over 75,000 miles, it is much larger than the Earth, as is shown in this diagram.

Distance from Earth *below*

Saturn is very remote. Its distance from the Earth is never less than 740 million miles. This is shown in the diagram. The synodical period is 378 days, Saturn reaching opposition on average two weeks later each year; thus an opposition occurred on 28 October 1969, and the following one will be on 11 November 1970.

Sidereal Period

Saturn takes 29·46 years to complete one revolution round the Sun. Its apparent diameter, as seen from Earth, ranges between 20″.9 and 15″.0.

Though Saturn has so long a "year", it has a very short axial rotation period (only 10h. 14m. at the planet's equator, though the polar period may be as much as 20 minutes longer).

Density *right*

Saturn is unique in that its density is, on average, less than that of water; the specific gravity is only 0·7. However, the escape velocity is high (22 miles per second, much greater than for any other planet apart from Jupiter). The surface gravity is not much greater than for Earth.

1	5·5	0·7

	Mars	Asteroids	Jupiter	Saturn
	141,600,000 mi		484,300,000 mi	886,100,000 mi
	227,800,000 km		779,100,000 km	1,426,000,000 km

S

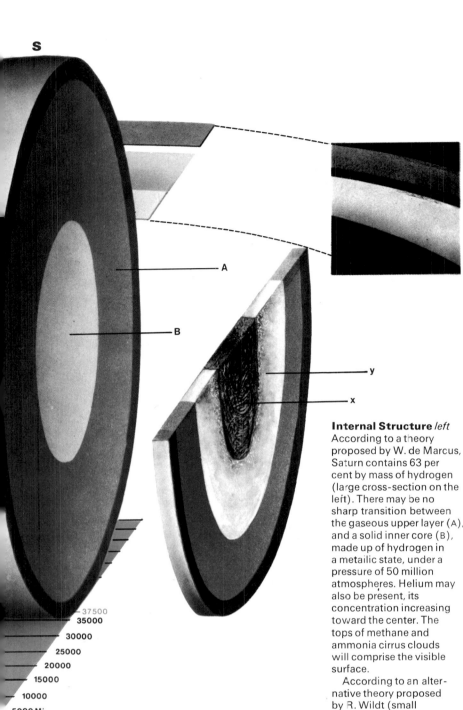

A

B

y

x

- 37500
- 35000
- 30000
- 25000
- 20000
- 15000
- 10000
- 5000 Mi

Internal Structure *left*

According to a theory proposed by W. de Marcus, Saturn contains 63 per cent by mass of hydrogen (large cross-section on the left). There may be no sharp transition between the gaseous upper layer (A), and a solid inner core (B), made up of hydrogen in a metallic state, under a pressure of 50 million atmospheres. Helium may also be present, its concentration increasing toward the center. The tops of methane and ammonia cirrus clouds will comprise the visible surface.

According to an alternative theory proposed by R. Wildt (small section at lower right of diagram), there is a rocky core (X), surrounded by a deep layer of ice (Y), which is surrounded in turn by the hydrogen-rich atmosphere. This model has now fallen out of favor. It has to be admitted that our knowledge of the internal constitution of Saturn is far from complete.

the surface of Saturn. It is transparent, and may be seen as a hazy darkish band where it crosses the disk. There are suggestions that it may have brightened up, but such an effect would in fact be very difficult to explain.

The Belts of Saturn

The belts of Saturn are not so conspicuous as those of Jupiter, but they are well-marked. The two which are most prominent are the equatorial belts, which lie to either side of the planet's equator.

Inclination of Saturn's Rings *below*

Though the ring system is very extensive, measuring 169,000 miles in diameter, it is also very thin ; the rings must be less than 10 miles thick, and when edge-on to the Earth, as in 1966, they are seen only as a slender line of light. When Saturn is observed with a small telescope at times of edgewise presentation, the rings cannot be seen at all, and the planet appears as a rather flattened globe.

At alternate intervals of 13 years 9 months and 15 years 9 months, the ring-plane passes through the Sun, since the axis of Saturn is inclined by 26 °.4 to the perpendicular to the planet's orbit (see diagram) and the rings lie in the plane of the equator.

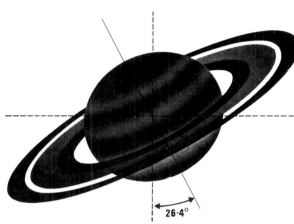

26·4°

During the 13¾ years between one edgewise presentation and the next, Saturn's south pole is tilted toward the Sun ; the southern ring-face is displayed, and the rings cover part of the northern hemisphere of the globe. During this interval, Saturn passes through perihelion, and will also at this time be moving at its greatest orbital velocity.

While the northern ring-face is displayed and the southern hemisphere partly hidden by the ring, Saturn passes through aphelion, and is moving more slowly ; this explains why the intervals between successive edgewise presentations are unequal.

When the rings lie exactly in the plane of the Sun or the Earth, they are invisible unless they are being observed through large telescopes.

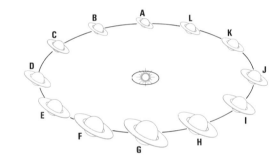

The Changing Aspects of the Rings *above* and *right*

The changing aspects of the rings are shown in these two series of drawings, which both cover a full cycle. As the sequence on the right shows, the rings were edge-on in 1966 (A) and will be so again in 1980 (G) ; the next edgewise presentation after that will occur in 1995 (A´). In 1972-3 (D), and again in 1987 (J), the rings are at their widest as seen from Earth.

A

B

Double Disappearance of the Rings *above*

When the rings lie in the plane of the Sun (A), they are illuminated only at the edge ; and as the ring-system is so thin, the amount of illumination is slight. This is why the rings become difficult to see at such times, as well as when the Earth lies in the ring-plane (B). However, a certain amount of light is reflected on to the rings from the globe, and the rings never vanish completely when observed through adequate telescopes.

The Rings and Their Shadows *right*

Because we always see Saturn from an angle, the rings appear elliptical. In fact the system is perfectly circular, as is shown in the "bird's-eye view" given here. The dimensions are listed in the large illustration at the top of the page. The rings lie entirely within the Roche limit for Saturn, but the innermost satellites lie outside this limit. The globe casts shadow on the rings, and the rings cast shadows on the globe — changing effects observable with a small telescope. Encke's Division, in the outer ring, is much less prominent than Cassini's.

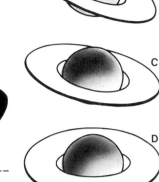 A

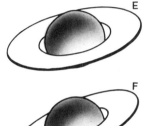 B

C

D

E

F

G

H

I

J

K

L

 A

 Pluto

Uranus

1,783,000,000 mi
2,870,000,000 km

2,793,000,000 mi
4,493,000,000 km

3,666,000,000 mi
5,898,000,000 km

Saturn 2

Divisions in the Rings

The only prominent division in Saturn's rings is Cassini's, so named because it was first seen in 1666 by G. D. Cassini. It is easily visible with a small telescope when the rings are open, and it is a true gap, due to the gravitational pulls of the inner satellites Mimas, Enceladus and Tethys. The effects of Janus, the satellite discovered by A. Dollfus in 1966, are not yet well known.

Encke's Division, in the outer rings is less conspicuous, and is normally well seen only at the ends or "ansæ" of the rings. Various other divisions have been reported in all three rings, but these have not been confirmed by observations made with large telescopes.

Cassini's Division

right A particle moving in the Cassini Division (C) will have a period half that of Mimas (M), one third that of Enceladus (E) and one quarter of that of Tethys (T).

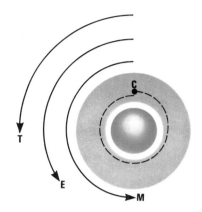

right After the particle has completed two revolutions round Saturn, Mimas will have completed one.

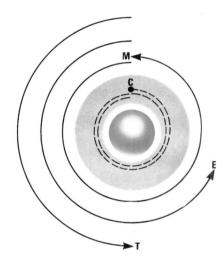

right When the particle in Cassini's Division has completed three revolutions, Enceladus has completed one.

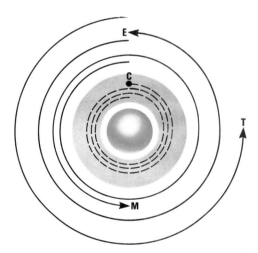

right When the particle has completed four revolutions, Tethys has completed one. Since the particle is subjected to regular perturbations from all three inner satellites, it is moved out of the Cassini Division area. In other words, the inner satellites keep the Cassini Division "swept clear" of ring particles.

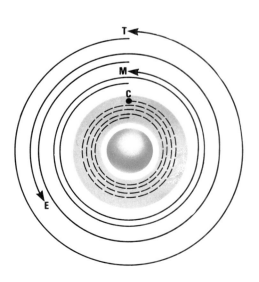

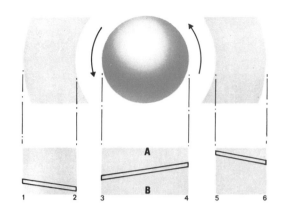

Doppler Effect *left*
Doppler shifts of light from the rings confirm their particles obey Kepler's Laws. The rings' velocity, and hence the displacement of a spectral line from its norm, is least at the outer edges (1 and 6), greatest at the inner (2 and 5). Displacement is toward the blue end of the spectrum (B) in light from the approaching side of the rings (1 and 2), and toward the red end (A) in that from the opposite side (5 and 6). A corresponding effect shows in light from the planet's surface (3 and 4).

Will Hay's Spot *left*
Conspicuous spots on Saturn are comparatively rare. The most prominent of modern times was discovered in 1933, in the equatorial region of the planet, by the British amateur W. T. Hay. At its brightest it was visible in a small telescope, and it persisted for some weeks.

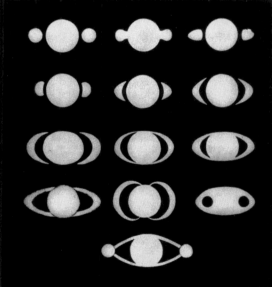

Historical Observations

When Saturn was first observed telescopically, its ring-system was at a narrow angle, and the feeble telescopes used by Galileo were not adequate to show its proper form.

Galileo believed Saturn to be a triple planet. Two years later, he found that the "companions" had vanished; this was because the rings were then edge-on to us. The original aspects returned in the years following 1613, but Galileo was never able to interpret it correctly.

Over the next few decades various curious theories were proposed; Hevelius, for example, considered that Saturn must be oval in shape, with two appendages attached to its surface.

The problem was solved by the Dutch astronomer C. Huygens, who was one of the best observers of his day. In 1659 he was satisfied that Saturn "is surrounded by a thin, flat ring, which nowhere touches the body" of the planet.

His theory met with a surprising amount of opposition, but it was confirmed by the observations of men such as Robert Hooke and G. D. Cassini; by 1665 it had been finally accepted.

The drawings given here were made during the 17th century, beginning with two by Galileo (1620) and ending with Hooke and Cassini.

The Satellites of Saturn

Of Saturn's 10 satellites, only one – Titan – is as large as the four Galilean satellites of Jupiter. Titan may, however, be the largest satellite in the Solar System. It is larger than the Moon, but its diameter is rather uncertain, and estimates range between 3500 miles and only 2700 miles. It is visible with a small telescope, and was discovered by Huygens in 1655.

The escape velocity of Titan is 1·73 miles per second. It is naturally very cold, and this has enabled it to retain an atmosphere, detected in 1943–4 by G. P. Kuiper and thought to be made up chiefly of methane. If the temperature of Titan were raised by as little as 38°C (100°F), the atmosphere would escape. Large telescopes can reveal some surface detail, and the colour of Titan is distinctly yellowish.

Mimas, Enceladus and Tethys, the three inner moons, are about as dense as water, and have been described as "snowballs"; the first two were discovered by Herschel in 1789, and Tethys by Cassini in 1684. The fourth satellite, Dione, may be no larger than Tethys, but is much denser and more massive; it also was discovered by Cassini in 1684. Rhea, farther from Saturn, is distinctly brighter. Its density is intermediate between that of Dione and those of the inner satellites.

Beyond Titan comes a small satellite, Hyperion, discovered by Bond in 1848. Far outside Hyperion lies Iapetus, another of Cassini's discoveries (1671), which is particularly interesting because it is variable. It is much brighter when west of Saturn than when to the east, so that it is either irregular in shape or else has a surface of unequal reflecting power. Its diameter is uncertain. Some authorities give a value of 700 miles; but recent observations indicate that at its brightest Iapetus may become as bright as the 9th magnitude, in which case it is as large as Rhea and possibly even larger.

Phœbe, the outermost satellite, is very small, and has a highly inclined retrograde orbit; it is probably a captured asteroid. Janus, the most recently discovered of the satellites, is actually the closest to the planet. It was found by A. Dollfus in December 1966.

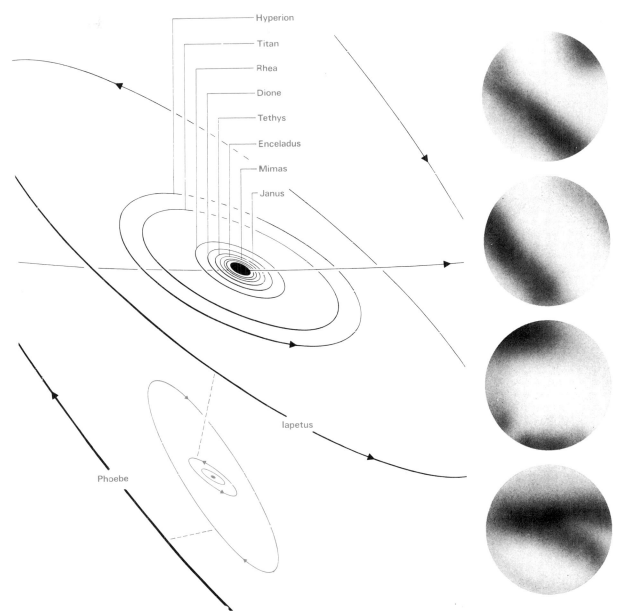

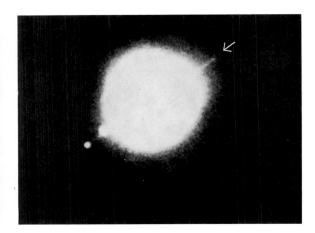

Discovery of Janus *above*
Janus, indicated by an arrow in this photograph of Saturn, is the planet's innermost satellite. At the time when the photograph was taken (December 1966) the rings were edge-on; when the ring system is wide open Janus cannot be seen, and it will not be possible to observe it again until 1980. This explains why it was not discovered until Dollfus at Pic du Midi detected it so recently.

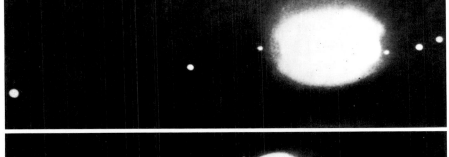

Titan's Surface *above*
Surface detail on Titan, drawn by Dollfus (24-in. refractor, Pic du Midi). It only shows some darkish patches and brighter zones.

The Inner Satellites *left*
upper left Saturn and its six inner satellites
lower left Same satellites on 24 March 1948; some stars also shown. Saturn itself is over-exposed, and Janus is not visible.
An extra satellite in orbit between Titan and Hyperion was reported by W. H. Pickering in 1904. It was named Themis, but has never been confirmed, and probably does not exist.

Data

Distance from Sun max. 937,600,000 mean 886,100,000 min. 834,600,000
Sidereal period 29·46 years
Synodical period 398·9 days
Orbital eccentricity 0·056 **Inclination** 2°29'
Mean orbital velocity 6·0 miles per second
Axial rotation period (equatorial) 10h. 14m.
Equatorial diameter 75,100 miles
Mass (Earth=1) 95 **Volume (Earth=1)** 763
Density (specific gravity) (water=1) 0·7
Surface gravity (Earth=1) 1·13
Escape velocity 22 miles per second
Albedo 0·41 **Oblateness** 0·10
Inclination of axis 26°.7
Apparent dia. seconds of arc max. 20·9, min. 15·0
Maximum stellar magnitude −0·4

Satellite Data

Name	Mean distance from centre of Saturn (miles)	Revolution Period d. h. m.	Orbital Eccentricity	Orbital inclination to Saturn's equator (degs)	Diameter (miles)	Magnitude
Janus	98,000	17 59	0·0	0·0	190(?)	14
Mimas	115,000	22 37	0·0202	1·5	300(?)	12·1
Enceladus	148,000	1 8 53	0·0045	0·0	350(?)	11·8
Tethys	183,000	1 21 19	0·0	1·1	600(?)	10·3
Dione	234,000	2 17 41	0·0022	0·0	600(?)	10·4
Rhea	327,000	4 12 25	0·0010	0·3	800(?)	9·0
Titan	758,000	15 22 41	0·0292	0·3	3000(?)	8·4
Hyperion	919,000	21 6 38	0·1042	0·6	300(?)	14·2
Iapetus	2,210,000	79 7 56	0·0283	11·7	1000(?)	var.
Phœbe	8,040,000	550 10	0·1633	150	130(?)	16

The diameters of all the satellites are very uncertain. Phœbe has retrograde motion.

Saturn 3

Appearance of Saturn

The photographs on this page demonstrate both the beauty of Saturn, and the difficulty of taking large-scale photographs of the planets with good definition.

Saturn is unique; there is nothing like it in the Solar System. When the rings are open, they hide part of the planet's disk. When they are edgewise-on the rings appear as a thin line of light, but the belts and other features of the globe are excellently displayed at this time.

Shadow Effects and Surface Features

The shadow effects are interesting. Except near opposition, the rings seem to be "cut off" by the shadow cast on to them by the globe, as is brought out by the large colour photograph taken at Palomar shown here. The rings themselves cast shadow on to the disk, and the Crêpe Ring is also seen as a dusky line bordering the ring seen against the globe. Photographs also show that there are changes in Saturn itself. Sometimes the poles may have dusky "hoods",

while at other times they are comparatively bright. The equatorial region, corresponding approximately to System I on Jupiter, is generally bright.

There are far fewer delicate surface details on Saturn than on Jupiter, and for this reason the rotation periods of the various belts and zones are not so well known. Every advantage is taken of the occasional bright spots, such as Hay's of 1933, but it usually happens that the belts are regular, with a total lack of spots or filaments.

Photographing Saturn

When photographing a planet, a time-exposure is essential. This means that there is bound to be interference from the Earth's unsteady atmosphere, so that the final picture is blurred. The colour photographs shown on this page are the best that can be obtained from Earth, since they were taken with giant telescopes under the most favourable conditions possible. The black-and-white photograph *below* was taken at the Lowell Observatory, Flagstaff.

Saturn among the Stars *below*
Saturn is reasonably bright, but there is nothing to distinguish it from other stars and it is not surprising that astronomers of pre-telescopic days had no idea of its real nature. It is quite impossible to see the ring-system, or any trace of it, with the naked eye. (Photograph by Patrick Moore, 1 October 1967.)

Occultation of Saturn by the Moon *below*
The rings are seen at a narrow angle. The limb of the Moon, necessarily somewhat over-exposed, is shown and the irregularities on the limb are very noticeable. When the photograph was taken, Saturn was approaching occultation. (Photograph by Commander H. R. Hatfield, 12-in. reflector, 1967.)

Aspects of Saturn

left Saturn with its rings at a narrow angle (U.S. Naval Observatory). The disk details are well shown. At this time the rings were not edgewise-on ; if they had been so, they would have shown up as nothing more than a very narrow line of light.

right Saturn, 14 October 1968 (G. P. Kuiper, 61-in. reflector). The southern hemisphere of the planet is displayed ; the bright equatorial zone and the Cassini Division are prominent. The rings were well opened out, as edgewise presentation had occurred in 1966.

below Saturn (60-in. reflector, Mount Wilson). The southern hemisphere of Saturn is displayed, and part of the northern hemisphere is covered by the rings. The difference in brightness between the outer ring (A) and the inner ring (B) is very marked. B is much the more brilliant.

Uranus, neptune and pluto

Uranus

Seven was the mystical number of the ancients, and it was natural for them to believe that there must be seven bodies in the Solar System – the five naked-eye planets, plus the Sun and the Moon. It came as a major surprise when, in 1781, a then-unknown, amateur astronomer named William Herschel discovered a new planet, well beyond the orbit of Saturn. Herschel had not been carrying out a deliberate search; he was interested mainly in the stars, and he was carrying out a systematic "review of the heavens", with a home-made reflector, when he found an object which showed a disk, and which moved perceptibly from one night to the next. Herschel did not at once recognize it as being a planet, and his official report was headed "An Account of a Comet", but as soon as the orbit could be calculated there could be no doubt as to the true nature of the object.

Uranus is extremely remote. Its distance is never less than 1600 million miles from the Earth, and its revolution period is 84 years. It is just visible to the naked eye when at its brightest, but not even large telescopes will show much detail on its pale, rather greenish disk. It is known to possess five satellites, though all of them are faint.

Neptune

With the discovery of Uranus, the Solar System was again regarded as complete, but Uranus refused to follow its predicted path. It persistently "wandered", and in 1834 an amateur, the Rev. T. J. Hussey, suggested that this wandering might be due to the pull of an unknown planet.

The whole problem was taken up by two astronomers, J. C. Adams in England and U. J. J. Le Verrier in France. Working quite independently, they computed the position of the unknown planet. Adams sent his results to the Astronomer Royal, Sir George Airy, but Airy took no immediate action. Le Verrier's calculations, finished in 1846, were sent to Berlin Observatory, where J. Galle and H. d'Arrest at once began to search. Almost immediately they located the new planet, very close to the position predicted by Le Verrier's calculations.

Neptune, as the new planet was named, is very similar to Uranus, but is much more remote. Its mean distance from the Sun is 2793 million miles, and its revolution period is 164·8 years. It is of magnitude 7·7, and so is too faint to be seen with the naked eye, but binoculars will show it. Its colour is slightly bluish; details on its disk are excessively hard to observe.

Pluto

Even after the discovery of Neptune, there was still something uncertain in the movements of the two outer planets. Percival Lowell, of Martian canal fame, took up the problem and computed the position of a hypothetical trans-Neptunian planet. He began searching, using the equipment at the Lowell Observatory in Arizona which he had founded, but the planet was still undiscovered at the time of Lowell's death in 1916.

Similar calculations made by W. H. Pickering led to a similar result, but a search carried out photographically by M. Humason at Mount Wilson, in 1918, also proved unsuccessful. For some years the search was suspended; but then Clyde Tombaugh, at the Lowell Observatory, returned to it. In March 1930 he discovered the long-expected planet, now called Pluto, very close to the position that had been given by Lowell. Pickering's calculations, too, were in close agreement. From Pluto the Sun would appear only as large as Jupiter appears from Earth, though it would still be extremely bright.

Pluto is too faint to be seen in small telescopes, and no detail can ever be seen on its surface. The axial rotation period seems to be 6·3 Earth-days, as has been estimated from short-term variations in the planet's magnitude.

Distances of the Outer Planets *right*
Pluto has an eccentric orbit; its next perihelion passage is due in 1989. At its next aphelion in A.D.2114, it will be too faint to be seen except with the world's largest telescopes

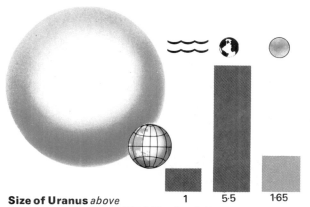

Size of Uranus *above*
Uranus has a diameter of 29,300 miles. It is considerably denser than Jupiter or Saturn, with a specific gravity of 1·65; the polar flattening is similar to that of Jupiter. The surface temperature can never rise above −190°C (−310°F), and all the ammonia has been frozen out of its atmosphere, though methane is detectable spectroscopically. The internal constitution is not known with any certainty, but Uranus resembles Jupiter and Saturn much more closely than the Earth. The most prominent surface feature is a bright zone to either side of the planet's equator.

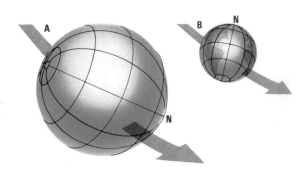

Rotation of Uranus *above*
Uranus has an axial rotation period of 10h. 48m. at the equator; near the poles the rotation period may be somewhat longer. The most remarkable characteristic is the tilt of the axis, which amounts to 98° – more than a right-angle as shown in the large diagram (A). This compares with Earth's axial tilt of 23½° shown in the smaller diagram (B), and means that the rotation is technically retrograde. Each pole of Uranus has a "night" lasting for 21 Earth-years, with a corresponding period of "daylight" at the opposite pole. The reason for this unusual axial inclination is unknown. From Earth, a pole may sometimes lie in the centre of the planet's disk.

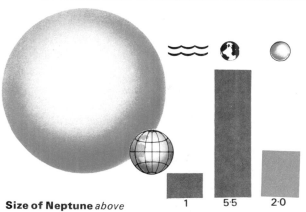

Size of Neptune *above*
The diameter of Neptune has been given as 27,700 miles, making it slightly smaller, though considerably more massive, than Uranus. However, in 1969 Neptune occulted a star, and this enabled G. E. Taylor, at Greenwich Observatory, to revise the diameter value; it now seems that Neptune is slightly larger than Uranus, and the diameter is thought to be 31,200 miles. The mass is 17 times that of the Earth, as against 15 Earth-masses for Uranus. The surface temperature is −220°C (−360°F). The internal composition is probably similar to that of Uranus.

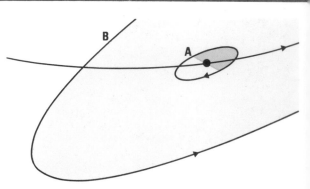

Satellites of Neptune *above*
Neptune has two satellites. The larger, Triton (A), was discovered by the English amateur astronomer William Lassell, and is brighter and larger than any of the satellites of Uranus. No atmosphere has been detected, but it is quite likely that Triton has a tenuous atmosphere made up of methane. The orbit is practically circular, but the orbital motion is retrograde.

The second satellite, Nereid (B), was discovered only in 1948. It is very small, and its orbit is more like that of a comet than a satellite; the motion is direct, and in all probability it is a captured asteroid rather than a satellite.

Size of Pluto *above*
Pluto has proved to be a most puzzling body. It is not a giant, but at first it was assumed to be considerably larger than the Earth. However, measurements of its diameter made by G. P. Kuiper with the 200-in. reflector at Palomar – the only telescope capable of showing a measurable disk – give a diameter of only 3700 miles, in which case Pluto is much smaller than the Earth and even smaller than Mars (represented by the inner circle above).

This raises problems. A small planet of this kind could not possibly exert any measurable influence on either Uranus or Neptune, and yet it was by this very influence that Pluto was tracked down! There have been various explanations, none of which is universally accepted. First, Pluto may be larger than the measures indicate, so that instead of measuring the full diameter we are measuring on a reflective part of the surface. Secondly, Pluto could be

exceptionally dense. Thirdly, the success of the calculations by Lowell and Pluto could have been purely fortuitous, in which case the real "Pluto" remains to be discovered.

Orbit of Pluto *above*
Pluto has an exceptional orbit, which is both relatively sharply inclined and decidedly eccentric (A); at perihelion it comes within the orbit of Neptune (B), though the inclination of 17° means that there is no fear of a collision. The next perihelion passage is due in 1989. There have been doubts as to whether Pluto should be ranked as a true planet. It is about the same size as Triton, the larger of Neptune's two satellites (assuming that Kuiper's measures are of the right order), and it is not impossible that Pluto may be a former satellite of Neptune which has moved away in an independent orbit. In any case, Pluto remains the most enigmatical member of the Sun's family of planets.

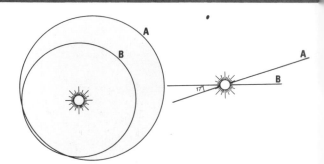

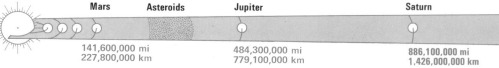

Mars	Asteroids	Jupiter	Saturn
141,600,000 mi		484,300,000 mi	886,100,000 mi
227,800,000 km		779,100,000 km	1,426,000,000 km

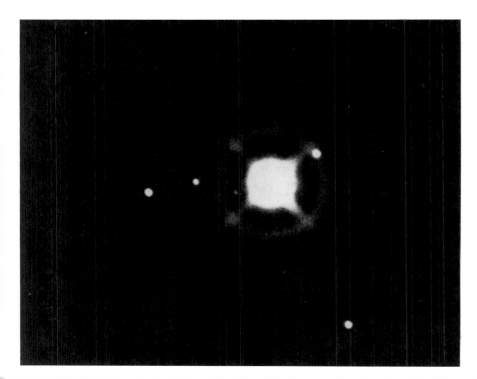

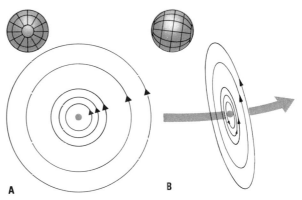

Data
Distance from Sun
max. 1,867,000,000 miles
mean 1,783,000,000 miles
min. 1,699,000,000 miles
Period of revolution 84·01 years
Synodic period 369·7 days
Orbital eccentricity 0·047
Orbital inclination 0° 46'
Mean orbital velocity
4·2 miles per second
**Axial rotation period at
equator** 10h. 48m.
Axial inclination 98°
Equatorial diameter
29,300 miles
Mass (Earth=1) 15
Volume (Earth=1) 50
Density (water=1) 1·65
Surface gravity (Earth=1) 1·1
Escape velocity
15 miles per second
Albedo 45 per cent
Oblateness 0·06
Max. surface temp.
−190 °C (−310 °F)

Satellites of Uranus *left and above*
Uranus has five satellites. All are faint ; the most recently discovered, Miranda, is visible only with giant telescopes. Their orbits lie in the plane of Uranus' equator ; when a pole is presented to the Earth (A) the orbits of the satellites appear circular, whereas when the equator is presented, as in 1945 (B), the orbits are virtually linear. The photograph to the left was taken by G. P. Kuiper with the 82-in. reflector at the McDonald Observatory, Texas. The satellites are shown, but the image of Uranus itself is necessarily over-exposed.

Neptune and Satellites *left*
Photograph of Neptune and its two satellites (G. P. Kuiper, 82-in. reflector, McDonald Observatory). Neptune is necessarily over-exposed.

Discovery of Neptune *below*
left A corner of the star-map used by Galle and d'Arrest during their search.
right The corresponding patch of sky, showing where Galle and d'Arrest found Neptune (marked by arrow) and where Le Verrier had calculated that the planet would be (marked by a cross). Adams' calculations were very similar. While Galle and d'Arrest were beginning their search, Challis, Professor of Astronomy at Cambridge, had started to look for the planet on the basis of Adams' work, and he actually saw Neptune twice without recognizing it as a planet rather than as a star.

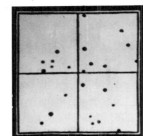 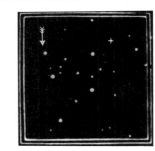

Data
Distance from Sun
max. 2,817,000,000 miles
mean 2,793,000,000 miles
min. 2,768,000,000 miles
Period of revolution
164·79 years
Synodic period 367·5 days
Orbital eccentricity 0·009
Orbital inclination 1°46'
Mean orbital velocity
3·4 miles per second
**Axial rotation period at
equator** 14 hours
Axial inclination 29°
Diameter 31,200 miles
Mass (Earth=1) 17
Volume (Earth=1) 61
Density (water=1) 2·0
Surface gravity (Earth=1) 1 1
Escape velocity
15 miles per second
Albedo 50 per cent
Oblateness 0·02
Max. surface temp.
−220°C (−360°F)

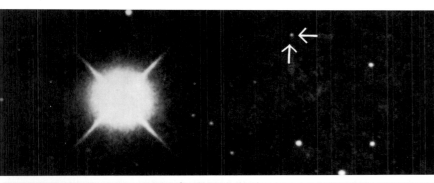

Discovery of Pluto *left*
The two photographs show the discovery of Pluto in March 1930. The upper photograph was taken on 2 March, and the lower on 5 March ; the large object to the left is the very over-exposed image of the 3rd-magnitude star Delta Geminorum. Pluto has moved perceptibly in the interval. Tombaugh was using a blink-microscope, a special device by which two photographic plates can be viewed alternately in rapid succession ; any object that has moved in the interval is seen to "jump". It was subsequently found that Humason, in his earlier search, had photographed Pluto twice, but on the first occasion the planet was masked by the image of a star, and on the second occasion Pluto's image fell upon a flaw in the photographic plate !

A Trans-Plutonian Planet ?
If Pluto is not the planet which Lowell sought, it is quite likely that a tenth planet exists. Unfortunately it will be very difficult to locate, because its effects upon the orbits of Uranus and Neptune will be very slight, while Pluto has not yet been followed for even half of a complete revolution (its period is almost 248 years, and it was discovered less than 50 years ago). Moreover, the hypothetical planet must be extremely faint. Only the world's largest telescopes could hope to photograph it, and they are fully occupied with stellar work.

Data
Distance from Sun
max. 4,566,000,000 miles
mean 3,666,000,000 miles
min. 2,766,000,000 miles
Period of revolution
247·70 years
Synodic period 366·7 days
Orbital eccentricity 0·246
Orbital inclination 17°10'
Mean orbital velocity
2·9 miles per second
**Axial rotation period at
equator** 6d. 9h.
Diameter 3700 miles (?)

Uranus
1,783,000,000 mi
2,870,000,000 km

Neptune
2,793,000,000 mi
4,493,000,000 km

Pluto
3,666,000,000 mi
5,898,000,000 km

Comets

Nature of Comets

Comets are the most erratic members of the Solar System. They have been observed from very early times, and records of them come from ancient China and Egypt; they were regarded as unlucky, and the fear of comets lasted until comparatively modern times. Yet a comet is not nearly so important as it may look. Its mass is very small compared with that of a planet, or even a satellite. Only the nucleus, which may be several miles in diameter, is at all substantial. The rest of the comet is made up of small particles (mainly ices) together with very tenuous gas.

Not all comets are brilliant, and by no means all of them have tails. Most comets appear as nothing more than rather blurred patches of light.

Comet Orbits

There are many short-period comets whose orbits are well known. These, however, are too faint to be seen with the naked eye. Most comets have orbits which are much more eccentric than those of the planets, and since a comet shines by reflected sunlight (plus a certain amount of self-luminosity when near perihelion), it can be seen only when reasonably close to the Sun and to the Earth. This means that even the short-period comets, whose orbits may extend out beyond that of Jupiter, cannot be followed continuously; when near their aphelion points, they are too faint to be seen. There are a few exceptional comets with much more circular paths, and these can be kept under observation, but comets of this type are faint and remote.

The more brilliant comets have orbits of very high eccentricity (more than 0.99 in many cases), and since their periods may amount to thousands or even millions of years they cannot be predicted. Only one bright comet, Halley's, has a period of less than several centuries.

Nomenclature of Comets

In general, a comet is named either after its discoverer or discoverers (as with Holmes' Comet), or after the mathematician who computed its orbit (as with Halley's Comet). According to the official nomenclature, the first comet discovered in a year is lettered a; thus 1970a was the first-discovered comet in 1970. A permanent nomenclature is subsequently adopted, in which Roman numbers are given in order of the perihelion passage; thus Comet 1973 II was the second comet to reach its closest point to the Sun in the year 1973.

A comet increases rapidly in brightness when nearing perihelion, because the Sun evaporates material from the ices in the nucleus, and a certain amount of self-luminosity is induced. Obviously, the comet suffers a steady wastage of material, and is short-lived on the astronomical scale. Several former periodical comets have been known to disintegrate. Thus Biela's Comet, which had a period of 6¾ years, was quite bright in 1826 and 1832; it was missed in 1839, because it was badly placed for observation; at its return in 1846, it split into two parts. The "twins" came back in 1852, but this was their last appearance – as comets. After being missed in 1866, a brilliant meteor shower was seen in 1872, unquestionably marking the debris of the dead comet.

Morehouse 1908 III *above*
Comet Morehouse 1908 III, as photographed by E. E. Barnard on 16 November. It was not brilliant, but was notable for its complicated tail structure, which showed rapid short-term variations.

Anatomy of Comet 1948 I *left*
Comet 1948 I (that is to say, the first comet to reach perihelion in 1948) was fairly bright. The tail was long : here the gaseous tail is shown, and there is considerable fine structure. The coma or head is made up of material evaporated from the nucleus. Various elements, such as sodium, have been identified, and in the tail there are hydrogen compounds such as ammonia, methane and cyanogen. The nucleus, which in many comets is almost stellar in aspect, is not shown on this particular photograph because of over-exposure of the coma so as to bring out the structure of the tail.

Comet Tails *right*
The tail of a comet always points away from the Sun. This is because the very small particles in the tail are affected by solar wind, and are "pushed away" from the nucleus.

There are two main types of comet tails. The gaseous tail is usually straight, while the dust tail is curved, because the dust particles are expelled from the nucleus less quickly. With many comets, both kinds of tails may be seen.

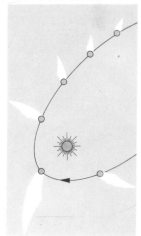

Holmes' Comet *left*
Holmes' Comet, seen in the region of the Andromeda Galaxy at its return of 1892. A short-period comet, it was lost for more than half a century after 1906 and rediscovered in the 1960s as an excessively faint object.

Donati's Comet *right*
Donati's Comet of 1858, as shown in an old woodcut. This was a brilliant naked-eye object. Both types of tail, the gaseous (straight) and the dust (curved), are excellently displayed. Donati's comet has a period amounting to many thousands of years.

Data

	Period (years)	Orbital inclination (degs.)	Distance from Sun, in a.u. perih.	aphel.	Orbital Eccentricity
Encke	3.3	12.5	0.3	4.1	0.846
De Vico-Swift	5.9	3.0	1.4	5.1	0.570
Tempel I	6.0	9.8	1.8	4.8	0.463
Forbes	6.4	4.6	1.5	5.3	0.556
Giacobini-Zinner	6.6	30.7	1.0	6.0	0.717
D'Arrest	6.7	18.0	1.4	5.7	0.611
Finlay	6.9	3.4	1.1	6.2	0.700
Holmes	6.9	20.8	2.1	5.1	0.420
Brooks 2	7.0	5.5	1.9	5.4	0.484
Whipple	7.5	10.2	2.5	5.2	0.349
Oterma 3	8.0	4.0	3.4	4.6	0.143
Comas Solá	8.5	13.7	1.8	6.6	0.576
Gale	11.0	11.7	1.2	8.7	0.761
Tuttle	13.6	54.7	1.0	10.3	0.821
Crommelin	27.9	28.9	0.7	18.0	0.930
Halley	76.0	162.2	0.6	35.3	0.967
Grigg-Mellish	164.3	109.8	0.9	59.1	0.969

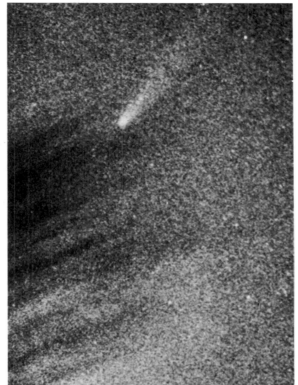

Arend-Roland 1957 III *above*
Comet Arend-Roland 1957 III, photographed by
E. M. Lindsay at Armagh on 25 April, 22.05 GMT.
The "spike" is due to material spread out along the orbit.
The comet was visible to the naked eye.

Kohoutek's Comet of 1973-4 *above*
Photographed on 17 November 1973 at the Royal Greenwich
Observatory. This comet is non-periodical, in that it will not
return for at least 10,000 years.

Humason's Comet *above*
Humason's Comet of 1961, photographed with
the 48-in. Schmidt telescope at Palomar. The comet
was a reasonably bright telescopic object; it has
an almost parabolic orbit.

**Three Main Types of
Comet Orbit** *right*
Comet orbits are of three
main kinds. The short-
period comets have
periods of a few years, and
are predictable (A);
their aphelia usually lie at
slightly greater than the
distance of the orbit of
Jupiter (B). Comets of
longer period, such as
Halley's (period 76 years),
have aphelia (C) beyond
the orbit of Neptune (D).
Some comets have inter-
mediate periods, such as
Crommelin's Comet
(period 27 years). Great
comets, however, have
orbits which are so
eccentric that the periods
are immensely long (E),
and so their appearances
cannot be calculated
with any accuracy.
Some comet orbits are
highly inclined to the
ecliptic (A in lower
diagram), and some
comets, including
Halley's among them, have
retrograde motion.

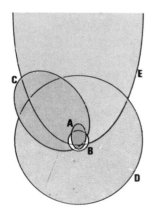

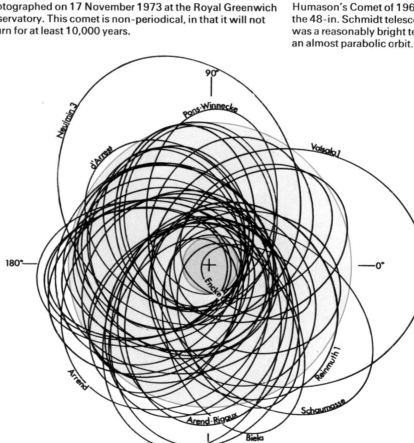

Families of Comets *left*
Orbits of the short-period
comets classed as
belonging to Jupiter's
"comet family". It used
to be thought that comets
came from outer space,
and were captured by
the pulls of the planets,
so that they were forced
into elliptical orbits, but
this theory is no longer
accepted.
  The first-discovered
member of the family,
Encke's Comet, has been
seen at 50 separate
returns, the last being that
of 1970. Its distance
from the Sun ranges
between 31,500,000 miles
and 381,300,000 miles.
It is never visible to the
naked eye, and there is
evidence that its brightness
is decreasing steadily, so
that by the end of the
century it may become a
very faint object. Few of the
comets in Jupiter's family
ever develop tails.

Development of the Tail of Halley's Comet
below Halley's Comet has been recorded at many returns,
and was seen in ancient times; it was also visible in 1066,
and is recorded in the Bayeux Tapestry. It was
visible in 1682, and Edmond Halley, afterwards

Astronomer Royal of Britain, calculated its orbit. He
found that the path was almost identical with those of
comets seen in 1531 and 1607, and came to the conclusion
that the comets were one and the same. He
predicted a return for 1758, and in this he was correct.

The comet came back again in 1835 and in 1910.
This series of photographs shows the development and
evolution of the tail in 1910. The next perihelion
passage is due in 1986. Halley's Comet is the only
bright comet to have a period of less than several centuries.

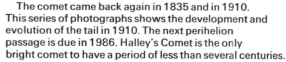

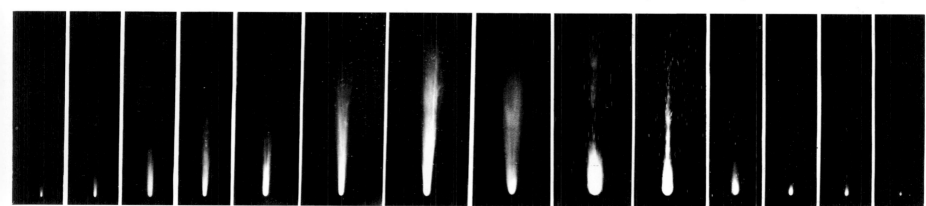

M t urs and M t urit s

Meteors

A meteor is a member of the Solar System. It is so small that when beyond the Earth's atmosphere it cannot be seen, but when it enters the upper air it becomes heated by friction, and destroys itself in the luminous streak that is usually called a shooting-star. The velocity of entry into the atmosphere, relative to the Earth, may be as high as 45 miles per second.

It is thought that more than 20 million meteors enter the atmosphere daily, but each is of very small mass, and few penetrate below a height of 50 miles above the ground before being destroyed. The average meteor visible with the naked eye is surprisingly small – smaller, indeed, than a grain of sand. Their composition has been studied spectroscopically. For obvious reasons meteor spectra are difficult to obtain, but several hundreds have now been secured, mainly by amateurs. It seems that there is a fundamental difference between a meteor and a meteorite. It is wrong to call a meteorite simply a "big meteor".

Extremely small particles, no more than 1/250 of an inch in diameter, are too small and friable to produce luminous effects when entering the atmosphere. These are known as micrometeorites. Since 1957 they have been extensively studied from space vehicles.

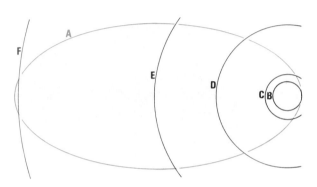

The Leonids *above*
Orbit of the Leonid meteor stream (A), associated with Tempel's Comet (1866). The main swarm encounters the Earth (B) only once in 33 years (the period of the shower), but some Leonids are seen annually. The orbits of Mars (C), Jupiter (D), Saturn (E), and Uranus (F) are also shown in the diagram.

Types of Meteors

Meteors are of two kinds: shower and sporadic. Shower meteors travel round the Sun in swarms, and are associated with comets. Every time the Earth passes through a swarm of meteors, the result is a shower of shooting-stars. This happens on many occasions annually, but the showers are not all equally rich. The Perseids of early August are very consistent. This sort of shower provides the best opportunity for trying to obtain meteor spectra; iron, calcium and other elements have now been identified by this means.

The Perseids appear each year (between 27 July and 15 August) because they are spread all round the orbit of their parent comet (1862 III). The Leonids, on the other hand, are concentrated in a bunch, and a major shower is seen only at intervals. The last was in 1966, when as seen from Arizona more than 100,000 meteors per hour were seen entering the atmosphere at the period of greatest activity.

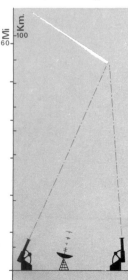

Computing Heights *left*
A meteor height can be computed only if the meteor trail is plotted by two observers some way apart, or if radar techniques are used (as is becoming increasingly common; by now, most useful meteor work, apart from spectra photography, is undertaken by radar). In general a meteor will burn out at a height of about 50 to 60 miles above sea-level. Very few penetrate below 40 miles, though some meteoritic material finally reaches the ground in the form of fine dust.

Meteor Trail *above*
Meteor trail, photographed by T. J. C. A. Moseley, at Armagh, 1968. The meteor was of about the 2nd magnitude, and lasted 1·5 seconds. The star-cluster near the middle of the photograph is Præsepe. As can be seen, the brightness of the meteor varied.

Meteor Showers *left and above*
Because the meteors in a shower are moving through space in parallel paths, as seen from Earth they seen to emanate from one particular point in the sky – just as parallel railway tracks will seem to converge toward the horizon. Each meteor shower thus has its own radiant – that distant point in the sky from which the shower appears to radiate (see diagram). The photograph was taken on 17 November by D. McLean from Kitt Peak, Arizona, while the Leonid shower was at its peak. Various meteors can be seen.
The Leonids gave brilliant displays in 1799, 1833 and 1866, but before the next return, that of 1899, the shower was perturbed by Jupiter, and the displays of 1899 and 1933 did not occur. However, in 1966 the Leonids again provided a magnificent display in some parts of the world. Beside the Leonids, other showers are relatively feeble.

Exploding Meteor *left*
An Andromedid meteor, exploding during its descent through the atmosphere; photograph by Butler, 23 November 1895, with a 2-in. lens.

Data

Name of Shower	Dates	Remarks
Quadrantids	1–4 January	Usually a sharp maximum on 3 Jan.
Lyrids	19–22 April	Swift meteors.
Aquarids	1–13 May	Long paths; swift meteors. Possibly associated with Halley's Comet.
Perseids	27 July–17 August	Rich, consistent shower.
Orionids	15–25 October	Swift meteors.
Leonids	17 November	Inconsistent; but a few Leonids can be seen round about 17 Nov.
Andromedids	26 November–4 December	Debris of Biela's Comet. Not now a rich shower.
Geminids	9–13 December	Rich, consistent shower.

Meteorites

A meteorite is a piece of material that comes from space, and lands on Earth without being completely burned away. It is different in nature from meteors, which do not reach the ground, being larger and much less friable. Most astronomers consider that there is no connection between meteorites and comets, but that there is a genuine link between meteorites and asteroids.

Many meteorites have been found and analysed. A few have even been observed to fall, so that they have been recovered almost immediately. They are of two main kinds, irons (siderites) and stones (aerolites), though the distinction is not clear-cut, and there are many intermediate types which are partly stony and partly nickel-iron. Rather surprisingly, it was only in 1802, after a fall of meteorites at the French village of L'Aigle, that scientists realized that material can come from the sky. Subsequent investigations revealed that there were meteorites in many countries; one, for instance, is the famous Sacred Stone in Mecca.

The largest known meteorite is at Hoba West, near Grootfontein in South-West Africa. It weighs over 60 tons, and is still lying where it fell in prehistoric times. The Hoba West Meteorite is very rich in nickel (16 per cent). Second to it in size is the Ahnighito Meteorite, found by the American explorer Peàry in Greenland in 1897, and now on show at the Hayden Planetarium in New York. The weight is not much inferior to that of the Hoba West Meteorite; it is mainly iron.

No unfamiliar elements have ever been found in meteorites. With siderites, the average composition is 90 per cent iron, 8 to 9 per cent nickel, rather less than 1 per cent cobalt, and smaller quantities of other elements such as phosphorus and sulphur. With aerolites, the average composition is 36 per cent oxygen, 24 per cent iron, 18 per cent silicon, 14 per cent magnesium and smaller quantities of other elements. These figures, of course, are only approximate, and different meteorites are of different composition.

Major meteorite falls are rare – and this is fortunate, since they can be highly destructive. A large meteorite may produce a crater, of which the best examples are the Barringer Crater in Arizona and the Wolf Creek Crater in Australia; many other craters, notably on the Canadian shield, have been attributed to meteoritic impact, but other authorities regard them as being of volcanic origin, and as yet there is considerable disagreement. (There is no evidence of a crater associated with the Hoba West Meteorite, though the fall took place so long ago that erosion must have been considerable.)

The most destructive fall of modern times was that of 1908, in the Tunguska region of Siberia. A meteorite came down in forested country, and flattened pine-trees for several miles around the point of impact; the effects of the fall can still be seen. The area is virtually uninhabited, and so far as is known there were no casualties, though many reindeer were killed. The Tunguska object may possibly have been the nucleus of a small comet, as is believed by some Russian astronomers; in any case, it broke up before impact and no crater was left. The

Tektite *left*
A small, glassy object, approximately to scale. Tektites, found notably in Australia, seem to have been heated twice; some are aerodynamically shaped.
It is not known whether they are of terrestrial or cosmic origin.

second large fall, that of 1947 – also in Siberia, in the Vladivostok area – also broke up, but many small pits were produced, and considerable meteoritic material was recovered.

There is no authentic record of anyone having been killed by a falling meteorite. It was formerly believed that meteoritic bodies would prove a major hazard in space research, but it is now clear that the danger is less than was originally thought, though it cannot be entirely ignored.

Orgueil Meteorite *left*
The Orgueil Meteorite, which fell in 1864; a carbonaceous chondrite, containing little metal, but numerous small pieces of graphite. It was suggested that it might contain the remains of organic material, but this has now been disproved.

Barwell Meteorite *left*
Part of the Barwell Meteorite, which fell in Leicestershire on 24 December 1965. During its descent it was seen from various parts of England, though observations were hindered by clouds. It broke up during the last stages of descent, and various fragments were scattered round Barwell village. Its original weight may have been 100 lb. The piece shown in the photograph, recovered by Patrick Moore, weighs 2½ lb. The Barwell Meteorite is a stone, made up chiefly of oxygen, iron and silicon.

Norton-Furnas Aerolite *left*
The Norton-Furnas Aerolite, which fell over Norton, Kansas, on 18 February 1948. The largest mass recovered weighs one ton. In this photograph it is seen at Albuquerque. It is the largest aerolite known.

Widmanstätten Patterns *left*
When an iron meteorite is cut and etched with acid, it may show what are called Widmanstätten patterns. They are due to the crystalline metallic structure which required unusual conditions for formation. Such crystals are unique to meteorites.

Hoba West Meteorite *left*
The Hoba West Meteorite near Grootfontein in South-West Africa – holder of the "heavyweight record" for meteorites. It is embedded in the limestone of the area, and is approximately rectangular. The weight of the mass is over 60 tons, though it is difficult to estimate accurately; it is thought that the original meteorite must have weighed over 80 tons before entering the atmosphere. The date of its fall is unknown, but it is certainly very old.

Impacts
above Destruction! Pine-trees, blown flat by the impact of the Siberian Meteorite of 1908.
left The impact crater near Winslow in Arizona; the diameter is almost 1 mile. It must be well over 10,000 years old.

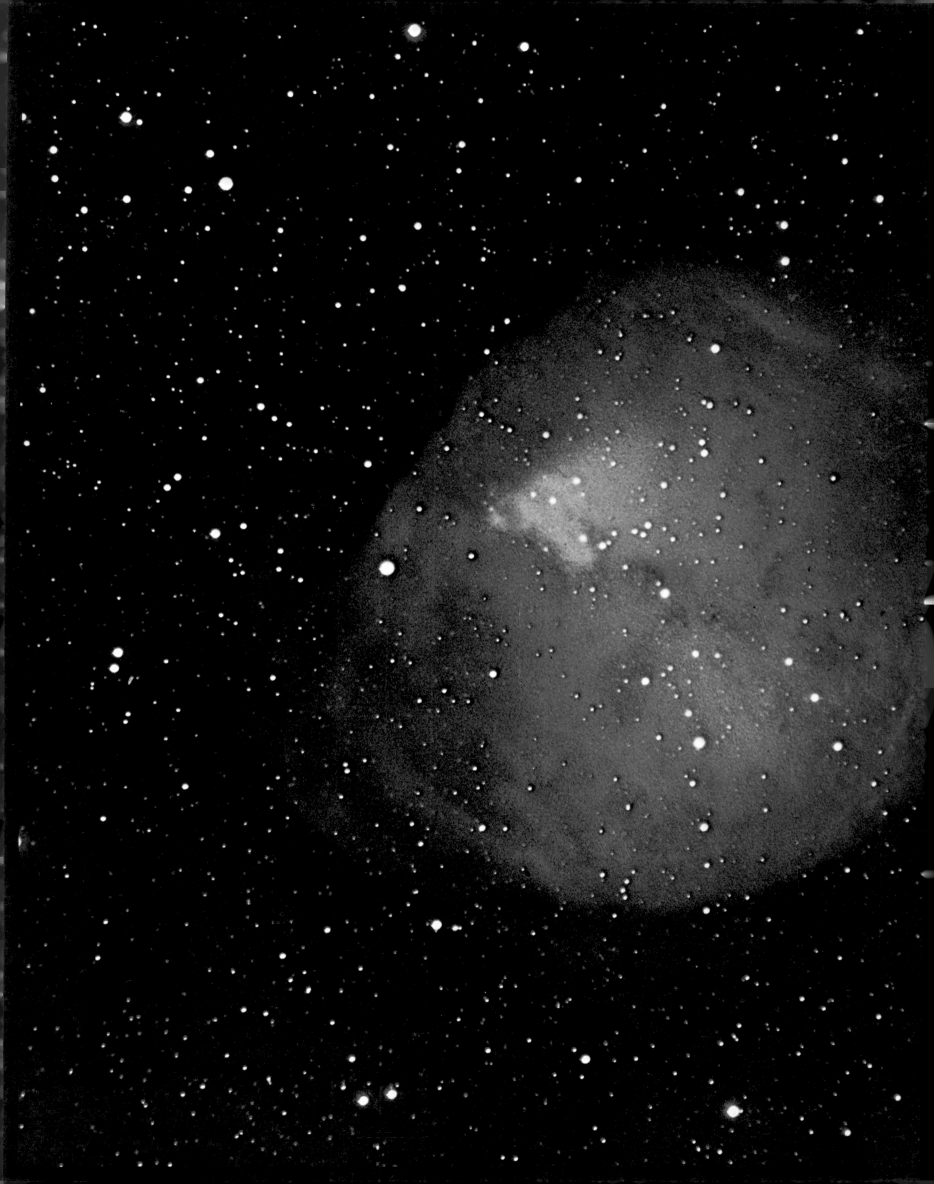

Atlas of the Stars

Early men regarded the stars as being small lamps fixed on to a solid background. This idea may sound absurd today, but it was not unreasonable in the days before astronomical science had begun; ancient peoples could have no idea that the stars are suns, and even the relatively enlightened Greeks could not bring themselves to believe that our Sun is nothing more than a star. All that could really be done was to catalogue the positions of the stars, and to divide them up into patterns or constellations. Without the help of modern-type instruments, the true nature of a star was bound to remain unknown.

Constellation patterns were drawn up by various races—probably by the Cretans, and certainly by the Chaldæans. The constellations shown by the Chinese and the Egyptians were quite different from those now in use; our sky-map follows basically upon the scheme given by Ptolemy of Alexandria, the last great astronomer of the Greek school, who worked between about A.D. 120 and 180. Ptolemy listed 48 constellations.

Dumbbell Nebula
The Dumbbell Nebula in Vulpecula, taken with the 200-in. reflector at Palomar in California. This object, one of the most interesting in the sky, was discovered by Messier in 1764, and is No. 27 in his catalogue of clusters and nebulæ.

Atlas of the Stars

What the ancients did not know was that the individual stars in a constellation are not necessarily associated with each other. A typical case is that of the Great Bear. In the tail of the Bear there are three stars, Alkaid, Mizar and Alioth. Alkaid and Mizar lie next to each other; but Alkaid is much the more remote. Indeed, it is farther away from Mizar than we are, and merely happens to lie in much the same direction as seen from Earth.

The astronomers of 2000 years ago believed, then, that all the stars lay at the same distance from us, and that the star-sphere rotated round the Earth once a day. The only real progress lay in measuring the positions of the individual stars, and this was done with considerable care. Hipparchus of Nicæa, about 130 B.C., produced an excellent catalogue, using

below Flamsteed House, the old observatory at Greenwich, founded 1675 and designed by Wren. All the scientific equipment has been moved to Herstmonceux and the observatory is now a museum.

right The 72-in. reflector constructed by the third Earl of Rosse, completed 1845, last used 1908. With it, Rosse discovered the spiral nature of the galaxies. This photograph, showing the tube and walls, was taken in 1969. *above* Old painting, showing Rosse preparing for a night's observation.

rudimentary measuring-instruments very skilfully. He was followed by Ptolemy, whose star catalogue has come down to us by way of its Arab translation (the *Almagest*). It was remarkably good, but even Ptolemy could not accept that a star might be a sun.

Beginnings of Modern Astronomy

After his death, astronomy languished for many centuries. The Arabs, between the 9th and 15th centuries, continued the work of star-cataloguing, but little theoretical progress could be made so long as the Ptolemaic system held sway. Ironically, the best star-catalogue of pre-telescopic times was produced, between 1576 and 1596, by the eccentric Danish astronomer Tycho Brahe, who was a violent opponent of the Copernican system, and who believed, for religious reasons, that the Earth must be the centre of the universe. Tycho's attention had been drawn to astronomy in 1572, when a brilliant new star flared out in the constellation Cassiopeia. Tycho was probably one of the first, if not the first, to notice it; it became bright enough to be visible in broad daylight, and we now know it to have been a supernova, involving the violent explosion of a formerly faint star. It was significant because it showed that the starry heavens were not unchanging, as the Greeks had taught. Tycho made a careful study of the star, and subsequently wrote a book about it.

Tycho was a superb observer, who would have made splendid use of the telescope, but it was only after his death in 1601 that Galileo first pointed one to the skies. Meantime the German astronomer J. Bayer had published a star-catalogue in which he allotted Greek letters to the stars in various constellations. The brightest star would become Alpha, the second Beta, and so on. In many cases the order is not strictly followed, but Bayer's nomenclature is still used, and his Greek letters have largely replaced the old proper names of the stars, some of which were tongue-twisting (Alkalurops, Azelfafage and Zubenelgenubi are examples) and most of which were Arabic. Nowadays, proper names are in common use only for the twenty or so brightest stars in the sky.

One of the important discoveries of the early telescopic period was made in 1612, when N. Peiresc observed the misty glow in Orion's sword that we now call the Great Nebula. However, the main emphasis of stellar work was still upon cataloguing, largely for navigational purposes. Greenwich Observatory was founded in 1675, by order of King Charles II of England, specifically so that a new star catalogue could be compiled for use by seamen; and after years of hard work by John Flamsteed, the first Astronomer Royal, the catalogue appeared.

After Newton's fundamental work of the late 17th century, the Ptolemaic System was finally rejected; the distance of the Sun could be measured with fair accuracy, and the idea that the stars are themselves suns took root. Determined efforts were made to measure their distances. James Bradley, the third Astronomer Royal, carried out some experiments in 1728, and although he failed to solve the problem his method was perfectly sound. There were also some speculations as to the distribution of the stars in space. It was even suggested that the star-system might be somewhat flattened in shape, though the first reliable scheme was not put forward before the work of Herschel in 1786.

Herschel's Contribution

It is worth pausing to write more about Herschel, because his researches were so fundamental. There was nothing random about them. He set out to survey the sky, using telescopes of his own construction, and in the course of his work he discovered many star-clusters, gaseous nebulæ, and also nebulæ which seemed to be made up of stars. In Herschel's view the star-system or Galaxy was shaped rather like "a cloven grindstone", with the Sun not far from the middle. More significantly still, Herschel wondered whether the starry or resolvable nebulæ, such as that in the constellation of Andromeda, might be separate star-systems well beyond our own. Again he was right (even though he never came to a final conclusion), but the problem was not finally cleared up until over a century after Herschel's death in 1822.

It was in 1838 that Bessel, in Germany, used the trigonometrical parallax method to establish the first reliable star-distance. The faint star 61 Cygni, in the constellation of the Swan, was found to lie at approximately 11 light-years. Light moves at 186,000 miles per second; its velocity had been measured by the Danish astronomer Ole Rømer in 1675. Therefore, in one year a ray of light can travel almost 6 million million miles; and this unit, the light-year, is a convenient one for astronomical purposes.

This triumph of Bessel's was followed by other distance measures; at almost the same time Thomas Henderson found that the brilliant southern star Alpha Centauri lies at only a little over 4 light-years, and Proxima, a faint companion of the Alpha Centauri, is still the nearest known star, at 4·2 light-years from us. However, the method of trigonometrical parallax is limited, and beyond a few hundred light-years the shifts become too small to be measured. Different lines of investigation have to be found, most of which depend upon instruments based upon the principle of the spectroscope.

Foundations of Spectroscopy

The pioneer work upon the analysis of light was carried out by Isaac Newton during the Plague years, 1665 and 1666, when he was working away at his quiet Lincolnshire home. He passed sunlight through a glass prism, and found that he had split the light up into a rainbow, with red at the long-wave end and violet at the short-wave end. Newton never really followed up this work, possibly because the prisms and lenses he had to use were of poor-quality glass, but the English scientist W. H. Wollaston returned to it in 1802, and found that the solar rainbow or spectrum was crossed by dark lines. Wollaston did not interpret the lines correctly, and so lost the chance of making a great discovery. Not many years later – in 1815 – a young German, Josef Fraunhofer, founded the science of solar spectroscopy, and mapped the dark lines which are still often called Fraunhofer lines. Unfortunately he died young, and it was not until 1859 that the solar spectrum was correctly interpreted by another German, Gustav Kirchhoff.

Kirchhoff laid down the laws of spectroscopy, which are of fundamental importance. An incandescent solid, liquid, or gas under high pressure will produce a continuous or rainbow spectrum. Incandescent gas under lower pressure will yield a bright-line or emission spectrum. The spectrum of the Sun is a combination of these two types, with the lines produced by the solar atmosphere being "reversed" into the dark absorption or Fraunhofer lines.

Of course, the stars were known to be suns; and so they too might be expected to yield spectra of the same kind. Pioneer work was carried out in Italy by Angelo Secchi, and in England by William Huggins, during the 19th century; star spectra were studied and interpreted, and it was found that although most of them showed bright backgrounds and absorption lines, there were considerable differences in detail. Hot, white stars did not produce the same spectra as yellow stars such as the Sun, or cooler orange-red stars such as Betelgeux in Orion. Secchi divided the stars up into various spectral types, but his preliminary classification was superseded by that drawn up at Harvard College Observatory, under the direction of E. C. Pickering, toward the end of the century.

According to modern classification, the stars are divided into nine types, which form a continuous sequence and which merge into each other. Meantime, astronomical photography had been developed, and eventually took the place of visual work at the eye-end of a telescope.

Clusters and Nebulæ

Star-clusters are not uncommon, and some, such as the Pleiades or Seven Sisters, are so conspicuous that they must have been known since early times. Nebulæ are fainter, and not many are visible without some optical aid.

In 1781 the French astronomer Messier had catalogued over 100 clusters and nebulæ (not because he was interested in them, but because he tended to confuse them with comets, in which he was very interested indeed). Messier or M-numbers are still used, though they have been officially superseded by the NGC or New General Catalogue numbers; thus the Orion Nebula is M.42, the Andromeda system M.31, and the Pleiades M.45.

Herschel, as we have noted, knew that the nebulæ are of two basic types: resolvable and irresolvable. Irresolvable nebulæ seemed to be made of "shining fluid", while the others were apparently composed of stars. In 1845 an Irish amateur, the third Earl of Rosse, constructed a huge reflector with a mirror 72 inches across, and discovered that many of the starry nebulæ are spiral in form, like Catherine-wheels.

Everything hinged upon the distances of the spirals, and for many years there seemed no prospect of making any reliable estimates. However, before the end of the 19th century spectroscopic work had established that the resolvable nebulæ really were starry; their spectra were somewhat confused, but the main absorption lines showed up plainly. On the other hand the irresolvable nebulæ yielded bright-line spectra, proving that they were composed of gas at low density.

The problem was finally solved in an ingenious way. It had long been known that there are some stars which do not shine steadily, but which brighten and fade over relatively short periods; these are the variable stars. Some variables, known as Cepheids, have periods of from a few days to a few weeks, and repeat their behaviour time and time again; they are perfectly regular, so that their brightness fluctuation can always be predicted. A few of them are visible to the naked eye, though most are telescopic objects.

Establishing Distances
In 1912 Miss Henrietta Leavitt, at Harvard, was studying photographs of the Small Magellanic Cloud, a star-system which lies in the southern sky. She found that there were many Cepheid variables, and that the brighter ones always had the longer periods. Because the Cloud stars are at approximately the same distance from us, this meant that the brighter Cepheids really were the more luminous; and eventually it became possible to work out a Cepheid's real luminosity merely by observing its behaviour. This meant that it also became possible for its distance to be worked out

In 1923 E. Hubble, working with the 100-in. reflector at Mount Wilson Observatory, detected Cepheids in some of the starry nebulæ, including M.31 in Andromeda. As soon as he found their periods, and measured their distances, the problem was solved. The Cepheids were so remote that they could not possibly be members of our Galaxy; and hence the spirals too were external systems. The modern estimate of the distance of the Andromeda Spiral is 2,200,000 light-years.

As soon as this fundamental step had been taken, other investigations could begin. In particular, there was the question of the distribution of the galaxies in space. The most important line of attack involved the Doppler principle, according to which a receding object is slightly reddened – because fewer light-waves per second enter the eye than would be the case if the object were at rest, so that the wavelength is apparently increased. It was already known that apart from a few of the closest systems, belonging to what is called the Local Group, all the galaxies showed red shifts in their spectra, so that they were moving away from us.

Hubble found that there was a relationship between the red shift and the distance; the farther away the galaxy, the faster its velocity of recession. It followed that the whole universe must be in a state of expansion, and this is still the theory accepted by most astronomers, though it is fair to add that there are a few notable dissentients.

Developments in Radio Astronomy
Not all the galaxies are spiral. Some are elliptical, some are spherical, and some – such as the southern Clouds of Magellan – are irregular in form. Their number is staggeringly great, as became evident when the new 200-in. reflector at Palomar came into commission in 1948. Also, it was found that some of

the myriads of galaxies were powerful emitters of radio waves.

In 1931 an American radio engineer, Karl Jansky, was investigating the causes of static when he found that his specially designed aerial was picking up long-wavelength radiation from the Milky Way. This was the start of the new science of radio astronomy. Jansky, strangely enough, never followed it through; but after the war large radio telescopes were constructed, and provided information which could never be obtained in any other way. Some of the radio sources were located inside our Galaxy; of these, many were due to the remnants of supernova explosions.

Tycho's star of 1572 is known to be one source, and another is the Crab Nebula, a patch of expanding gas which represents the débris of a supernova observed by Chinese and Japanese astronomers as long ago as 1054. However, most of the radio sources are extra-galactic, and there are some systems which are surprisingly energetic in the radio range.

During the past few years there have been more developments, totally unexpected, and which complicate the picture still further. In 1963 it was found that some radio sources are associated with objects which look rather like faint blue stars, but which have spectra showing tremendous red shifts. These objects, now known as quasars, are very much of a problem. If the distance-estimates based upon their Doppler shifts are correct, they are very remote – in some cases over 8,000 million light-years – and are incredibly powerful.

For many centuries astronomy was a static science. So far as the stars were concerned, almost no theoretical progress was made between A.D.200 and 1600, and it is fair to say that we have learned more since 1945 than we did during the preceding three centuries. The 1960s were particularly productive, and we may expect fundamental advances in the near future, even though we may be no nearer to solving the greatest mystery of all: that of the creation of the great universe in which we all live.

right Dome of the 200-in. reflector at Palomar. The Hale telescope, as it is known, remains the most powerful in the world, and its tremendous light-grasp enables it to photograph objects so faint that no other instrument can record them.

St llar sp ctra and st llar vulutiun

Since the stars are suns, they might be expected to yield spectra of the same basic type. When it became possible to study stellar spectra, it was found that most stars showed continuous rainbow backgrounds and dark absorption lines, as with the Sun, but there were differences in detail. These differences were linked with surface temperature. Hot bluish or white stars showed spectra different from those of red or orange stars.

Following the Harvard classification, the stars are divided into main types W, O, B, A, F, G, K, M, R, N and S, each of which is further subdivided. Types W, O, R, N and S are comparatively rare; most stars are included in the sequence B to M, which is also a sequence of decreasing surface temperature.

The H-R Diagram

In 1908 the Danish astronomer E. J. Hertzsprung drew up a diagram in which he plotted the luminosities of the stars against their spectral types. Similar researches were being carried on in America by H. N. Russell, and diagrams of this kind are now known as Hertzsprung-Russell or H-R Diagrams. Most of the stars lie along a well-defined band or Main Sequence from the top left (very luminous white stars) down to the lower right (feeble red stars). Red stars seem to come in two kinds: luminous giants and dim dwarfs. The division into giant and dwarf branches is also evident, though less marked, for the orange and yellow stars, but not for the white.

The Energy of the Stars

It was tempting to regard the H-R Diagram as marking a strict evolutionary sequence. At that time the source of stellar energy was not known. It had been assumed that the stars were radiating because they were contracting under gravitational forces; but this would not provide enough energy for many millions of years, and there was strong evidence that the age of the Sun must be of the order of 5000 million years. Instead, Russell proposed that the stars might keep shining because of the mutual annihilation of protons and electrons.

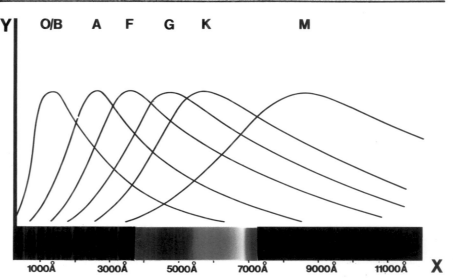

Protons and electrons are fundamental particles; a proton carries unit positive charge, while an electron, with only 1/1837 the mass of a proton, carries unit negative charge. On Russell's theory, a proton and an electron could collide and annihilate each other, releasing energy. However, the time-scale was again wrong; it was calculated that by the process of annihilation, a star could shine for millions of millions of years. It is now known that the atom is too complicated for straightforward annihilation of the type proposed by Russell to occur.

The Modern Theory

The correct solution was derived from the work of H. Bethe and C. von Weizsäcker from 1939. Hydrogen is the most abundant element in the universe, and a normal star contains a great deal of it. Inside the star, where the pressures and temperatures are tremendous, hydrogen nuclei are combining to form nuclei of helium. It takes 4 hydrogen nuclei to make one nucleus of helium; in the process, a little mass is lost,

and energy is released. It is this energy which keeps the stars radiating.

Stellar Evolution

On the old Russell theory, a star was born in a nebula, and became a red giant; it then contracted, joining the Main Sequence at the top left, and passed down to the lower right, ending its career as a red dwarf. On the modern theory, a star begins inside a nebula and contracts, joining the Main Sequence at a point determined by its initial mass; massive stars join nearer the top left. The star remains on the Main Sequence for most of its brilliant career. When its hydrogen runs short, it uses different nuclear processes to maintain its energy; it becomes first a red giant, and then either suffers a violent supernova outburst or else collapses into a very small, dense white dwarf star. A star may even end its career as a pulsar, emitting radio waves but little visible light. It is believed that a pulsar is a neutron star, millions of times denser than a normal star such as the Sun.

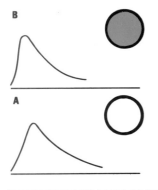

B-type stars. White or bluish, with surface temperatures of 25,000 °C (type B0). Helium lines prominent, so that B-stars are sometimes called "helium stars".

Spica (B1), Rigel (B8), Regulus (B7), Achernar (B5), Alpheratz (B9), Alcyone (B7), Alkaid (B3), Beta Centauri (B1), Beta Crucis (B0), Gamma Orionis (B2).

A-type stars. White, with surface temperatures of 11,000 °C (type 0); hydrogen lines prominent. Sometimes called Sirian stars. Colour index = 0 (type A0).

Altair (A7), Vega (A0), Fomalhaut (A3), Alioth (A0), Sirius (A1), Rasalhague (A5), Deneb (A2), Alpha Coronæ (A0), Beta Aurigæ (A2), Beta Ursæ Majoris (A1).

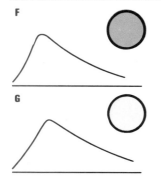

F-type stars. Yellowish; temperatures 7500 °C. Hydrogen and helium lines less prominent; calcium lines very conspicuous ("calcium stars").

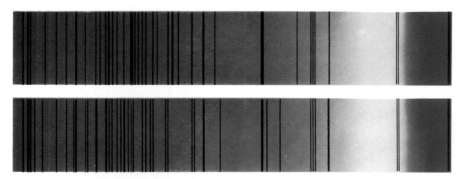

Canopus (F0), Procyon (F5), Polaris (F8), Alpha Persei (F5), Beta Cassiopeiæ (F2), Rho Puppis (F6), Theta Scorpii (F0), Alpha Hydri (F0).

G-type stars. Yellow; surface temperatures 5300 °C (G0 giant), 5800 °C (G0 dwarf). Giant and dwarf division very clear-cut. Metallic lines numerous.

The Sun (G2), Capella (G0), Zeta Herculis (G0), Epsilon Geminorum (G8), Gamma Persei (G8), Eta Boötis (G0), Beta Hydri (G1), Beta Corvi (G5).

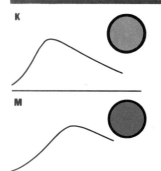

K-type stars. Orange; surface temperatures 4000 °C (K0 giant), 4900 °C (K0 dwarf). Hydrocarbon bands appear. Sometimes called Arcturian stars.

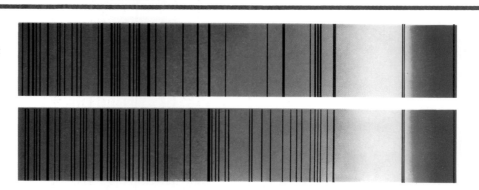

Arcturus (K2), Pollux (K0), Aldebaran (K5), Alpha Trianguli Australe (K2), Alpha Arietis (K2), Alpha Cassiopeiæ (K0), Gamma Leonis (K0).

M-type stars. Surface temperatures 3000 °C (giants), 3400 °C (dwarfs). All M-stars are red or orange-red, and many are variable. Broad titanium oxide and calcium lines

Betelgeux (M2), Antares (M1), Gamma Crucis (M3), Beta Pegasi (M2), Mu Geminorum (M3), Mira (M6), Rasalgethi (M2), Beta Andromedæ (M0).

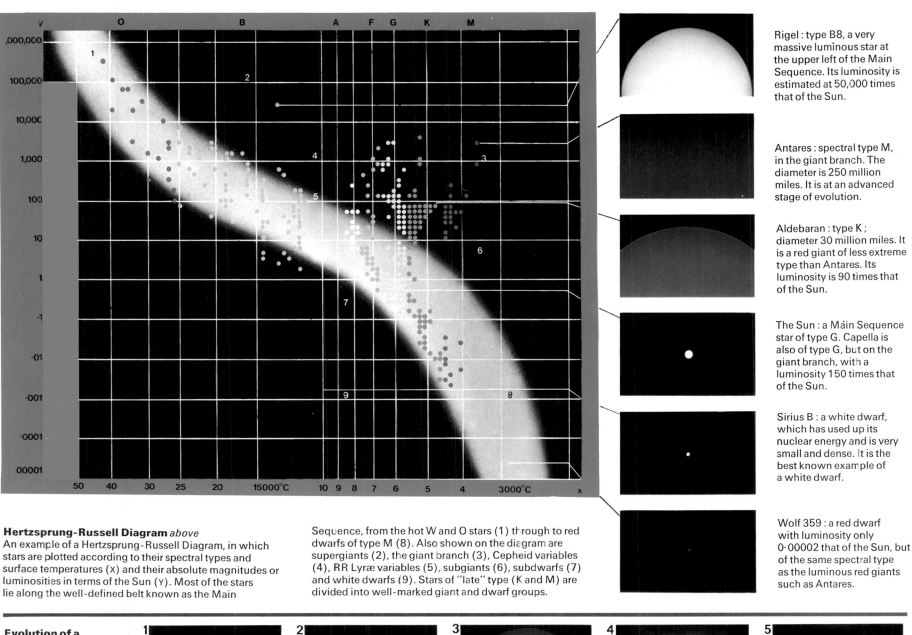

Hertzsprung-Russell Diagram *above*
An example of a Hertzsprung-Russell Diagram, in which stars are plotted according to their spectral types and surface temperatures (x) and their absolute magnitudes or luminosities in terms of the Sun (Y). Most of the stars lie along the well-defined belt known as the Main

Sequence, from the hot W and O stars (1) through to red dwarfs of type M (8). Also shown on the diagram are supergiants (2), the giant branch (3), Cepheid variables (4), RR Lyræ variables (5), subgiants (6), subdwarfs (7) and white dwarfs (9). Stars of "late" type (K and M) are divided into well-marked giant and dwarf groups.

Rigel : type B8, a very massive luminous star at the upper left of the Main Sequence. Its luminosity is estimated at 50,000 times that of the Sun.

Antares : spectral type M, in the giant branch. The diameter is 250 million miles. It is at an advanced stage of evolution.

Aldebaran : type K ; diameter 30 million miles. It is a red giant of less extreme type than Antares. Its luminosity is 90 times that of the Sun.

The Sun : a Main Sequence star of type G. Capella is also of type G, but on the giant branch, with a luminosity 150 times that of the Sun.

Sirius B : a white dwarf, which has used up its nuclear energy and is very small and dense. It is the best known example of a white dwarf.

Wolf 359 : a red dwarf with luminosity only 0·00002 that of the Sun, but of the same spectral type as the luminous red giants such as Antares.

Evolution of a Massive Star
A massive star (for instance, of 15 solar masses) will contract out of the interstellar material (1) and will join the Main Sequence after only 100,000 years (2). After approximately 150 million years – the exact period will depend on the mass – it will move into the giant region (3), "burning" first helium and then heavier elements. It may suffer a supernova explosion (4) and will send most of its material into space, leaving a neutron star of very high density and low luminosity (5).

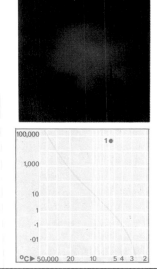 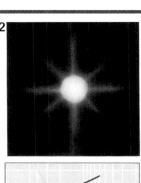 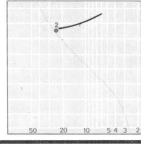 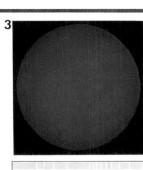 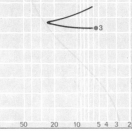   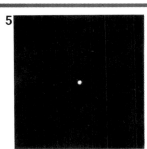 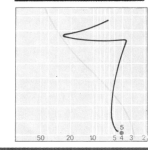

Collapsars and Black Holes

It has been well established that the final fate of a star depends upon its initial mass. If the mass is low (by stellar standards), the evolution will be comparatively gentle, since in extreme cases the nuclear processes may be unable to start; if the temperature is not high enough, the hydrogen-into-helium process cannot begin. With a normal star, of mass comparable with the Sun, the evolution takes place as described on page 120; when the supply of available hydrogen becomes exhausted, the star will pass through the Red Giant stage and will collapse into a White Dwarf. If, however, the mass is much greater, the end result may be a supernova outburst. Much of the star's material will be blown away into space; all that will be left will be a patch of expanding gas, together with a neutron star or pulsar. The Crab Nebula provides the best-known example of this, though other supernova remnants have been identified. The Crab, at a distance of 6,000 light-years, is much the brightest example.

Yet even a neutron star may not be the ultimate in stellar density, and it has been suggested that there may be even more remarkable objects – which we can never actually see. If the density progresses far beyond even the neutron star stage (presumably because of gravitational forces, about which our knowledge is still very slight) we may have a very small, super-dense object or "collapsar", whose surface gravity is so great that nothing – not even light – can escape from it. It will, in fact, be the centre of a "forbidden area" or Black Hole.

Obviously, such an extraordinary object would be unobservable, but it is just possible that it could be located. In the constellation of Auriga there is an innocent-looking star, Epsilon Aurigæ, which is known to be an eclipsing system. Normally it shines as a star of the 3rd magnitude, but every 27 years it is eclipsed by an invisible companion which radiates only in the infra-red. Until recently it was always assumed that this invisible component must be a very young star, still shrinking towards the Main Sequence

and not yet hot enough to shine in the visible range. But according to another theory, the companion is not young; it is extremely old, and has evolved into the collapsar stage, so that what we are really seeing is the hiding of the bright primary by the Black Hole, surrounding the collapsar. In this case, the infra-red radiation which we receive comes from clouds of dust and gas which are spiralling inward; when they enter the Black Hole, and are captured by the collapsar, they will become unobservable.

This sounds a strange theory, but it is becoming widely accepted, even though many authorities still have marked reservations about it. Research into these extreme features of the universe is still at a very early stage; but our whole outlook has altered during the past decade. Collapsars are not proved, but they may exist; pulsars are undoubtedly genuine – and there may be other features of the Galaxy which remain, as yet, unsuspected.

The Motions and Distances of the stars

The constellations, or star-patterns, do not appear to alter perceptibly from year to year. The groups that we see today are the same as those that must have been seen in the time of Julius Cæsar or the builders of the Pyramids; the old term "fixed stars" was quite logical. The apparent daily east-to-west movement of the sky is due entirely to the rotation of the Earth, and has nothing to do with the stars themselves.

Measuring Proper Motions

Yet the stars do have individual or "proper" motions. They are not genuinely fixed in space, but are moving about at high velocities. Over sufficiently long periods, the proper motions become large enough to be measured, as was first done by Edmond Halley in the 17th century; he compared the positions of three brilliant stars (Sirius, Procyon and Arcturus) with the positions as given in the old catalogues, and concluded, quite correctly, that there had been perceptible motion.

The next step was to measure the distance of a star, but this proved to be a very difficult problem. It was attempted by James Bradley, in 1728, and also by Sir William Herschel, who was among the greatest of all observational astronomers; it was Herschel who put forward the first reasonably accurate prediction of the shape of the Galaxy. However, it was only in the 19th century, in 1838, that the first star-distance was successfully measured.

Measuring Distance

Friedrich Bessel, Director of the Königsberg Observatory in Germany, decided to attack the problem in the same way that Bradley and Herschel had done – by using the method of parallax. Basically, this is the same method as that used by a surveyor to measure the distance of some inaccessible object, such as a mountain-top. If the object is observed from two different positions, it will seem to shift against the more distant objects in the background, and the amount of angular shift will allow its distance to be calculated. Bessel selected a star, 61 Cygni, which he thought must be close, because it is a wide binary and has also a relatively large proper motion (5″·25 per year).

He made careful observations at an interval of six months, during which time the Earth moved from one side of its orbit to the other; Bessel was therefore using a baseline of twice the radius of the Earth's orbit (twice 93 million miles, or 186 million miles). The parallax shift amounted to 0″·30. In 1838, Bessel was able to announce that 61 Cygni must lie at a distance of 11 light-years. This was very near the truth; the modern value for the distance of the star is 10·7 light-years.

The Nearest Stars

At about the same time T. Henderson, at the Cape of Good Hope, measured the distance of the brilliant southern star Alpha Centauri, which, like 61 Cygni, is a visual binary with a comparatively great annual proper motion, and found it to be only 4·3 light-years away. Proxima, a faint companion of the Alpha Centauri system, is slightly closer; it is 4·2 light-years away, and is the nearest of all stars with, of course, the exception of the Sun.

A third distance-measurement was undertaken by F. G. W. Struve, at Dorpat (Estonia). The star selected by Struve was Vega, in Lyra. Since Vega is 27 light-years away, much farther than 61 Cygni or Proxima, Struve's result was less accurate, but it was of the right order.

Only 8 stars lie within ten light-years of the Sun. Most stars are much more remote; for instance, the distance of Arcturus is 41 light-years, of Acrux 230 light-years, of Rigel 900 light-years – so that we are seeing Rigel not as it is now, but as it used to be 900 years ago. The method of trigonometrical parallax, used by Bessel and Henderson, is practicable only for relatively near stars. Beyond 150 light-years the shifts become too small to be measured accurately and less direct methods have to be used. Most of these depend upon measurements of the real luminosity of the star.

Once this and the apparent magnitude are known, the distance can be worked out, provided that due allowance is made for factors such as the absorption of light in space.

Even for the nearest stars, the parallax shifts are very small. That for Proxima amounts to 0″·79.

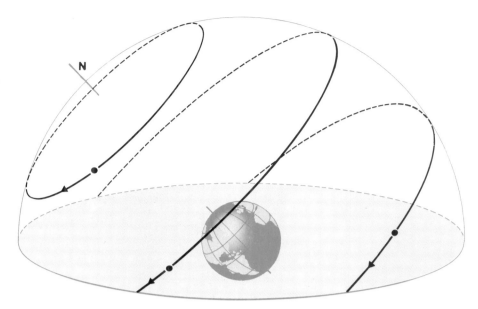

The Diurnal Motions of the Stars left

The Earth has a rotation period, relative to the stars, of 23h. 56m. 4s.·091 ; this is the length of the sidereal day. All the stars seem to describe circles round the celestial pole in this period, as shown in the diagram.

The Sun, Moon and planets share in this daily or diurnal motion, but since they are so much closer to us than the stars they also shift perceptibly against the constellations over short periods. The celestial poles, on the other hand, show no diurnal motion.

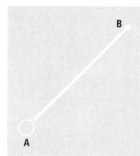

Actual Motion *above*
The diagram shows the actual motion (A–B) of a star in space ; a combination of *radial* and *transverse* motion.

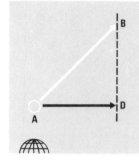

Radial Motion *above*
Radial motion (A–C) is velocity toward or away from Earth ; positive if the star is receding, negative if approaching.

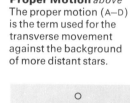

Proper Motion *above*
The proper motion (A–D) is the term used for the transverse movement against the background of more distant stars.

Radial and Transverse Velocity

The stars are so far away from us that their proper motions are slight. Radial velocity does not produce a shift in the star's position against the background of more remote stars, and can be measured only by spectroscopic methods when all the spectral lines can be seen to move toward one end of the spectrum (Doppler effect). Once both the radial and transverse velocities are known, the real motion of the star in space can be found.

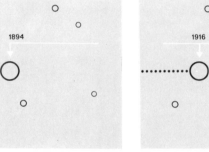
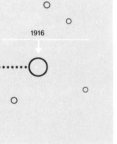
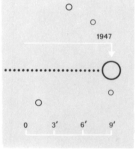

Barnard's Star *left*

The star with the greatest known proper motion is Barnard's Star ; its official designation is Munich 15040. The proper motion is 10″.29 per year, so that it takes the star 90 years to cover a distance across the sky equal to half the apparent diameter of the full moon. The star is 6 light-years away, and is a red dwarf of spectral type M with a luminosity 0·0005 that of the Sun. Slight irregularities in its proper motion have led P. van de Kamp, in America, to conclude that it is attended by a non-luminous body (or bodies) presumably of planetary nature.

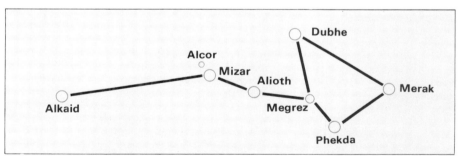
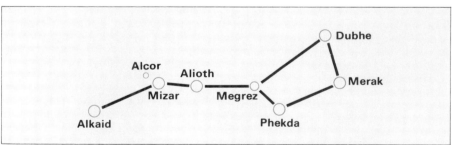
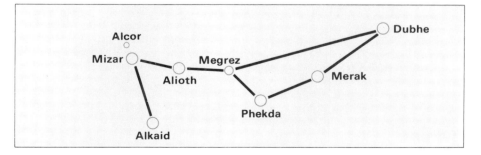

Effects of Proper Motion *left*

The long-term effects of proper motion are shown in this series of three diagrams. Of the seven main stars in Ursa Major (the Great Bear), five are moving through space in the same direction; the remaining two, Alkaid and Dubhe, are moving in a different direction. The upper diagram shows the Bear as it appeared 100,000 years ago ; the central diagram gives the pattern as it is today; and the lower shows how the Bear will appear in 100,000 years' time. The proper motions of the stars are, however, too slight to become noticeable with the naked eye, except over very long periods of time.

 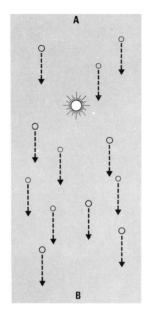

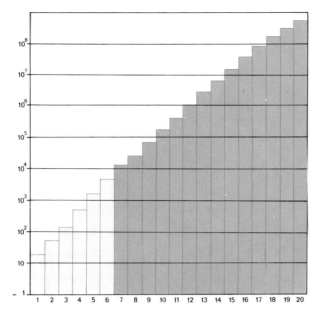

Solar Apex and Antapex *above*

Like all other stars which are contained in our system, the Sun is in motion round the centre of the Galaxy; it takes 225 million years to complete just one such journey.

At present the motion of the Sun is directed toward a point in the constellation Hercules, known as the apex of the Sun's way (A). The effect of the Sun's motion in space leads to an overall shift of the stars in that part of the sky away from the apex and toward the opposite point in the sky, which is known as the antapex (B)

Apparent and Absolute Magnitudes of the Stars

A star may appear brilliant either because it is comparatively near (as with Sirius) or because it is really very luminous (as with Rigel).

There is no direct connection between a star's apparent magnitude (that is to say, its brightness as seen from the Earth) and its actual luminosity. The brighter the object, visually, the lower is the designated apparent magnitude (see table *right*).

The absolute magnitude is the apparent magnitude that the star would have if it could be seen from a standard distance of 32·6 light-years (10 parsecs). Thus the Sun would appear of magnitude +4·7, so that it would be a dim naked-eye object; Sirius would be of magnitude +1·3; the Pole Star would be a brilliant object of magnitude −4·6; Rigel would appear of magnitude −7, so that it would cast notable shadows. Absolute magnitude is therefore a measure of the star's real luminosity.

Stars of Different Apparent Magnitudes *above*

The apparent magnitude of a star is its brightness as seen from Earth. The brightest object in the night sky, the full moon, is of magnitude −12·5 (on the same scale, the magnitude of the Sun is −26·6); Venus at its brightest attains −4·4. Of the stars, Sirius, at −1·4, is much the brightest; the magnitude of Polaris is +2·0, while the faintest stars normally visible to the naked eye are of magnitude +6. The Palomar reflector is capable of recording stars down to photographic magnitude +23.

The graph above shows the number of stars of different magnitude. Conventionally "1st-magnitude stars" are those from magnitude −1·4 (Sirius) down to +1·4 (Regulus); there are 21 stars which are known to be of the 1st magnitude.

The Apparently Brightest Stars

	Apparent mag.	Absolute mag.	Spectral class	Parallax	Dist. (l.y.)
Sirius	−1·4	+1·3	A1	0·37	8·6
Canopus	−0·7	−7·4?	F0	0·0	900?
Alpha Centauri	−0·3	+4·7, +6·1	G2,K	0·76	4·3
Arcturus	−0·1	−0·2	K2	0·07	41
Vega	0·0	+0·6	A0	0·12	26
Capella	0·1	−0·6	G0	0·07	47
Rigel	0·1	−7	B8	0·0	900?
Procyon	0·4	+3·0	F5	0·31	10
Achernar	0·5	−0·9	B3	0·05	66
Betelgeux	var.	−3 (var)	M2	0·02	190
Beta Centauri	0·7	−3·9	B0	0·01	300

(Alpha Centauri is a wide binary system. The distances and absolute magnitudes of the more remote stars, such as Canopus and Rigel, are uncertain.)

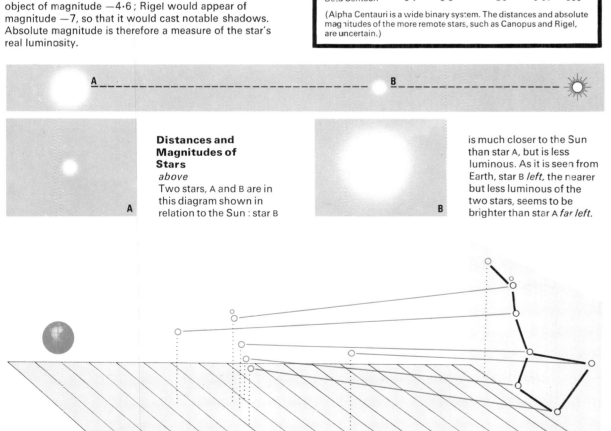

Distances and Magnitudes of Stars *above*

Two stars, A and B are in this diagram shown in relation to the Sun: star B is much closer to the Sun than star A, but is less luminous. As it is seen from Earth, star B *left*, the nearer but less luminous of the two stars, seems to be brighter than star A *far left*.

Distances of the Stars in Ursa Major

The diagram shows the seven chief stars of the Great Bear pattern, also seen on the opposite page, drawn to their correct relative distances from the Earth. Of the seven,

Alkaid, at 210 light-years, is much the most remote. It seems to lie next to Mizar in the sky, but the distance of Mizar is only 88 light-years, so that Alkaid is farther away from Mizar than we are.

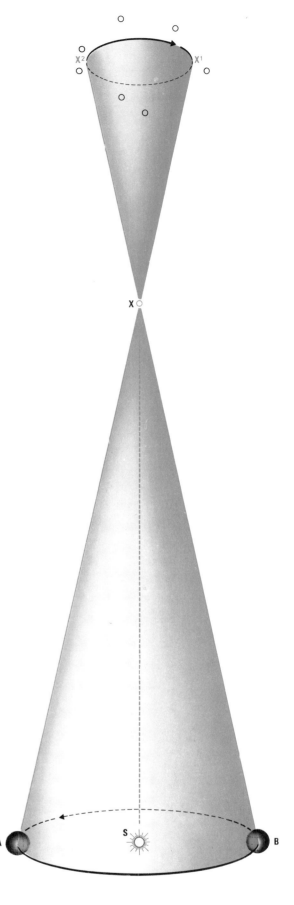

Method of Trigonometrical Parallax *above*

A represents the Earth in its position in January; the nearby star X, measured against the background of more remote stars, appears at X¹. Six months later, by the month of July, the Earth has moved to position B; as the Earth is 93 million miles from the Sun, the distance A–B is twice 93 = 186 million miles. Star X now appears at X². The angle AXS can therefore be found, and this is known as the parallax.

Since the length of the baseline A–B is known, the triangle can be solved, and the distance (X–S) of the star can be found.

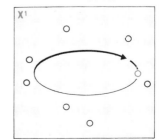 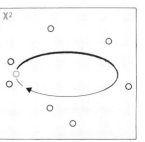

Unstable Stars

To astronomers, variable stars are among the most interesting of all stellar objects. They are of many kinds. Some are regular in behaviour, so that their light-curves are repeated time and time again; others are erratic, so that their changes in brightness cannot be predicted in advance. We also have the so-called 'novæ', or new stars, which flare up unexpectedly from insignificance to brilliance, and remain bright for a relatively brief period of a few days, weeks or months before fading away.

The Cepheids

The Cepheids, named after the prototype star Delta Cephei, are regular variables of short period. They are very common in the Galaxy; and because they are highly luminous F- or G-type supergiants they are visible over great distances.

The most important feature of Cepheids is that their real luminosities are closely related to their periods of variation. The longer the period, the more luminous the star. In fact, it is possible to find out the real brightness of a Cepheid merely by observing its fluctuations; and this in turn gives a key to the star's distance. Cepheids act as the astronomer's "standard candles" in space. It was by studying the Cepheids contained in M.31, the Andromeda Spiral, that E. Hubble, in 1923, was able to prove that the galaxies are independent, external systems.

Long-period Variables

The long-period variables are quite different. Their periods range up to over 400 days, and they are much less regular than the Cepheids; neither is there any link between their luminosities and their changes in brightness. The most famous star of this kind is Mira in Cetus (the Whale), which can sometimes become as bright as the Pole Star; its period is 331 days on average, and when at its faintest it is too dim to be seen even with binoculars, though of course a small telescope will always show it. Long-period variables are often called Mira stars. They are red giants, of vast diameter but relatively low surface temperature.

Irregular Variables

Other stars are irregular in behaviour. Some, such as Betelgeux in Orion, fluctuate in a more or less random manner with a small range of magnitude; others are much more violent. For instance, Eta Carinæ, unfortunately too far south to be seen from Sweden, is now too faint to be visible with the naked eye; but from 1835 to 1845 it outshone every star in the sky apart from Sirius!

What Makes a Star Vary?

Variable stars are regarded as unstable stars, most of which have left the Main Sequence and are well advanced in their evolution. (Eclipsing binaries, such as Algol and Epsilon and Zeta Aurigæ, are not intrinsic.) It is uncertain whether every star becomes a variable during some stage in its career, or whether it represents a condition which occurs only in particular instances.

Novæ

There may be a link between the U Geminorum or SS Cygni stars and the more spectacular novæ. A nova is not a 'new' star. What happens is that a formerly obscure star suffers a violent outburst which causes it to flare up abruptly, sometimes in a few hours; after a sudden rise there is a more gentle decline back to the original brightness. For instance, Nova Persei 1901, which at its best rivalled Capella, is now a faint telescopic object. The last really brilliant nova to be visible from Sweden was DQ Herculis of 1934, which reached the first magnitude.

Supernovæ

Much more spectacular than ordinary novæ are the supernovæ, which represent true stellar explosions; the star blows most of its material away into space. Only four have been seen in our Galaxy in recorded times: those of 1006, 1054, 1572 and 1604, though many have been recorded in outer galaxies. It seems that only a very massive star can undergo a supernova outburst; an ordinary star, such as the Sun, is unlikely to suffer so violent a fate.

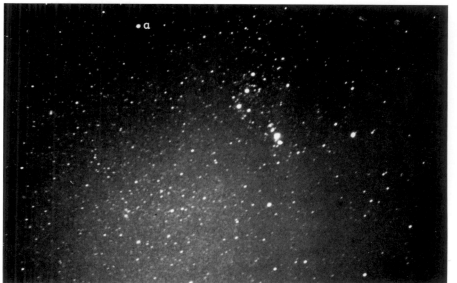

Perseus *left*
Two variable stars, Beta (β) and Rho (ρ). Algol, a common name for Beta, is not truly variable; it is the prototype eclipsing binary, ranging between magnitudes 2·2 and 3·5. Eclipses take place every 2½ days. Rho Persei, spectral type M, is an intrinsic variable; its extreme range is between 3·3 and 4·2, and the fluctuations seem to be completely irregular. A good comparison star is Kappa (κ), on the far side of Algol. When this photograph was taken, Algol was at maximum, with Rho appearing as slightly brighter than Kappa.

Auriga *left*
Auriga, with the two famous eclipsing binaries Epsilon (ε) and Zeta (ζ). Epsilon has a range of between magnitude 3·3 and 4·6, with a period of 9898·5 days – 27½ years. The brighter star is an F-type supergiant 60,000 times as luminous as the Sun; the fainter star radiates chiefly in the infra red, and is the largest star known (diameter over 2000 million miles). Zeta has a range of from 5·0 to 5·7; period 972 days. The brighter star is of type B9, with a diameter of 2,400,000 miles; the fainter is a K5 supergiant, diameter 180,000.000 miles

Cassiopeia *left*
Two interesting stars, Gamma (γ) and Alpha (α, or Shedir). Gamma is an irregular variable, with a spectral type of B0 (peculiar). It is normally of about magnitude 2·3, but suffers occasional outbursts to 1·6 (as happened in 1936), and can sink to as low as 3. Alpha is of type K0, and was listed in the variable star catalogue by Kukarkin as being "constant", but other observations indicate that it fluctuates slowly and irregularly between magnitudes 2·1 and 2·6. Useful comparison stars are Beta (β) (2·3) and Delta (δ) (2·6).

Orion *left*
Orion, showing the famous red giant Betelgeux. In 1840 Sir John Herschel found that Betelgeux is variable.
It has been claimed that there is a rough period of about 2070 days, but there are marked irregularities. The range is officially listed as between magnitude 0·3 and 1·3, but it is seldom that Betelgeux becomes as faint as Aldebaran (0·9), and past observations have recorded it as being fully equal to Rigel (0·15). Aldebaran makes a useful comparison star, owing to the similarity in colour.

Crab Nebula *right*

The Crab Nebula, M.1. Tauri ; U.S. Naval Observatory photograph. There can be no doubt that this nebula is the débris of the supernova seen by Chinese and Japanese astronomers in 1054 ; it became brighter than Venus, and remained visible to the naked eye for a year.

The Crab Nebula is of special importance to astronomers, as it emits radio waves and X-rays as well as visible light. It also contains one of the remarkable radio sources known as pulsars ; this is the only pulsar that has so far been identified with a definite optical object.

Supernovæ in NGC 7331 and NGC 4725
above and below

The top pair of illustrations shows a supernova in the spiral galaxy NGC 7331 in Pegasus (R.A. 22h.34m.8., decl. +34°10'). The integrated apparent magnitude of the galaxy is 9·7 ; the apparent dimensions are 10' x 2'.3. These two photographs were taken with the Lick Observatory 120-in. reflector. In the upper photograph the supernova is not visible ; in the lower, the supernova (indicated by arrow) is at its maximum.

A similar event can be seen in the bottom pair of illustrations in NGC 4725, a spiral galaxy in Coma Berenices (R.A. 12h.48m.1., decl. +25°46'). Again the supernova is indicated by an arrow.

Bright Novæ

Year and Constellation	Maximum magnitude	Discoverer	Notes
1572 Cassiopeia	−4	Tycho Brahe	Tycho's Star. Now a radio source. Supernova.
1600 Cygnus	3	Blaeu	P Cygni ; shell star ; now variable between magnitude 4½ and 5¼.
1604 Ophiuchus	−2·3	?	Kepler's Star ; a supernova.
1670 Vulpecula	3	Anthelm	Not now identifiable.
1783 Sagitta	6	D'Agelet	
1848 Ophiuchus	4	Hind	
1866 Corona Borealis	2	Birmingham	T Coronæ ; 91 recurrent nova ; second outburst ; in 1946.
1876 Cygnus	3	Schmidt	Q Cygni.
1891 Auriga	4·2	Anderson	
1898 Sagittarius	4·9	Fleming	
1901 Perseus	0·0	Anderson	GK Persei. Now a close binary.
1903 Gemini	5·0	Turner	
1910 Ara	6·0	Fleming	
1910 Lacerta	4·6	Espin	
1912 Gemini	3·3	Enebo	
1918 Aquila	−1·1	Bower	Now a close binary.
1918 Monoceros	5·7	Wolf	
1920 Cygnus	2·0	Denning	

Year and Constellation	Maximum magnitude	Discoverer	Notes
1925 Pictor	2·0	Watson	RR Pictoris. Fine slow nova. Now a close binary.
1927 Taurus	6·0	Schwassmann & Wachmann	
1934 Hercules	1·2	Prentice	DQ Herculis. Fine slow nova. Now a close binary.
1936 Aquila	5·4	Tamm	
1936 Lacerta	1·9	Gomi	CP Lacertæ. Fast nova ; rapid decline.
1936 Sagittarius	4·5	Okayabasi	
1939 Monoceros	4·3	Whipple & Wachmann	
1942 Puppis	0·4	Dawson	Fast nova
1950 Lacerta	6·0	Bertaud	
1960 Hercules	5·0	Hassell	
1963 Hercules	5·7	Hassell	
1967 Delphinus	3·7	Alcock	HR Delphini. Exceptionally slow nova. Pre-outburst magnitude, 11½.
1968 Vulpecula	4·3	Alcock	Fast nova ; magnitude 12 at end of 1969.
1970 Serpens	4·6	Honda	
1970 Aquila	6	Honda	

The supernova S Andromedæ (1885) in the Andromeda Galaxy, M.31, was just visible with the naked eye when at maximum.

Origin of Supernovæ

According to modern theories of stellar evolution, very massive stars may end their brilliant careers by becoming supernovæ ; certainly the Crab Nebula is a supernova remnant, and many galactic radio sources are of the same type. There have been suggestions that ordinary novæ may evolve into planetary nebulæ, similar to the Ring Nebula (M.57 in Lyra *right* or NGC 6543 *far right*), where the central star is very faint and hot ; but this is a matter for debate, and as yet there is no general agreement.

It has also been suggested that pulsars may be representative of the last stages of a supernova, and it is true that the only pulsar so far optically identified lies in the Crab Nebula. A pulsar is a radio source emitting sharp, regular and extremely short pulses ; the first to be discovered in 1967 (in Vulpecula), emits a pulse every 1.3728 seconds. It is now thought that pulsars are neutron stars, millions of times denser even than white dwarfs. Over 60 pulsars are now known, but, although studies of these are in hand it would be premature to assert that they are associated with supernovæ.

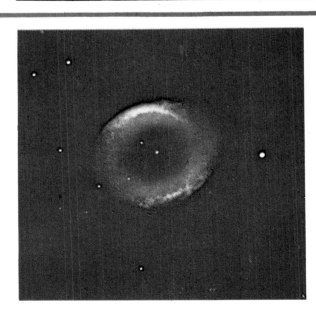

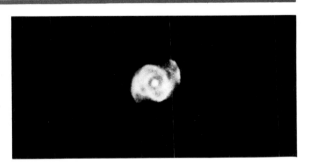

Ring Nebula *left*

The Ring Nebula in Lyra, M.57 (NGC 6720) ; R.A. 18h.51m.7s., decl. +32°58' ; distance 1410 light-years. The integrated magnitude is 9·3 ; magnitude of the central star, 15. M.57 lies directly between Beta and Gamma Lyræ.

Planetary Nebula *above*

NGC 6543, in Draco ; R.A. 17h.58m.7., decl. +66°38', integrated magnitude 8.8 (magnitude of central star, 11·1).

Double and Multiple Stars

Binary Systems

In 1767 the English clergyman John Michell suggested that most double stars were likely to be physically associated or *binary* systems. By 1804 the great observer William Herschel was able to show that this suggestion was correct, and in 1827 F. Savary showed that in the case of the binary system Xi Ursæ Majoris the two components revolved round their common centre of gravity in a period of 60 years.

Most double stars are binary pairs; optical doubles are comparatively rare. In some cases, as with Gamma Arietis in the Ram, the components of a binary are equal in luminosity and mass; in others, as with Mizar, there is an obvious difference between the primary and the secondary; and in many cases, as with Sirius, the primary is very much larger and more luminous than the companion. Yet although the stars differ so widely in size and in luminosity, there is a smaller range in mass, because a small star is always much denser than a large one. Therefore, the centre of mass of a binary system is always located between the two components; it is not a case of the fainter star moving round the brighter. Binary stars have been very useful to theoretical astrophysicists, because the orbits of the components provide information from which the combined mass of the system can be worked out.

Some double stars, such as Albireo (Beta Cygni), show superb contrasting colours; in other cases, as with Gamma Arietis, both components are of the same spectral type. Other fine pairs are Castor, Alpha Centauri and Alpha Crucis in the far south of the sky, and Gamma Virginis, which however is "closing up" and will be difficult to see by the end of the century. The revolution periods, too, have a wide range. With widely separated pairs there is no detectable orbital motion, and the revolution period may be many millions of years, so that all we can really say is that the components are moving through space in company. With close pairs, the revolution period is short. Zeta Herculis, a fine visual binary, has a period of 34 years, so that both the separation and the position angle alter quite rapidly; other binaries have periods which are even shorter.

Spectroscopic and Eclipsing Binaries

If the components are too close to be seen individually they may still be detected by spectroscopic means. There are binaries, too, in which the fainter star periodically passes in front of the brighter, and cuts off some or all of its light. These eclipsing binaries (sometimes termed eclipsing variables) are of various kinds; the best-known are Algol, in Perseus, and Beta Lyræ, which lies near Vega.

Double stars, both optical and binary, are favourite objects of study for the amateur observer. Many of them are within range of small telescopes, and some are really spectacular. A list of double stars for observation is given on page 127.

Double and Multiple Stars

In 1650 the Italian astronomer Riccioli turned his telescope toward Mizar, the second star in the tail of the Great Bear, and saw that it was made up of two separate components, so close together that to the naked eye they appeared as one. The distance between them is 14·5 seconds of arc; the position angle is 150°. Mizar was the first known telescopic double star. One component is of magnitude 2·2, the other 4·0, so that they are decidedly unequal. The position angle (P.A.) is the direction of the fainter star as reckoned from the brighter, from 0° (north) round by east (90°), south (180°) and west (270°).

Naked-Eye Doubles

Mizar is particularly interesting, because it forms a naked-eye pair with the 5th-magnitude Alcor (80 Ursæ Majoris). There are various other naked-eye doubles, such as Alpha Capricorni, in the Sea-Goat, and Epsilon Lyræ, near Vega. Yet at first it was not realized that the components of a double star might be physically associated. If two stars happen to lie in the same approximate direction, one will be seen "behind" the other; they will appear as though side by side in the sky, even though there is no real connection between the two apart from the visual one. Pairs of this kind are termed optical doubles.

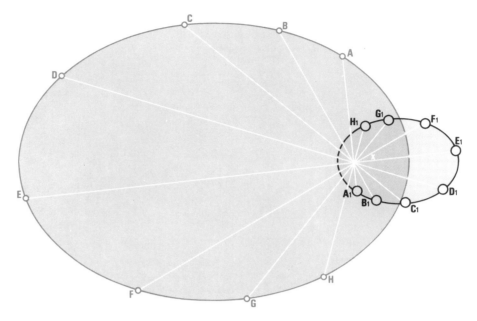

Orbits *left*
Though the stars have so wide a range in size and in luminosity, the mass-range is not nearly so great. In the diagram, X represents the centre of gravity of an unequal binary. The more massive component has the smaller orbit (A1, B1 ... H1); the less massive component has the larger orbit (A, B ... H). It is seen that the components are always at opposite ends of an imaginary line joining them and passing through X. With equal components, the point X is midway between the two stars, and the orbits are of equal size. (For clarity this diagram is somewhat over-simplified.)

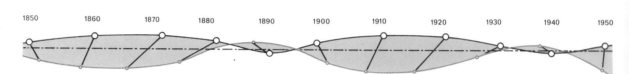

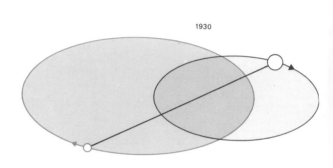

Sirius *above and right*
In 1834, F. W. Bessel found that the proper motion of Sirius was not regular, and he deduced the presence of a faint binary companion. In 1862, Clark at Washington found the Companion in the predicted position.
It is a white dwarf, with a diameter of only 26,000 miles, but a mass equal to that of the Sun. The diagram above shows the proper motions of the two components between 1850 and 1950; the diagram to the right relates this to the actual orbital positions in 1930.
The Companion is of magnitude 8·6, but is not easy to observe, as it is overpowered by the brilliance of the primary; the ratio is 10,000 to 1.

Examples of Double Stars

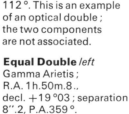

Optical Double *left*
Alpha Tauri (Aldebaran):
R.A. 04h.33m.0., decl.
+16°25'. Aldebaran has an 11th-magnitude companion at 31''·4, P.A. 112°. This is an example of an optical double; the two components are not associated.

Equal Double *left*
Gamma Arietis;
R.A. 1h.50m.8.,
decl. +19°03; separation 8''·2, P.A.359°.
This is a very wide, easy double. The components are equal at magnitude 4·4, and both are of spectral type A.

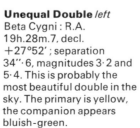

Unequal Double *left*
Beta Cygni: R.A. 19h.28m.7, decl. +27°52'; separation 34''·6, magnitudes 3·2 and 5·4. This is probably the most beautiful double in the sky. The primary is yellow, the companion appears bluish-green.

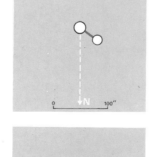

Very Unequal Double *left*
Alpha Herculis (Rasalgethi); R.A. 17h.12m.4.; decl. +14°27'; separation 4''·6, P.A.109°. The primary is an M-type red giant, variable from magnitude 3 to 4; the companion appears strong green by contrast.

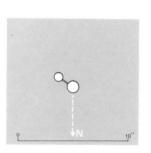

Rapid Binary *left*
Zeta Herculis: R.A. 16h.39m.4., decl. +31°42'. The maximum separation is 1''·6; period 34 years. The magnitudes of the components are 3·1 and 5·6. The primary is of spectral type G, and is obviously yellowish.

Triple *left*
Alpha Centauri: R.A. 14h.36m.2., decl. −60°38'. The separation and position angle alter fairly rapidly, as the period is only 80·1 years. Two of the components of Alpha Centauri are of magnitudes 0·3 and 1·7; types G4 and K5.

Multiple *left*
Epsilon Lyræ: R.A. 18h.42m.7., decl. +39°37'. The main pair is separable with the naked eye (magnitudes 4·7, 4·5; 207''·8); each is again double. Several unconnected stars lie in the same telescopic field.

Multiple *left*
The Trapezium (Theta Orionis), involved in the great nebula Messier 42, is a fine example of a multiple star. All the components are of spectral type O, and presumably had a common origin. The four main stars are visible with a small telescope.

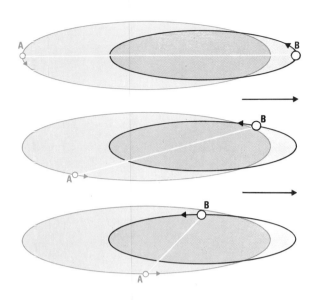

Variations in Algol *right*

In 1669 G. Montanari, at Bologna, found that the bright star Algol (Beta Persei) showed variations in brightness; every 2 days 11 hours it faded from magnitude 2 down to magnitude 3½, taking 5 hours to do so, after which it remained at minimum for 20 minutes and then took a further 5 hours to regain maximum. In 1783 J. Goodricke suggested that this behaviour must be due to the fact that Algol is a binary. When the fainter star passes in front of the brighter (1 and 3 in the diagram), a marked increase in brightness occurs shown in the light-curve. When the bright star passes in front of the fainter (2 and 4) there is a slight secondary minimum.

In the Algol system, the brighter component is of type B8, surface temperature 12,000 °C; the diameter is 2·5 million miles. The fainter star is of type K. The eclipses are not total.

Algol is the prototype eclipsing binary (often known, misleadingly, as an eclipsing variable). Many others are now known. There is no fundamental difference between an eclipsing binary and an ordinary spectroscopic binary; it simply happens that the orbit is at an angle suitable for eclipses to occur as seen from the Earth.

Beta Lyræ *below*

With the eclipsing binary Beta Lyræ, the two components are almost in contact; as the components are not so unequal as is the case with Algol, light variations are always going on, and there are alternate deep and shallow minima, as are displayed by the light curve below.

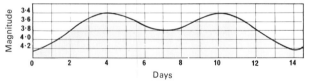

Epsilon Aurigæ and Zeta Aurigæ *below*

Two remarkable eclipsing binaries lie close to the brilliant star Capella, in Auriga. Epsilon Aurigæ is made up of a supergiant, 60,000 times as luminous as the Sun, together with an immense diffuse star radiating almost entirely in the infra-red; it is 2000 million miles in diameter, and may possibly be a "Black Hole". Zeta Aurigæ is equally remarkable. The difference in size between the components is so great that when the smaller star crosses the disk of the supergiant, as shown here, it should really be termed a transit *below left*.

The period is 972 days; the diameter of the orange K9-type supergiant is 180 million miles, while that of the B7-type companion is only 2·5 million miles. When the B7 star is eclipsed its light shines through the atmosphere of the supergiant for up to 3 weeks *below right*, and the spectral changes give invaluable information about the composition of the atmosphere of the larger star.

Spectroscopic Analysis

If the components of a binary are too close together to be seen individually, the system may be studied spectroscopically.

top Both components of the binary, A and B, are moving transversely to the line of sight, so that there are no Doppler shifts.

middle Star A is now moving toward the Earth, so that its spectral lines show a violet shift; B is receding, and will show a red shift—so in the combined spectrum, the lines are doubled.

lower Maximum splitting of the lines is produced when the velocities are entirely radial.

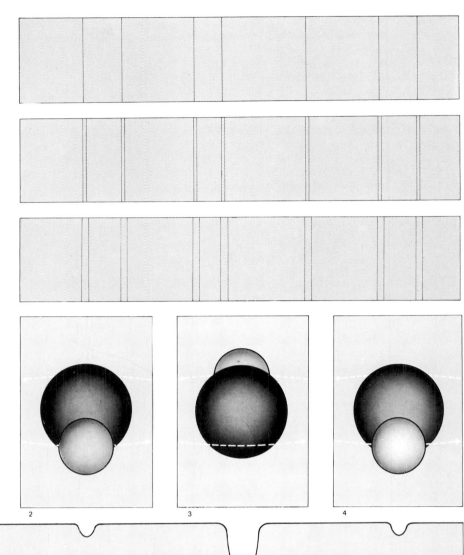

Eclipsing Binaries

The Algol stars and the Beta Lyræ variables are the best-known of the eclipsing binaries, but there are other types also. With the W Ursæ Majoris stars, both components are usually comparable in size and mass with the Sun, and are so close together that their revolution periods are short—sometimes less than 5 hours. The minima are of equal depth, since the two components are of the same luminosity, and there is continuous variation in light, so that the stars are often regarded as a sub-group of the Beta Lyræ type. W Ursæ Majoris stars are not common, and few of them become bright enough to be seen in binoculars. W Ursæ Majoris itself varies between magnitude 8·3 and 9·1 in a period of 0·33 days; each component is of spectral type F8.

An eclipsing binary is in fact a double star whose orbital plane is edgewise to the Earth, and whose components occult each other in turn. In other words, all double stars would appear to be eclipsing binaries if viewed from a certain angle. During the eclipse there are two periods of minimum luminosity—the primary minimum, when the darker star passes in front of the lighter one, and the secondary minimum, when the lighter star occults the darker. The light curve from an eclipsing binary will normally show a large dip at the primary minimum and a shallow dip at the secondary.

The UX Ursæ Majoris stars have Algol-like characteristics, but have extremely short periods, and differ sufficiently from the Algol class to justify their being placed in a separate category. UX Ursæ Majoris, RW Trianguli, and DQ Herculis belong to this class; DQ Herculis is the 1934 nova which for a time exceeded the 2nd magnitude. Then there are a few eclipsing binaries, such as CV Serpentis, in which one component is an extremely hot star of spectral type W (Wolf-Rayet); and with W Serpentis the two components

of the system are approximately equal in size, but differ in surface brightness and therefore in total luminosity.

With an eclipsing binary, the characteristics of the light-curve provide information which can lead to the accurate determination of the real orbit. As soon as this is known, important conclusions can be drawn. Since the components move round their common centre of gravity, the more massive star will move in the smaller orbit; and it becomes possible to work out the mass ratio between the two stars.

This cannot be done nearly so accurately with non-eclipsing binaries. For instance, with the Algol-type star U Cephei, it has been found that the masses of the components are respectively 4·7 and 1·9 times that of the Sun; the spectral types are B8 and G0; the diameters are 5·8 and 9·2 times that of the Sun respectively; and the surfaces of the stars are only 4,500,000 km. apart.

List of Eclipsing Binaries

This list includes eclipsing binaries with a magnitude range of more than 0.5, and a maximum magnitude of 8·0 or above.

Algol type

Star	Maximum	Primary Min.	Secondary Min.	Period	Spectrum
Algol	2·2	3·5	2·3	2·87	B8
U Cephei	6·7	9·8	6·8	2·49	B8+G2
Delta Libræ	4·8	5·9	4·9	2·33	A1
U Sagittæ	6·4	9·0	6·5	3·38	B9+G2
RZ Scuti	7·7	8·9	7·8	15·1	B2
Lambda Tauri	3·3	4·2	3·6	4·0	B3
Z Vulpeculæ	7·0	8·6	7·1	2·45	B3+A3
VV Cephei	6·6	7·4	6·6	7430	M2+B0
Epsilon Aurigæ	3·3	4·6	—	9889	F0
Zeta Aurigæ	5·0	5·7	—	972	K5+B7

Beta Lyræ Type

Star	Maximum	Primary Min.	Secondary Min.	Period	Spectrum
Beta Lyræ	3·4	4·3	3·8	12·91	B8
TT Aurigæ	8·1	9·0	8·4	1·33	B3+B3
V Crateris	9·5	10·2	9·9	0·70	A6
V Puppis	4·5	5·1	5·0	1·45	B1+B3

Clusters and Nebula

Star Clusters and Nebulæ

Look up at the Pleiades or Seven Sisters, in the constellation of Taurus, and you will see what seems at first sight to be a misty haze. Closer inspection will show several separate stars; people with normal eyesight can see at least seven. Binoculars reveal many more, and altogether the group contains over 400 members, most of them hot and white. The Pleiades group makes up a typical open or galactic star cluster. Photographs taken with large telescopes show that there is a beautiful gas-cloud spread between its stars.

Loose Clusters

Loose clusters are of no definite shape. Some of them contain many hundreds of stars; others may include only a few dozens. Though the individual stars may seem closely crowded together, they are really widely spread; there is no danger of mutual collisions.

The Pleiades lie at 1400 light-years from us. Their brightest star, Alcyone, is of the 3rd magnitude, and the cluster is a familiar feature of the Scandinavian night sky all through the winter. Also in Taurus is another loose cluster, the Hyades, surrounding the brilliant red star Aldebaran. However, Aldebaran is not a true member of the Hyades. It simply happens to lie almost midway between the Hyades and ourselves. Other prominent loose clusters are Praesepe or the Beehive, in Cancer, and Kappa Crucis in the Southern Cross, too far south to be seen from Sweden.

Most loose clusters consist mainly of hot young stars, known as Population I; however, all loose clusters are being disrupted by gravitational forces, and eventually their stars will be scattered.

Globular Clusters

This is not the case with the globular clusters, which are of different type. They are symmetrical, and may contain as many as half a million stars within a radius of 150 light-years. Even in the centre of a globular cluster there is little danger of mutual collision, but if our Sun were in such a region the night sky would be glorious indeed, with many stars brilliant enough to cast shadows. There could be no true darkness at all.

Altogether there are 119 known globular clusters associated with our Galaxy. They are all very distant, and form what may be called an "outer framework". The two brightest globulars (Omega Centauri and 47 Tucanæ) are too far south to be seen from Europe, but we can observe Messier 13 in Hercules, which is just visible with the naked eye, and is conspicuous through binoculars.

Messier's Catalogue

The first really useful list of clusters was drawn up in 1781 by the French astronomer Charles Messier. His catalogue numbers are still used; thus the Pleiades cluster is officially known as Messier 45. However, the catalogue contained other objects besides clusters, Messier also included the gas-clouds which we now call galactic nebulæ, as well as some external galaxies.

Galactic Nebulæ

Galactic or gaseous nebulæ are made up of tenuous gas together with small "grains", usually termed dust. The most spectacular example is Messier 42, which is easily visible with the naked eye; it lies in the Sword of Orion, below the three bright stars of the Hunter's Belt. It contains a very hot multiple star, Theta Orionis, nicknamed the Trapezium because of the arrangement of its four main members. The very thin material of the nebula is excited by the short-wave radiations sent out by the hot stars embedded in it, and is made self-luminous; this is an emission nebula. Many others are known, though most of them are more remote than Messier 42 (rather over 1000 light-years from us).

If the stars in a galactic nebula are less hot and energetic, they will be unable to make the nebular material self-luminous. The nebula will then shine only by reflection. The nebulosity in the Pleiades is of this kind. If there are no suitable stars in the material, the nebula will be dark, and detectable only because it blots out the light of stars beyond. There are dark nebulæ in Cygnus, visible from Europe; but the best example of all is the so-called Coal Sack in the Southern Cross.

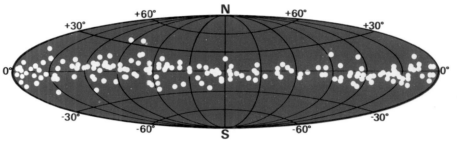

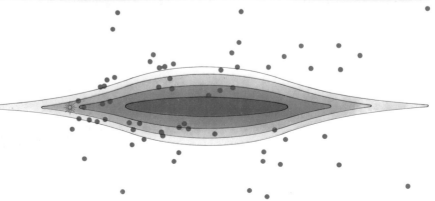

The Pleiades *left*
The photograph shows part of the constellation Taurus, including the famous open cluster M.45, known officially as the Pleiades but often called the Seven Sisters. The cluster lies at a distance of 410 light-years, and contains 400 stars in a spherical volume of space 25 light-years in radius. The brightest star, Alcyone, is of the 3rd magnitude, and normal-sighted people can see at least 6 others without optical aid. The leading stars of the cluster are of early spectral type. There is also a beautiful reflection nebula, well shown only by long-exposure photographs taken with large telescopes; it does not appear on this picture, which is taken on a much smaller scale. Photograph by Alan Williams, of Portsmouth.

Distribution of Galactic Clusters *left*
Galactic clusters are of Population I, and lie near the main plane of the Galaxy, though there are a few exceptional clusters which are well away from it; of these, the best example is the old cluster M.67 Cancri.

Cluster M.5 *left*
The globular cluster M.5 (NGC 5904), in Serpens; R.A. 15h.16m., decl. +2°16'. M.5 was discovered by G. Kirch in 1702, and is a bright telescopic object with an integrated magnitude of 6·2; its distance is 27,000 light-years. More than 100 RR Lyræ variables have been discovered in it. Its diameter is 130 light-years, according to Bečvár. It is in the same field as the 5th-magnitude star 5 Serpentis, and is easily found.

Distribution of Globular Clusters *left*
The diagram shows the distribution of globular clusters, which belong to the halo of the Galaxy. Globular clusters, like all objects of the halo, are entirely of Population II, and have very eccentric orbits round the galactic nucleus. The main Galaxy is shown, together with the position of the Sun.

It was by studies of the distribution of the globular clusters that Shapley was able to draw up the first reliable picture of the size of the Galaxy. His final value for the diameter of the Galaxy was of the right order, though it has since been amended to the present figure of 100,000 light-years.

Emission Nebula *right*

If nebulosity is excited to self-luminosity by a suitable star (A), the light emitted (B) is characteristic of the substance making up the nebula.

far right The Lagoon Nebula, M8 Sagittarii, is a typical emission nebula, 4850 light-years away. The "lagoon" is produced by the foreground dust-clouds. It is thought that the star producing the H-II region is of type O, and is so deeply embedded in the nebula that it cannot be observed. Globules have been detected, indicating that star formation is in process. The Lagoon is a dense nebula; there are 10^3 to 10^4 atoms per cubic centimetre in the central region – even though this density still corresponds to what we would normally term a laboratory vacuum.

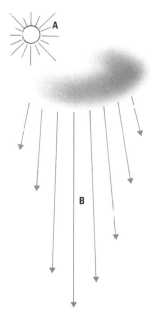

In the accompanying spectrum bright lines are a result of the presence of nebulous material in the H-II region. The hydrogen atoms are ionized by the short-wave radiations sent out by the hot star, and the gas is then excited to luminosity. The spectrum of a galactic nebula is of totally different type from that of a galaxy; a galaxy is of course made up of stars, and can therefore be expected to yield absorption lines against a continuous background.

Dark Nebula *right*

A dark nebula (B) cuts out the light of stars beyond (A), but is essentially similar to a bright nebula; the only difference is that it is not illuminated or excited by a suitable star.

far right The dark nebulosity in Monoceros was photographed in red light with the Palomar 200-in. reflector; the light of more distant stars is completely obscured, and it is easy to understand why Herschel at first described dark nebulæ as "holes in the heavens". The absorption is an effect caused by the solid particles, not by the interstellar gas. The nebulosity has two effects upon the light passing through it; a general dimming in all wavelengths, and a reddening resulting from the greater absorption of the shorter wavelengths. In the left-hand

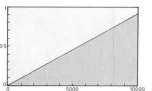

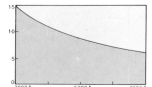

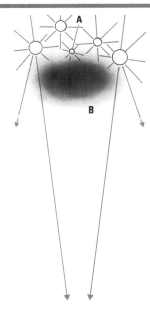

graph, the loss of brightness, in magnitudes, is plotted against the distance (in light years) traversed through a typical nebula.

In the right-hand graph, the loss in magnitudes over a fixed distance is plotted against the wavelength in Ångström units. The absorption is greater at the shorter wavelengths.

Reflection Nebula *right*

There are some gaseous nebulæ which shine only because it so happens that light is being reflected from a suitable star (A in the diagram). Such nebulæ usually contain a high dust content, but in fact we know far less about their constitution than we do in the cases of emission nebulæ.

The nebulosity which occurs in the Pleiades Cluster in Taurus *far right* is of the reflection type; the photograph shows the region of the star Merope, one of the brightest stars in the cluster. It was taken with the Palomar 60-in. telescope; the nebulosity in the Pleiades cannot be well observed visually, and it is necessary to take long-exposure photographs for a proper examination of this formation.

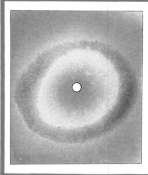

Real and Apparent Size of Nebulosity *left*

Nebulosity can be seen directly only if it is either excited to self-luminosity (emission nebula) or is lit up by a suitable star (reflection nebula). Almost always there will be more extensive nebulosity which is not luminous.

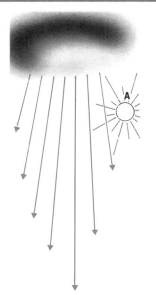

Mapping the Galaxy

Rotation
The Galaxy is rotating; the Sun takes 225 million years to complete one journey round the centre. Stars close to the centre have shorter periods, while those farther out take longer to make a full revolution.

Structure of the Galaxy
Radio astronomy studies have confirmed that the Galaxy is a loose spiral. This had been strongly suspected beforehand, from studies of the distribution of bright, hot stars of Population I, which tend to be most frequent in the spiral arms. It is not easy to obtain an overall impression from visual observation – simply because we are situated inside it.

Surrounding the main system is the spherical galactic halo, containing globular clusters as well as individual stars; these are predominantly of Population II, and are relatively old objects.

Mapping the Galaxy
Our Galaxy contains 100,000 million stars. It has an overall diameter of 100,000 light-years; the maximum breadth is 20,000 light-years, and the Sun, with its planets, lies close to the main plane, 32,000 light-years from the galactic centre.

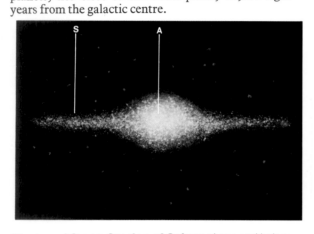

Shape and Cross-Section of Galaxy *above and below*
The diagram above shows the edge-on view, with the well-marked nucleus (A). The lower diagram shows the plan view, with rather loose spiral arms. The approximate position of the Sun (S) is marked.

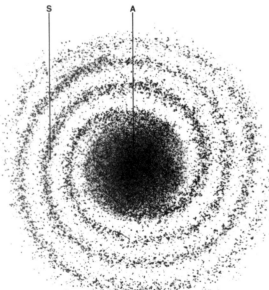

The Milky Way
The Milky Way effect is caused by the fact that when we see it we are looking along the main plane of the Galaxy (S-A in the above diagrams.) We cannot see through to the galactic centre, because of the obscuring material scattered in space which intervenes. The centre of the Galaxy lies toward the rich star-clouds in Sagittarius.

There is a very definite distinction to be made between the meaning of the terms "Milky Way" and "Galaxy". "Milky Way" is the name used to describe the luminous appearance in the sky with which we are all familiar, while "Galaxy" is the name given to the entire star-system – though it may be considered somewhat confusing that the system is often referred to as the Milky Way Galaxy.

Though the Galaxy is not exceptionally large by the standards of the universe, it is above average. In the Local Group there are five important galaxies; in order of size and mass, these are M.31 (the Andromeda Spiral); our Galaxy; M.33 (the Triangulum Spiral); and the Large and Small Clouds of Magellan.

A Galaxy Similar to Our Own *left*
The photograph, taken with the Palomar 200-in. telescope, shows the spiral galaxy NGC 7331, in Pegasus. In size and mass this galaxy is comparable with our own. Its position is R.A. 22h. 34m. 8., decl. +34°10'; the visual magnitude is 9·7.

The arms of NGC 7331 are more tightly wound than those of our own Galaxy. The view that we have of them is almost edge-on; if the angle of tilt were different, the galaxy would appear very imposing, and would resemble the 'Whirlpool' galaxy M.51 in Canes Venatici (which will be found illustrated on page 154).

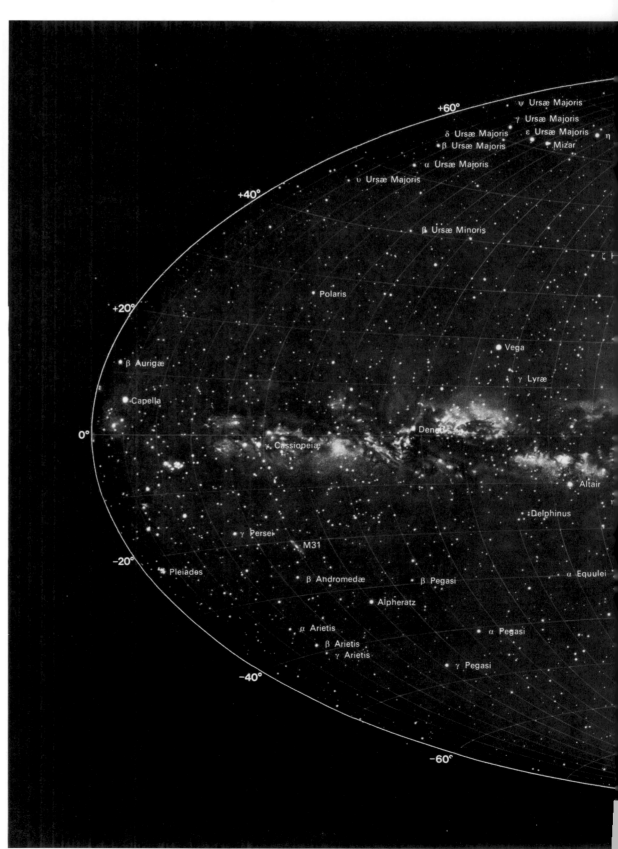

Map of the Milky Way *below*

This map of the Milky Way was drawn by Martin and Tatiana Teskūla at the Lund Observatory, Sweden. The co-ordinates refer to galactic latitude and longitude, measured from the galactic plane or mean plane of the Milky Way; the zero point for longitude is the intersection of the galactic plane and the celestial equator, at R.A. 18h.40., on the border of Aquila and Serpens.

The north galactic pole lies in Coma Berenices; the south galactic pole in Sculptor. In these areas the interstellar absorption is at its least, and it is possible to obtain a good view of external galaxies. Near the plane of the Milky Way, almost no galaxies are to be seen – not because they do not exist in these directions, but because their light is obscured by the material which is spread throughout our own Galaxy. Astronomers often referred to this area of apparent emptiness as the "zone of avoidance".

The Milky Way itself is shown extending along the galactic equator; there are many bright stars, as well as gaseous nebulæ. Various objects away from the Milky Way are identifiable on the maps, but appear to be somewhat distorted compared with their appearance on most charts, since they have been plotted on an unusual projection.

Radio Map of Galaxy *right*

The radio map shows the distribution of the clouds of neutral hydrogen in the plane of the Galaxy. Each point has been assigned the maximum density which would be seen in projection against the plane; contours have then been drawn in accordance with the density scale to the lower left, giving the average number of atoms per cubic centimetre. Distances were inferred from the radial velocities of the clouds, measured from the Doppler shift in their radio frequency, with assumptions regarding galactic motions. In the region left blank, radial components of motion are so small the method fails.

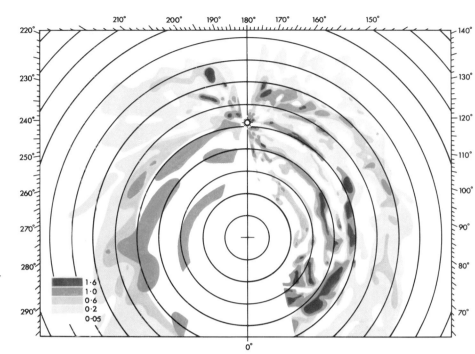

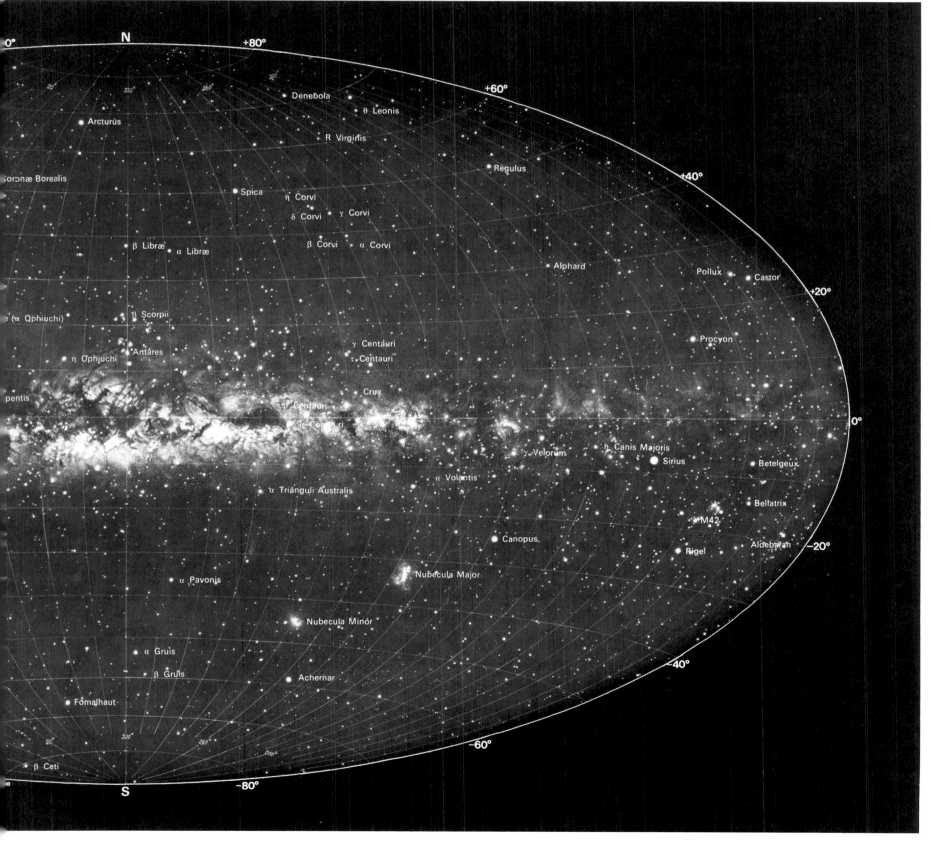

Galaxies Beyond Our Own

Until 1923 it was uncertain whether the objects we now call galaxies were contained within our own system, or whether they were external. The problem was solved by the American astronomer E. Hubble, who detected Cepheid variables in some of the spirals, notably M.31 in Andromeda. Since the period-luminosity law enabled the distances of these Cepheids to be measured, the distances of the galaxies could also be found. For many years the distance of M.31 was given as 750,000 light-years, but in 1952 W.

Baade, at Palomar, found that there had been an error in the Cepheid scale, and that all the galaxies were farther away than had been thought. The modern value for the distance of M.31 is 2·2 million light-years.

Cepheids are highly luminous stars (up to absolute magnitude —6) and the method can be used for galaxies out to 20 million light-years. Bright super-giant stars can be used out to greater distances, and globular clusters can extend the range out to 80

million light-years – on the assumption that globular clusters in any galaxy, including our own, are likely to be of about the same dimensions and luminosity. For more remote galaxies, distance-measurements have to be carried out by means of spectroscopic analysis. Hubble's Constant relates the recessional velocity of a galaxy (determined from its red-shift), to its distance. In this Atlas, the Constant is taken as 100 km./sec. per megaparsec for all galactic distances based on red-shift.

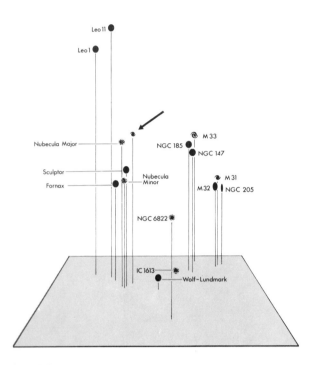

Local Group of Galaxies *above*
Our Local Group is a small cluster, having less than 30 known members: the spirals M31, M.33 and our Galaxy; the Clouds of Magellan, and smaller elliptical and irregular systems. These galaxies are so close to us that their own peculiar motions mask the effect of the red shift/distance relation.

Irregular and Regular Galaxies *left and right*
left The Large Cloud of Magellan (Nubecula Major) is a good example of an irregular galaxy. Relatively few galaxies are totally irregular.
right NGC 6946, an example of a regular galaxy, is a loose spiral.

Cluster of Galaxies *above*
Cluster of galaxies in the constellation of Hercules. photographed with the Palomar 200-in. reflector. Spirals, barred spirals and elliptical galaxies are seen. The clustering is genuine, and the individual galaxies are not receding from each other, though the entire group is receding from our own Galaxy.

Stars in the Arms of the Andromeda Galaxy *above*
Part of the spiral structure of the Andromeda Galaxy (M.31), taken in blue light with the Palomar 200-in. telescope. The hazy patch at the upper left is made up from unresolved Population II stars. The Andromeda Galaxy lies at a distance of 2·2 million light-years from our own Galaxy.

Interacting Galaxies *above*
Interacting spirals, NGC 5432 and 5435, photographed with the Lick Observatory 120-in. reflector. The spirals are at different angles to us, but are of the same type, and their mutual gravitational effects can be clearly seen in the way their patterns have developed; faint luminous matter forms a link between the two systems.

The Hubble Classification of Galaxies *right*
The system of classification of galaxies introduced by Hubble is still in use. There are elliptical galaxies (E0 to E7), spirals (Sa, Sb and Sc), barred spirals (SBa, SBb and SBc) as well as irregular systems which are not shown on the diagram given here. It is not now generally thought that the Hubble diagram represents an evolutionary sequence. There are many refinements ; for instance, Seyfert galaxies (many of which are radio sources) have very bright, condensed nuclei. The photographs below show examples of different types.

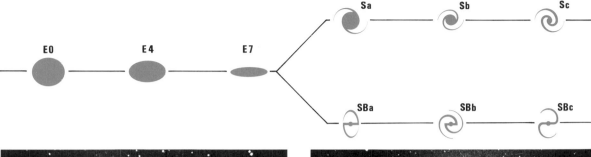

 **Type E0:** M.87 in Virgo, magnitude 9·2, distance 41 million light-years (Lick 120-in. reflector). M.87 is a giant system, and since it is symmetrical it resembles a globular cluster. It is a powerful source of radio emission.

Type E4: the dwarf galaxy NGC 147 in Cassiopeia, magnitude 12·1. Typical of the relatively small systems, and made up entirely of Population II, so there are no very luminous Main Sequence stars, and star formation has ceased.

Type E6: NGC 205, photographed in red light (200-in. reflector, Palomar). Its system appreciably more elongated than with NGC 147, it is the smaller companion of the Andromeda Galaxy and is made up of Population II objects.

Type Sa: NGC 7217, spiral galaxy in Pegasus (200-in. Palomar reflector.) The nucleus is well-defined, and the arms are symmetrical and tightly wound. It has now been established that a galaxy rotates with its arms "trailing".

Type Sb: M.81 (NGC 3031) in Ursa Major (200-in. Palomar reflector). Seen at a narrower angle than NGC 7217, its arms are "looser" ; magnitude is 7·9. Like its fainter companion, the peculiar M.82, M.81 is a radio source.

Type Sc: M.33 (NGC 598), the Triangulum Spiral (48-in. Schmidt, Palomar). The nucleus is less defined, and the spiral arms not so clear. M.33 is the most distant known member of our Local Group.

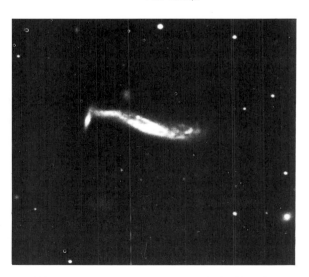

Type SBa: NGC 3504 in Leo Minor (Palomar 200-in. reflector). The "bar" through the centre of the system is noticeable, and the spiral arms extend from the ends. The reason for the formation of a bar of this kind is not known at the present time.

Type SBb: NGC 7479 in Pegasus (120-in. Lick reflector). Magnitude 11·6. The bar formation is much more pronounced, and there are only two noticeable arms, which extend from the ends of the bar. The other arms are very weakly developed.

Type SBc: a galaxy in the Hercules cluster, photographed with the Palomar 200-in. reflector. Here the bar formation is dominant, and the arms are reduced to little more than extensions of it. A second spiral, farther away, appears adjacent to the barred spiral.

The Unfolding Univers

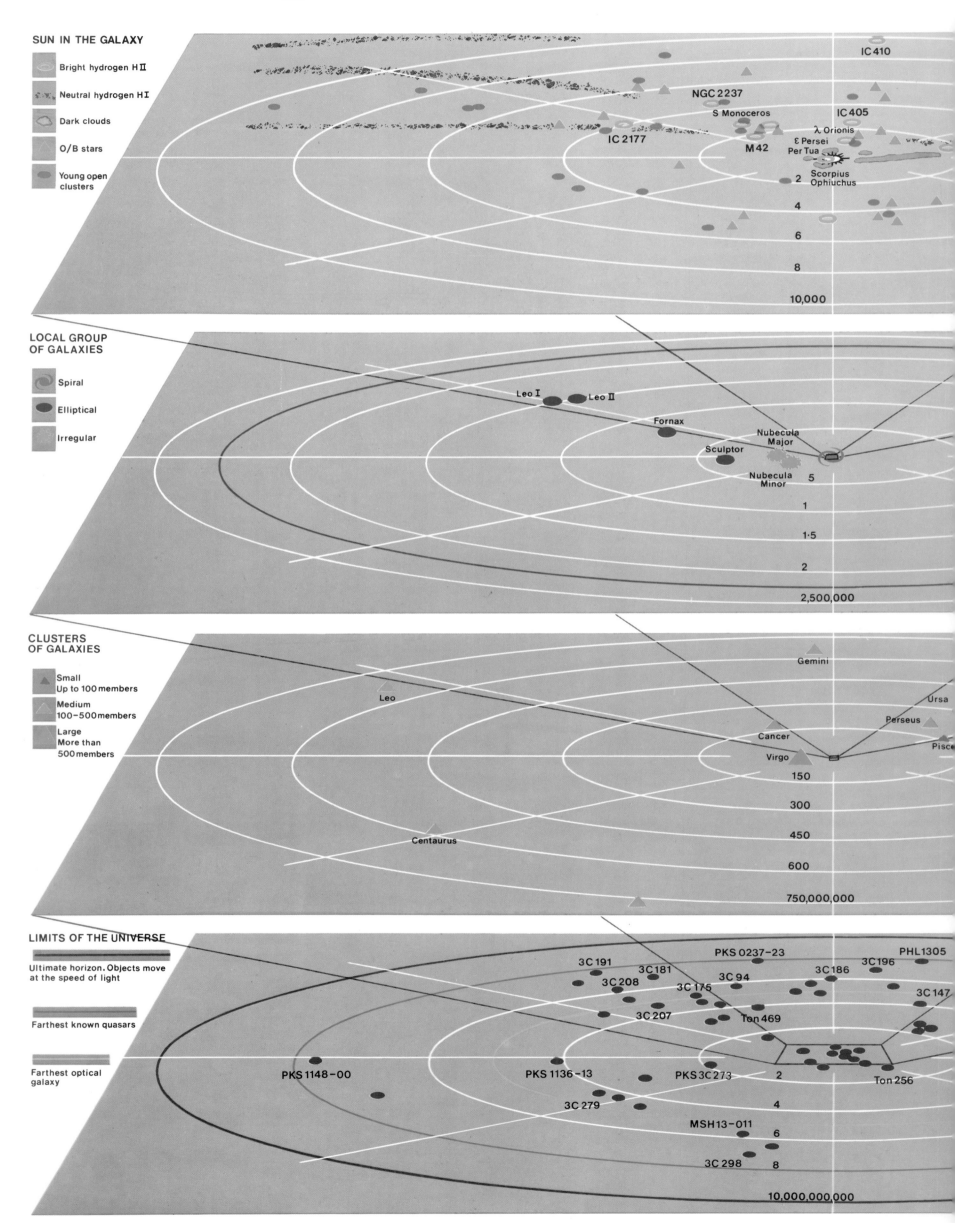

SUN IN THE GALAXY

- Bright hydrogen H II
- Neutral hydrogen H I
- Dark clouds
- O/B stars
- Young open clusters

IC 410

NGC 2237

S Monoceros

IC 405

λ Orionis

ε Persei
Per Tua

IC 2177

M 42

Scorpius
Ophiuchus

2

4

6

8

10,000

LOCAL GROUP OF GALAXIES

- Spiral
- Elliptical
- Irregular

Leo I   Leo II

Fornax

Nubecula Major

Sculptor

Nubecula Minor

5

1

1·5

2

2,500,000

CLUSTERS OF GALAXIES

- Small
  Up to 100 members
- Medium
  100–500 members
- Large
  More than 500 members

Gemini

Leo

Ursa

Perseus

Cancer

Pisce

Virgo

Centaurus

150

300

450

600

750,000,000

LIMITS OF THE UNIVERSE

- Ultimate horizon. Objects move at the speed of light
- Farthest known quasars
- Farthest optical galaxy

3C 191   PKS 0237-23   PHL1305

3C 181   3C 186   3C 196

3C 208   3C 175   3C 94

3C 147

3C 207   Ton 469

PKS 1148-00   PKS 1136-13   PKS 3C 273   2   Ton 256

3C 279   4

MSH 13-011   6

3C 298   8

10,000,000,000

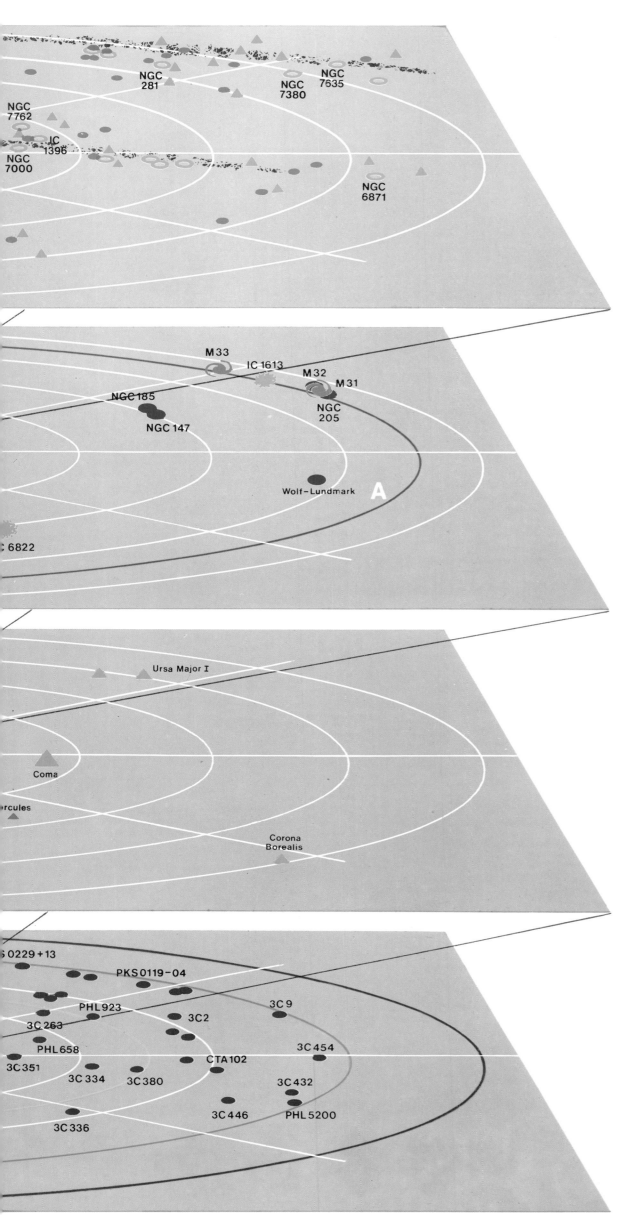

These diagrams show the immense scale of the universe. The Sun belongs to our Galaxy; our Galaxy is contained in the local group of galaxies, and other systems are farther away. Everything here has been projected on to the plane of the Milky Way.

The Region of the Sun *left*
The diagram to the left shows the region of our Galaxy which we can examine optically; it is impossible to see through to the centre of the system, because of the obscuring interstellar material. On this scale the entire Solar System would be a microscopic dot. Proxima Centauri, the nearest star, excluding the Sun, is 4·2 light-years away from us (too close to be shown on this scale). In addition to the visible stars there are the clusters, galactic nebulæ such as the Rosette Nebula NGC 2237, and the Orion Nebula, M.42 and stellar associations.

On this scale *left*, the area covered in the upper diagram is shown by the small rectangle at the centre. The Local Group of galaxies, of which our own is a member, has a radius of 2·3 million light-years — the distance of the Triangulum Spiral M.33.

The Local Group of Galaxies *left*
There are about 27 known galaxies in the Local Group, of which the most important are the Andromeda Spiral (M.31), our Galaxy, the Triangulum Spiral (M.33), and the two Clouds of Magellan, which may be regarded as companions of our Galaxy. M.31, at a distance of 2·2 million light-years, is the most remote object visible with the naked eye. This distance is shown by the line (A) in the diagram. The Local Group is an isolated unit, and the random motions of its systems outweigh the overall recession of the universe as a whole.

A further increase in scale; the diagram above is now represented by the central rectangle. The area contains many clusters of galaxies receding from each other at velocities which become greater as the distance increases, according to Hubble's Law.

Clusters of Galaxies *left*
Though isolated galaxies occur, there is a general tendency toward clustering; our Local Group forms such a cluster, though not an extensive one on the universal scale. Some clusters, such as that in Virgo, contain thousands of systems, but these are exceptional. With very remote galaxies and clusters of galaxies, individual stars cannot be studied clearly enough to yield reliable distance-estimates, and less direct methods have to be used — for example, after identifying the type of galaxy, an average brightness may be assumed for it, and its distance estimated.

Again the area above is in the central rectangle. Optical studies show galaxies up to 5000 million light-years; the most remote known galaxies are radio sources. If the red shifts of quasars are cosmological, their distances are certainly still greater.

The Limits of the Universe *left*
The inner line shows the distance of the most remote known galaxies. The second line indicates the distance of the farthest known quasars, assuming that their red shifts provide a reasonably accurate method of measuring their remoteness. The outer line represents the boundary of the observable universe, at which distance a galaxy would be receding at the velocity of light and would thus be unobservable. It must, however, be stressed that there is no proof that Hubble's Law holds good out to this limit — although measurements of velocity by the red shift are probably reliable, measurements of distance are certainly less so — and so the diagram is uncertain.

The Radio Sky

Visible light makes up only a very small part of the total electromagnetic spectrum. In 1931 an American research worker, Karl Jansky, was experimenting with a home-made radio aerial which he had built to investigate static when, to his surprise, he found that he was detecting long-wavelength radiations from the Milky Way. His results were published in 1933, but caused very little general interest. Further experiments were conducted in 1937 by an amateur, Grote Reber, who built the first true radio telescope, but it was only after the end of World War II that radio astronomy became recognized as an important branch of science.

The Jodrell Bank 250-foot paraboloid was completed in 1957, and has been responsible for fundamental advances in radio astronomy; its rôle may be compared with that of the Hale 200-in. reflector in optical astronomy.

Radio Telescopes

A radio telescope does not produce a visual picture of the object under study, and the familiar hissing noises are produced in the equipment, so that the term "radio noise" is somewhat misleading. Moreover, radio telescopes are of various designs; some of them consist of long chains of aerials. So far, the Jodrell Bank 250-foot "dish" remains the largest fully steerable radio telescope in operation.

Radio emission from the quiet Sun was detected before World War II by Southworth in the United States. Flare emission was detected by British researchers during the war. (For a time it was not recognized, and was thought to be due to German transmissions.) The planet Jupiter is also a radio source. However, most of the radio sources lie far beyond the Solar System. Some are within our Galaxy; others are more remote.

Galactic Radio Sources

Of galactic radio sources, one of the most important is M.1, the Crab Nebula, which is the wreck of the supernova seen in 1054. Much of the radio radiation is synchroton emission (that is to say, radiation from electrons moving in a magnetic field at velocities near that of light); the gas is still spreading outward from the old explosion-centre. Other galactic sources also are thought to be due to old supernovæ; one radio source corresponds to the position of the supernova studied by Kepler in 1604, and other source, in Cassiopeia, is due to Tycho's Star, a supernova observed in 1572.

Extragalactic Sources

The radio sources beyond our Galaxy are of various kinds. There are some galaxies which are unexpectedly powerful in the radio range; one is M.87, a giant elliptical in Virgo, from which issues a curious "jet".

Then there are galaxies such as M.82, in Ursa Major, which give evidence of having suffered some violent explosion near the centre, so that gas is still moving outward at high velocity. The quasars, discovered in 1963, are much smaller, but are powerful radio emitters. These "quasi-stellar" objects are still an enigma, as their energy appears to be very much greater than their size would suggest.

Colliding Galaxies

Many theories have been advanced to explain why some galaxies are strong radio sources while others are not.

It used to be thought that in some cases, as with the powerful source Cygnus A (550 million light-years away) the explanation might be the collision of two separate galaxies. The individual stars would seldom collide, but the interstellar material would be in constant collision, and it is believed that this might explain the radio waves; the two galaxies would be "passing through" each other in the manner of two orderly crowds moving in opposite directions past the same point. However, it has now been established that the process would be quite inadequate to explain the strength of the radio emission which is observed.

Little is yet known of the precise cause of the emission, and neither is it certain whether every galaxy becomes a "radio galaxy" at some stage in its evolutionary career.

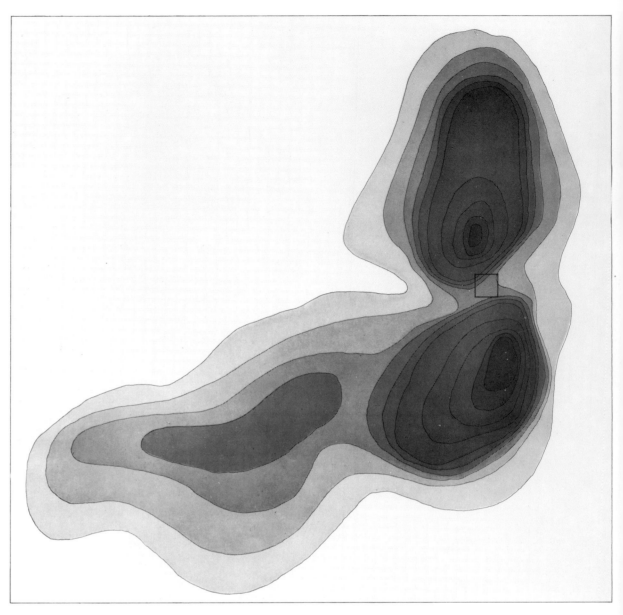

Radio Map of Centaurus A *above*
In some cases a radio telescope can detect features which are much more extensive than their optical counterparts. This is so with Centaurus A (NGC 5128).

The photograph to the left shows the visible object, but the radio map covers a much wider area – the region of the photograph is represented by the small rectangle in the middle! Also, the radio source is of a double nature, and the main centres of emission are not coincident with the optical object, as is shown by the arrangement of the contours in the diagram.

The reason for this curious situation is unknown. It used to be thought that Centaurus A must be made up of two galaxies in collision, but this theory has had to be abandoned, and at present the mystery remains.

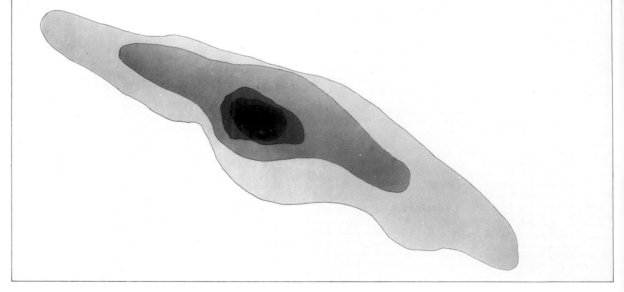

Radio Map of M.82 *above*
With the irregular galaxy M.82 the situation is different, inasmuch as the optical object *left* is larger than the radio source *above*; the radio emission comes from the central part of the galaxy, as shown by the contour lines.

In M.82 there are intricate hydrogen gas structures of immense size, and velocities up to 600 miles per second indicate that a tremendous outburst has taken place in the middle of the system. The outburst occurred between 1 and 2 million years before our present view; since M.82 is 8·5 million light-years from us, the outburst must date back over at least 9·5 million years.

M.82 contains few hot blue stars, and the radio radiation which it emits may be largely put down to synchrotron emission.

Quasar 3C-47 *right*
Distribution of radio intensity in the quasar 3C-47. The optical object (marked +) is the quasar; the two main radio emitting regions are 62'' of arc apart (700,000 light-years).

Quasar 3C-273 *above*
(Palomar 200-in. reflector). This quasar is notable as being a source of X-rays as well as of visible light and radio radiation.

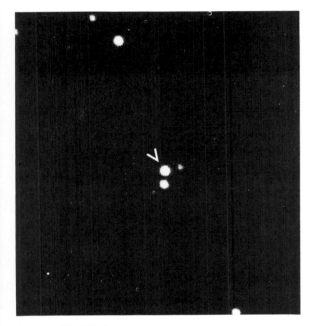

Quasar 3C-147 *above*
Photographed with the 200-in. Palomar telescope; it is indicated by the pointer, and at first glance it looks very like a faint star.

Quasars
During the compilation of a catalogue of radio sources, radio astronomers at Cambridge tried to identify the sources with optical objects. Some of the sources seemed to be associated with objects which did not look like galaxies. One of these, 3C-273 in Virgo, is of the 13th magnitude, and at first glance there is nothing to single it out from a star. Early in 1963, M. Schmidt at Palomar studied the spectrum of 3C-273, and discovered that there were emission lines, shifted toward the red end of the spectrum by 16 per cent. This indicated a high velocity of recession, and also a very great distance.

The discovery was quite unexpected; it showed that 3C-273 could not be a star. Other objects of the same kind were studied, and nearly all were found to have greater shifts. Originally the objects were called QSOs or quasi-stellar objects; they are now known as quasars.

If the red shifts in the spectra are Doppler effects, the quasars are thousands of millions of light-years away; yet they are much smaller than galaxies, and must be immensely luminous. A powerful quasar may be the equal of 100 normal galaxies. As yet there is no satisfactory explanation of the way in which a quasar produces its energy.

Whatever may be their true nature, there can be little doubt that quasars are of fundamental importance. Until 1963 they were entirely unsuspected, which shows that our knowledge of the universe may still be more limited than we like to think.

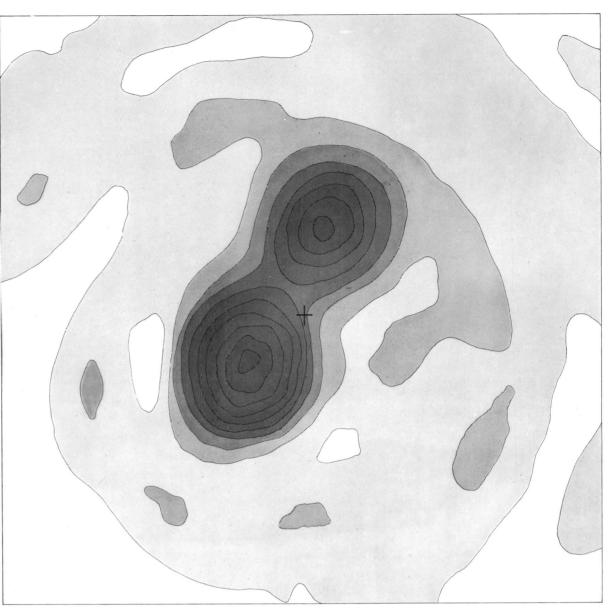

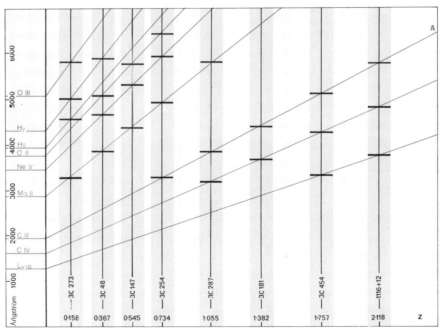

Red Shifts in the Spectra of Quasars *left*
In the diagram, the spectra have been arranged so that their emission lines fall on straight, inclined lines; as the red shift (z) increases, a given emission line changes colour. For example, the CIII line (A) is normally at 1909 Ångströms, in the ultraviolet. It is blue in the spectrum of 3C-287, green in 3C-454, and orange in 1116+12, the most distant quasar whose spectrum is shown here.

Most authorities regard the red shifts as Doppler effects, but a minority view holds that they are not cosmological so that quasars would be neither so remote nor so luminous as is generally thought.

NGC 2623 *left*
An Sc-type galaxy in Cancer, shown here in a photograph taken with the Palomar 200-in. reflector. The galaxy is notable as being a strong source of radio radiation, and was once thought to be made up of two systems in collision, though this theory has now been rejected. The spiral arms are not well defined, but are strong enough to be traced.

Mapping the Constellations

The origin of the constellation patterns is not certainly known. The Chinese and the Egyptians drew up fanciful sky-maps (two of the Egyptian constellations, for instance, were the Cat and the Hippopotamus), and so, in all probability, did the Cretans. The pattern followed today is Greek, and all the 48 constellations given by Ptolemy of Alexandria in his book the *Almagest*, written about A.D.150, are still in use.

Ptolemy's list contains most of the important constellations visible from the latitude of Alexandria. Among them are the two Bears, Cygnus, Hercules, Hydra and Aquila, as well as the twelve Zodiacal groups. There are also some small, obscure constellations such as Equuleus (the Foal) and Sagitta (the Arrow), which seem surprisingly faint and ill-defined to be included in the original 48.

Star Legends

It has been said that the sky is a mythological picture-book and certainly most of the famous old stories are commemorated there. All the characters of the Perseus tale are to be seen – including the sea-monster, though nowadays it is better known as Cetus, a harmless whale! Orion, the Hunter, sinks below the horizon as his killer, the Scorpion, rises; Hercules lies in the north, together with his victim the Nemæan lion (Leo). The largest of all constel-

lations, Argo Navis – the Ship which carried Jason and his companions in quest of the Golden Fleece – has been unceremoniously cut up into its keel (Carina), poop (Puppis) and sails (Vela), as it was thought unwieldy.

Modern Constellations

Ptolemy's constellations did not cover the entire sky. There were gaps between them, and inevitably these were filled. Later astronomers added new constellations, sometimes modifying the original boundaries. Later still, the stars of the far south had to be divided into constellations, and some of the names have a very modern flavour. The Telescope, the Microscope and the Airpump are three of the more recent groups. The Southern Cross, Crux Australis, is a 17th-century constellation. It was formed by Royer in 1679, and so has no claim to antiquity. Many additional constellations have been proposed from time to time, but these have not been adopted, though one of the rejected groups (Quadrans, the Quadrant) is remembered in the name of the annual Quadrantid meteor shower. There have also been occasional attempts to revise the entire nomenclature, but it is unlikely that any radical change will now be made. The present-day constellations have been accepted for too long to be altered.

Janssonius' Map *below*
The map below was drawn by Joannes Janssonius in 1660, and is constructed with reference to the pole of the ecliptic. Here too the constellation figures are shown, and it is not surprising that Sir John Herschel, the great 19th-century astronomer, commented that the patterns had been drawn up "so as to cause as much inconvenience as possible"! The constellation boundaries were modified and laid down by the International Astronomical Union in 1933.

Eimmart's Maps
The maps *right* were drawn by G. C. Eimmart, and show the ancient groups as well as some of the more modern ones. Around the main charts are given diagrams to show the phases of the Moon *lower left*, the seasons *lower right*, and the tides *upper right*, as well as three of the

"systems of the world". The centre top drawing shows the Ptolemaic theory, with the Earth in the middle of the universe and all the other bodies moving round it. To the bottom centre is the chart of the Copernican system, with the Sun occupying the focal position and the planets revolving round it; the planetary orbits are circular, since Copernicus himself still regarded the circle as the "perfect" form. The diagram to the top *left* shows the Tychonic system, supported by the Danish astronomer Tycho Brahe. Here the Earth is central, and the Sun, Moon and stars move round it, but the planets are in orbit round the Sun. Tycho was one of the best observers in astronomical history, and between 1576 and 1596 he drew up the star-catalogue which enabled Kepler to demonstrate the truth of the heliocentric system. Yet Tycho was too conventional to dethrone the Earth from its central position, and he was also a firm believer in the pseudo-science of astrology.

Ancient astronomers believed the sky to be solid, and that the stars were fixed on to it. The idea of a "celestial sphere" is still convenient for the purposes of explanation; the sphere is assumed to be concentric with the surface of the Earth, as shown in the diagram to the right.

The direction of the Earth's axis indicates the north and south celestial poles, while the celestial equator is the projection of the Earth's equator on to the celestial sphere. Just as the terrestrial equator divides the Earth into two halves (north and south), so the celestial equator divides the sky into two halves. The equator runs through Orion, and passes very close to Mintaka (Delta Orionis), the northern star in the Hunter's belt. It will be seen that the celestial equator is a great circle on the celestial sphere. (A great circle is defined as a circle whose plane passes through the centre of the sphere; it therefore divides the sphere into two halves.)

The Ecliptic

The ecliptic is defined as the projection of the plane of the Earth's orbit on to the celestial sphere, but it may also be described as being the apparent yearly path of the Sun among the stars; it passes through the twelve constellations of the Zodiac. Because the Earth's axis is inclined to the perpendicular at an angle of $23\frac{1}{2}°$, so the ecliptic is inclined to the celestial equator by the same amount, $23\frac{1}{2}°$.

About 21 March each year—the date is not quite constant, owing to the inconsistencies of our calendar —the Sun crosses the celestial equator, moving from the southern to the northern hemisphere of the sky; this is termed the Vernal Equinox, or First Point of Aries (Υ). Six months later, about 23 September, the Sun again crosses the equator, moving from north to south; this is the Autumnal Equinox, or First Point of Libra ($\libra$).

The Vernal Equinox is used as the zero point for right ascension (see diagram, *lower right*). It used to be in Aries, but precession has now shifted it out into the adjoining constellation of Pisces, the Fishes.

Precession

Precession is due to the gravitational effects of the Sun and Moon on the Earth's equatorial bulge. The Earth's axis is not constant in direction, but describes a circle 47° in diameter in a period of 25,800 years. This means that the position of the celestial pole alters, which in turn affects the position of the celestial equator and of the equinoxes. The First Point of Aries moves along the ecliptic, from west to east, by 50 seconds of arc annually. To the ancient Egyptians, the pole star was not Polaris, but Alpha Draconis (Thuban), which is of magnitude $3\frac{1}{2}$. In 12,000 years' time the north celestial pole will lie near the brilliant blue star Vega (Alpha Lyræ).

In consequence of precession, all values of right ascension must be qualified by a statement of the year to which they refer. In this atlas all figures relate to 1950 unless otherwise stated.

At present there is no bright south polar star. The nearest naked-eye star to the pole is Sigma Octantis, of the 5th magnitude.

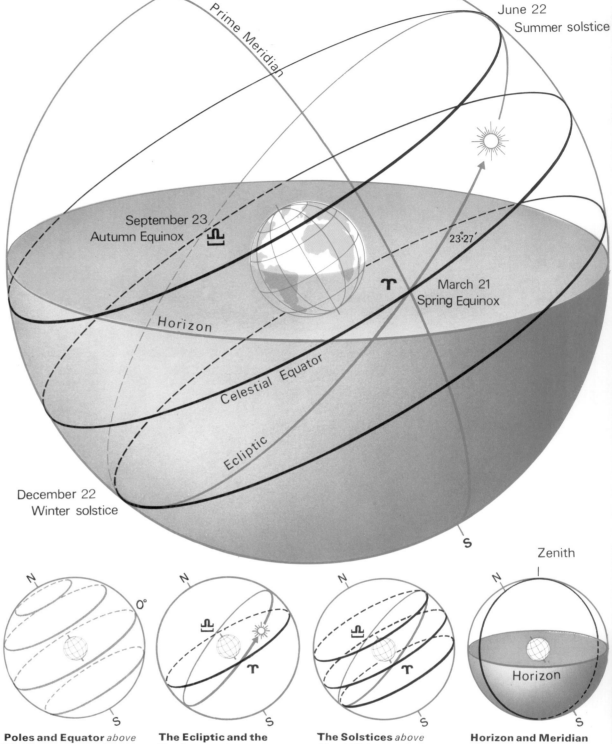

Poles and Equator *above*
The diagram shows the celestial poles (the points on the celestial sphere indicated by the direction of the Earth's axis) and the celestial equator (the projection of the Earth's equator on to the celestial sphere).

The Ecliptic and the Equinoxes *above*
The ecliptic (the projection of the plane of the Earth's orbit on to the celestial sphere) and the equinoxes (points of intersection between the ecliptic and the celestial equator) are shown.

The Solstices *above*
About 22 June and 22 December the Sun respectively reaches its northernmost and southernmost positions – the summer and winter solstices. The declination of the Sun is then 23 °27′ from the celestial equator.

Horizon and Meridian *above*
The observer's horizon and meridian are shown. The meridian is the great circle on the celestial sphere which passes through both poles and the zenith (observer's overhead point).

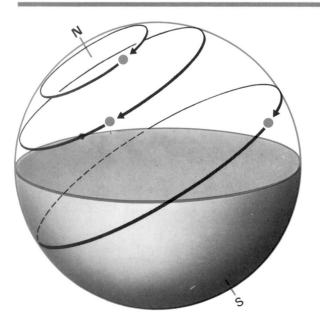

Diurnal Motions *left*
A star close to the celestial pole never sets, but remains permanently above the horizon; it is termed a circumpolar star. Stars farther from the pole will drop below the horizon at their lowest. From central Europe and the northern United States, constellations such as Ursa Major and Cassiopeia are circumpolar, but not from more southerly countries. From Australia, the northernmost constellations never rise above the horizon, but the Southern Cross, never visible from Europe, is circumpolar. Orion, which is cut by the celestial equator, is visible from every continent.

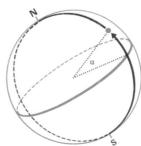

Declination *above*
The declination of a celestial body is its angular distance north or south of the celestial equator (α in the diagram). Objects on the celestial equator have declination 0 °; the celestial poles are at 90 °. The altitude of the visible celestial pole must always be equal to the observer's latitude, so that, for instance, from the latitude 51 °N (approximately that of London) the north celestial pole will be 51 ° above the horizon.

Right Ascension *right*
The right ascension of a celestial body is measured by its angular distance from the First Point of Aries (Υ). The prime meridian (A) is defined as the great circle running north–south through the First Point of Aries; the right ascension is measured eastward. Thus a star which culminates 4h.33m. (B) after the First Point of Aries has done so, has an R.A. of 4h. 33m.

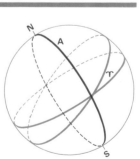

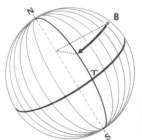

The Sky Through the Year *right*

Due to the orbital motion of the Earth, the Sun appears to move along the ecliptic, passing through the 12 constellations of the Zodiac as well as part of a 13th constellation (Ophiuchus, between Scorpius and Sagittarius). Obviously, it is not possible to see the Sun and the stars at the same time – except during a total solar eclipse – and as the diagram will show, an observer on Earth will see the Sun in the area of Aries and Pisces during March; these constellations will therefore be above the horizon only during daylight. In June, during the course of the summer solstice, the Sun will be in Gemini, so that the whole of this region (including Orion) will be unobservable.

Precession *right*

The precession circle, 47° in diameter, showing the shift in position of the north celestial pole around the pole of the ecliptic (A). In Egyptian times (c. 3000 B.C.) the polar point lay near Thuban or Alpha Draconis; it is now near Polaris in Ursa Minor (decl. +89°02'); in A.D. 12,000 it will be near Vega. The south celestial pole describes an analogous precession circle.

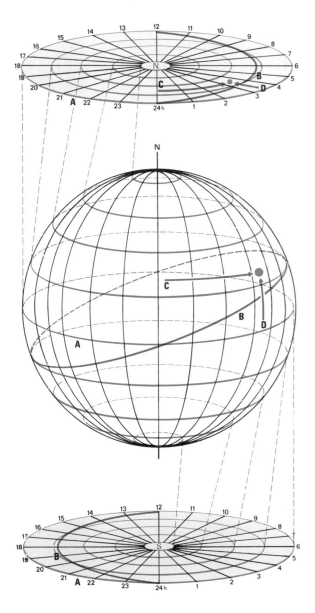

Star Map Projection *above*

The diagram above shows the method of construction that has been used to draw up the star maps given in this atlas. The co-ordinates of the celestial sphere are plotted so as to give an easily understood and convenient survey of the entire sky.

On both northern and southern sky projections. A shows the celestial equator projected on to the map as the outer (0°) concentric circle. B shows the ecliptic, C and D the projection of right ascension and declination respectively. In the maps given in this atlas, the two hemispheres of the sky *below* are each divided into two, and each hemisphere is shown in two sections. In each case the celestial equator runs round the periphery of the map. Orion is divided by the equator, and thus appears in both sections; it is easy to find the corresponding position for other groups. For instance, Virgo also is divided by the equator; the northern part of the constellation is shown on the right-hand map (pages 148-9, 156-7) and the southern part on the left-hand map (pages 164-5, 172-3).

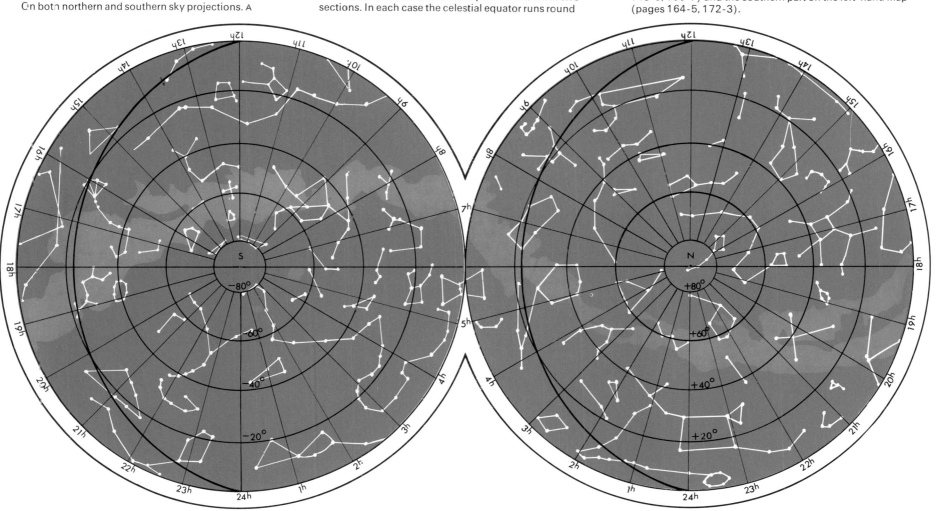

Seasonal Star Maps: North

Latitudes *left*
Latitudes of the major cities of the Northern Hemisphere. For the observer in the northern hemisphere all the stars of the northern sky are visible in the course of a year, but he can see only a limited distance south of the celestial equator. If his latitude is α °N the most southerly point he can see in the sky is 90 ° — α °S. Thus for example to an observer at latitude 50 °N, only the sky north of 90 °—50 ° (40 °S) is ever visible.

The charts given on this page are suitable for observers who live in the northern hemisphere, between latitudes 50° and 30°N. The horizon is given by the latitude marks near the bottom of the charts. Thus for an observer who lives at latitude 30°N, the northern horizon in the first map will pass just above Deneb, which will be invisible.

A star rises earlier, on average, by 2 hours a month; thus the chart for 20.00 hours on 1 January will be valid for 18.00 hours on 1 February, or for 22.00 hours on 1 December, and so on.

The limiting visibility of a star for an observer at any latitude can be worked out from its declination. (The latitudes of the major cities of the world are shown in the key maps, and a list of the positions of bright stars is given on page 147). To an observer in the northern hemisphere, a star is at its lowest point in the sky when it is due north; a star which is "below" the pole by the amount of one's latitude will touch the horizon when at its lowest point. If it is closer to the pole than that it will be circumpolar. From latitude 51°N, for example, a star is circumpolar if its declination is 90°—51°=39°N, or greater. Thus Capella, decl. +45° 57' is circumpolar to an observer in London, Cologne or Calgary (latitude 51°N). A minor allowance must be made for atmospheric refraction.

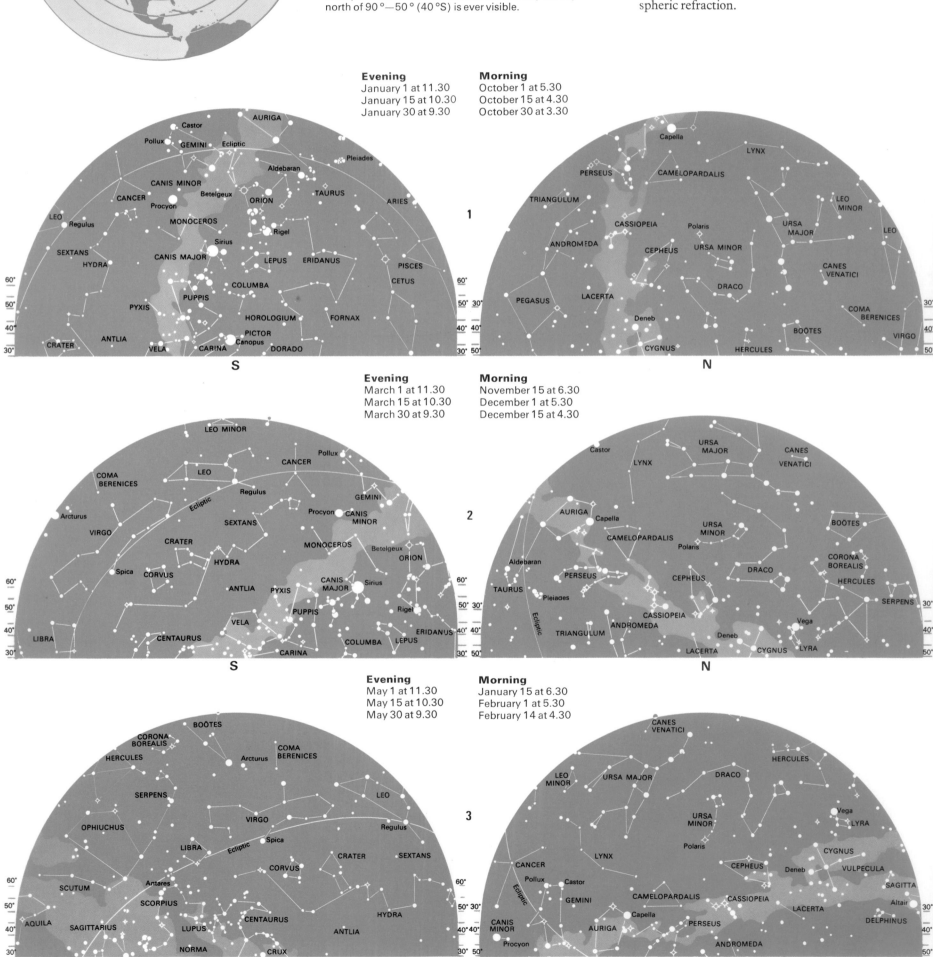

Evening
January 1 at 11.30
January 15 at 10.30
January 30 at 9.30

Morning
October 1 at 5.30
October 15 at 4.30
October 30 at 3.30

Evening
March 1 at 11.30
March 15 at 10.30
March 30 at 9.30

Morning
November 15 at 6.30
December 1 at 5.30
December 15 at 4.30

Evening
May 1 at 11.30
May 15 at 10.30
May 30 at 9.30

Morning
January 15 at 6.30
February 1 at 5.30
February 14 at 4.30

Similarly, to an observer at latitude 51°N, a star with declination south of —39° will never rise. Canopus lies at declination —52°40′, therefore it is invisible from London (latitude 51°N) but can be seen from any latitude south of latitude 37°20′N (that is to say, 90°—52°40′), again neglecting the effects of refraction.

Alkaid or Eta Ursæ Majoris, the southernmost of the seven stars of the Great Bear, has declination +49°33′44″, which for this sort of calculation may be rounded off to +50°. It will therefore be circumpolar from latitudes north of 40°N, and will be permanently invisible from latitudes south of 40°S.

The charts given here show the northern (right) and the southern (left) aspects of the sky from the viewpoint of an observer in northern latitudes. They are self-explanatory; the descriptions given below apply in each case to the late evening, but more accurate calculations can easily be made by consulting the notes given by the side of each chart.

Chart 1

In winter, the southern aspect is dominated by Orion and its retinue. Capella is almost at the zenith or over-head point, and Sirius is at its best. Observers in Britain can see part of Puppis, but, as has been noted, Canopus is too far south to be seen from any part of Europe. The Sickle of Leo is very prominent in the east; Ursa Major is to the north-east, while Vega is at its lowest in the north. It is circumpolar from Britain, but not New York, and it is not on the first chart.

Chart 2

In spring, Orion is still above the horizon until past midnight; Leo is high up, with Virgo to the east. Capella is descending in the north-west, Vega rising in the north-east; these two stars are so nearly equal in apparent magnitude (0·1 and 0·0 respectively) that, in general, whichever is the higher will also seem the brighter. In the west, Aldebaran and the Pleiades are still visible.

Charts 3-6

In early summer (Chart 3) Orion has set, and to British observers the southern aspect is relatively barren, but observers in more southerly latitudes can see Centaurus and its neighbours. During summer evenings (Chart 4) Vega is at the zenith, Capella low in the north; Antares is at its highest in the south. By early autumn (Chart 5) Aldebaran and the cluster of the Pleiades have reappeared, and the Square of Pegasus is conspicuous in the south, with Fomalhaut well placed. And by early winter (Chart 6) Orion is back in view, with Ursa Major lying low in the northern sky.

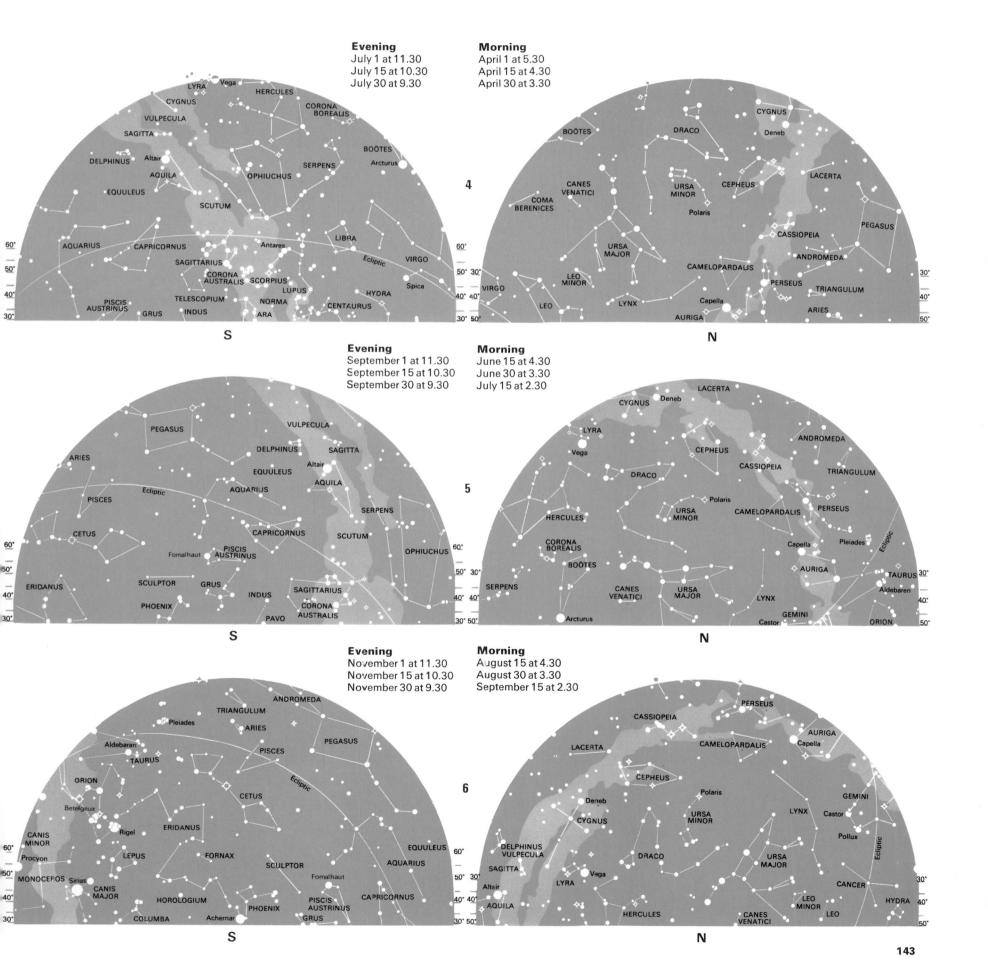

Seasonal Star Maps: South

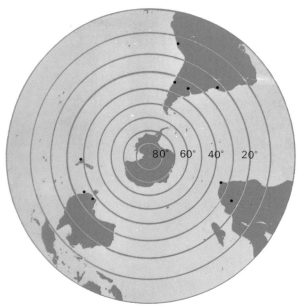

On the whole, the stars in the south polar area of the sky are brighter than those of the far north, even though the actual pole lies in a barren region, and there is no pattern of stars so distinctive as the Great Bear – apart from the Southern Cross, which covers a much smaller area. Canopus, the brightest star in the sky apart from Sirius, has a declination of −37° (to the nearest degree), and is not visible from Europe, but it rises well above the horizon from Mexico, and from Australia and New Zealand it is visible for much of the year.

Latitudes *left*

For the observer in the Southern Hemisphere all the stars of the southern sky are visible in the course of a year, but he can see only a limited distance north of the celestial equator. If his latitude is ∝ °S, the most northerly point he can see in the sky is 90°−∝ °N. Thus, for example, to an observer at latitude 50 °S only the sky south of 90°−50°=40 °N is ever visible.

In the far south, too, there are the clouds of Magellan, which are of such great importance in modern astrophysical research. They are prominent naked-eye objects, and the Large Cloud can be seen without optical aid even under conditions of full moonlight.

An observer at one of the Earth's poles would see one hemisphere of the sky only, and all the visible stars would be circumpolar. It is not even strictly correct to say that Orion is visible from the entire surface of the Earth.

An observer at the south pole would never see Betelgeux, whose declination is +7°. From latitudes south of S.83° (that is to say, 90°−7°), Betelgeux would never rise; but this applies only to a small part of the Antarctic continent.

The charts given here show the northern (left-hand chart) and southern (right-hand chart) aspects of the sky from latitudes between S.15° and S.35°. As with

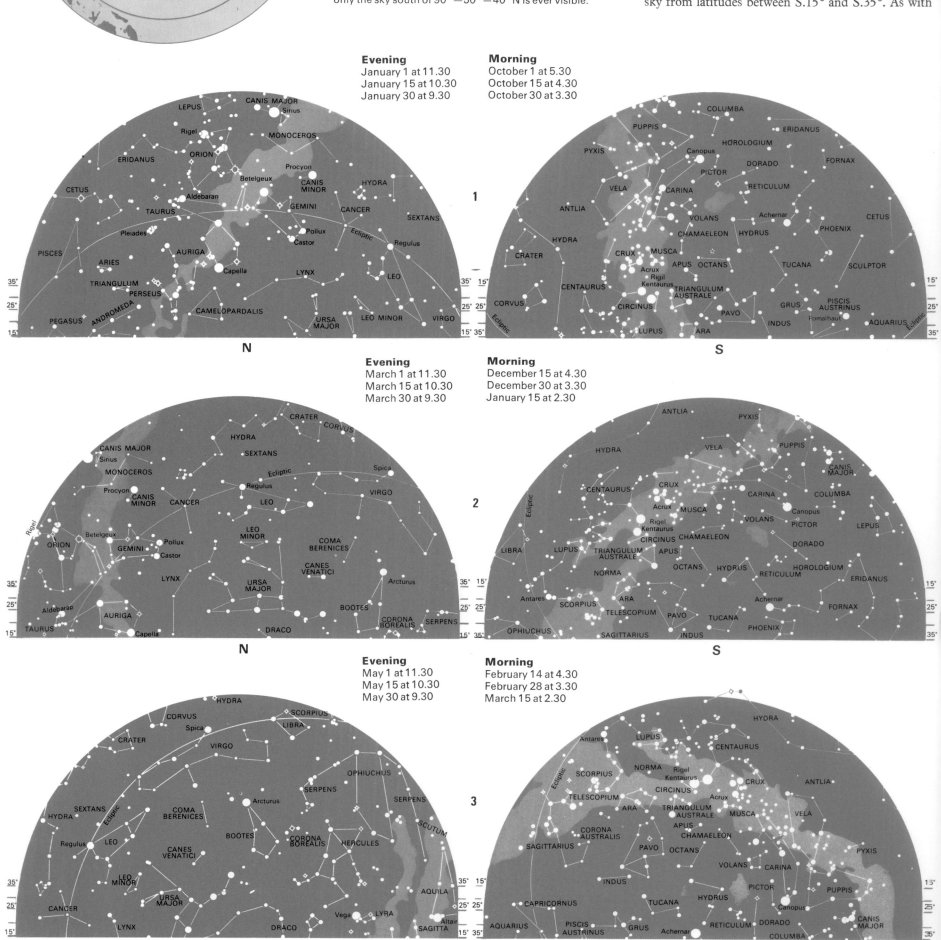

the northern charts, the descriptions are given for late evenings, though each chart is valid for all the dates and times shown by each.

Chart 1

In January, the two most brilliant stars, Sirius and Canopus, are high up. Sirius seems appreciably the brighter of the two (magnitude —1.4, as against —0.8), but its eminence is due to its closeness rather than its real luminosity. It is an A-type Main Sequence star, only 26 times as luminous as the Sun; Canopus is an F-type supergiant, whose luminosity is 80,000 times that of the Sun according to one estimate. This means that it is many times more powerful than Sirius, pure white, and also Canopus, which appears yellow. Lower down, the Southern Cross is a prominent feature, with the brilliant pair of stars Alpha and Beta Centauri in the same area.

In the north, Capella is well above the horizon;

Orion is not far from the zenith, and if the sky is clear a few stars of Ursa Major may be made out low over the northern horizon.

Chart 2

In March, Canopus is descending in the south-west, and Crux rising to its greatest altitude; the south-east is dominated by the brilliant groups of Scorpius and Centaurus. (Scorpius is a magnificent constellation. Its leading star, Antares, is well visible from Europe, but the "tail" is too far south to be seen to advantage.) To the north, the Great Bear is seen; Orion is descending in the west.

Charts 3-4

The May aspect (Chart 3) shows Alpha and Beta Centauri very high up, and Canopus in the south-west; Sirius and Orion have set, but Scorpius is

brilliant in the south-east. In the north, Arcturus is prominent, with Spica in Virgo near the zenith. By July (Chart 4) Vega, Altair and Deneb are all conspicuous in the northern part of the sky, and Arcturus is still high above the north-western horizon; Antares is not far from the zenith. The Milky Way is glorious, stretching from the southern horizon to the far north.

Charts 5-6

The September view (Chart 5) shows Pegasus in the north, and the "W" of Cassiopeia is above the horizon. In September the Southern Cross is almost at its lowest.

By November (Chart 6) Sirius and Canopus are back in view; Alpha and Beta Centauri graze the horizon, and the region of the zenith is occupied by large, comparatively barren groups such as Cetus and Eridanus.

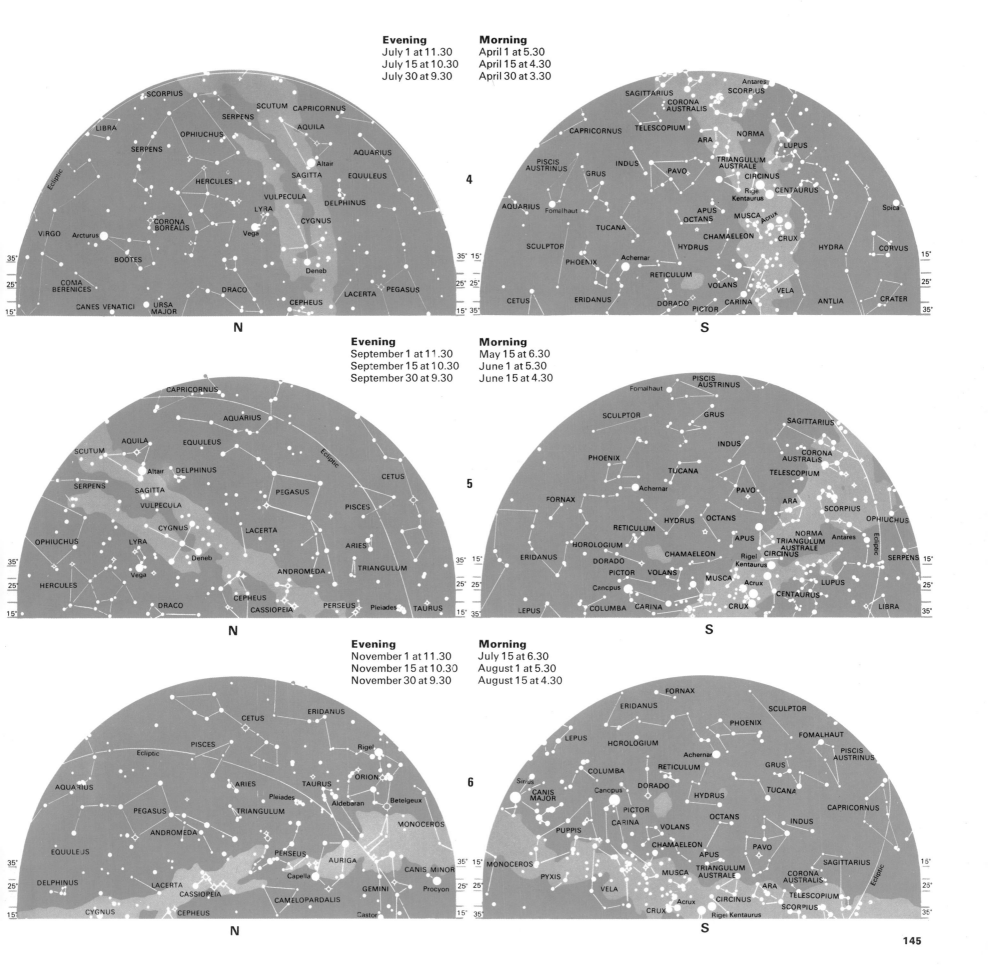

Key to the Star Maps

Star recognition is not nearly so difficult as it might appear at first glance. The best method is to select a few distinctive constellations which can be found without difficulty, and use these as guides to identify the rest.

To observers in the central latitudes of Europe or the northern United States, the most convenient groups are Ursa Major and Orion. Using these as "starting-points", all the other constellations may be learned. Southern observers can use Centaurus and Crux similarly. Orion is crossed by the celestial equator, and is visible from every inhabited area.

The charts given on this page are intended as guides only; as soon as the leading stars in a constellation have been identified, reference to the detailed maps given on pages 148-9, 156-7, 164-5 and 172-3 will locate the rest. The list given opposite indicates the area in which each constellation is to be found.

All the stars named in these key maps are bright enough to be seen even when the sky is bright, during periods of moonlight.

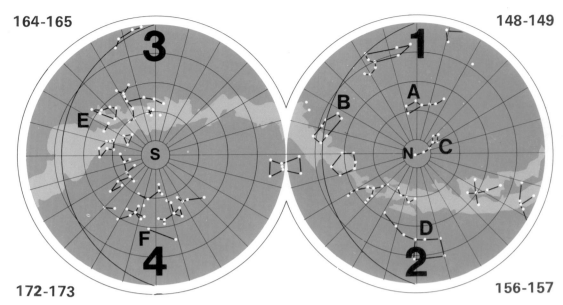

A The Great Bear
Ursa Major (the Great Bear) is circumpolar from Britain and the New York area. The "curve" of the tail leads to Arcturus, the brilliant orange-red star in Boötes (the Herdsman), and thence to Spica; Polaris is indicated by the two pointers, Merak and Dubhe; Leo (the Lion) and Cassiopeia can also be found. Castor and Pollux, in Gemini, and Capella, in Auriga, are shown on this key map and also on that of Orion. Though Ursa Major contains no star of the 1st magnitude, its distinctive shape makes it quite unmistakable.

B Orion
Orion is crossed by the celestial equator and is thus visible from everywhere on the Earth. The Belt points in one direction to Aldebaran, and in the other to Sirius; Procyon, Castor and Pollux, and Capella can also be found easily. The Belt is made up of three bright stars, Delta Orionis (Mintaka), Epsilon (Alnilam) and Zeta (Alnitak); the equator passes very close to Mintaka. Orion is particularly suitable as a "guide" because of its distinctive shape and its brightness. All the leading stars apart from Betelgeux are of early-spectral type, and are very hot, white and luminous.

C Cygnus
During summer evenings in Europe and the northern United States, the night sky is dominated by what is often nicknamed the Summer Triangle – Vega in Lyra, Deneb in Cygnus, and Altair in Aquila. Deneb may be found from Ursa Major, as shown in this key map. The cross of Cygnus is very distinctive, while Altair in Aquila has a star to either side of it. Of the three 1st-magnitude stars in the "Summer Triangle", Deneb appears the faintest, but is actually much more remote and luminous than Vega or Altair.

D Pegasus
The southern sky during autumn evenings in Europe and the northern United States is dominated by Pegasus, whose four chief stars make up a square. The "W" of Cassiopeia may be found from Ursa Major (first key map); Cassiopeia itself may then be used to locate Pegasus, which is very distinctive, even though its individual stars are not brilliant. Alpheratz or Sirrah, in the Square, is officially included in the adjacent constellation, Andromeda. Well south of Pegasus is the 1st-magnitude star Fomalhaut, in a somewhat isolated position.

E The Southern Cross
Crux Australis (the Southern Cross), never seen from Europe, is the most famous of all constellations in the southern part of the sky. Its position is indicated by the two 1st-magnitude stars Alpha and Beta Centauri; it stands out because of its four bright stars close together, of which two (Alpha and Beta Crucis) are of the 1st magnitude. The "longer arm" of Crux indicates the position of the south celestial pole, in Octans, which is not marked by any bright star.

F The Crane
Well south of Fomalhaut, permanently below the British horizon, is the constellation of Grus (the Crane); it is more distinctive than the other "Southern Birds", Pavo, Phœnix and Tucana. Its leading star, Alnäir, is of the 2nd magnitude, and the outline of the constellation is easy to recognize. Adjoining Phœnix is the 1st-magnitude Achernar, at the southernmost point of Eridanus, the River. Part of Eridanus is visible from Europe (Beta Eridani, or Kursa, lies near Rigel in Orion) but Achernar never rises above the European horizon.

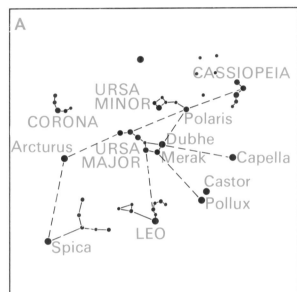

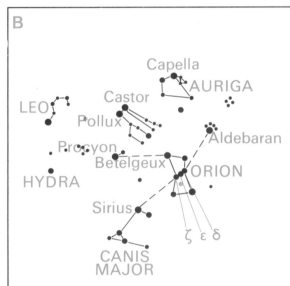

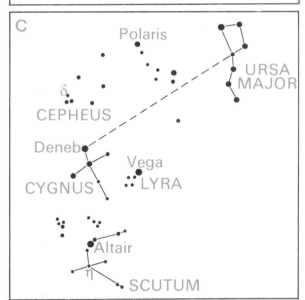

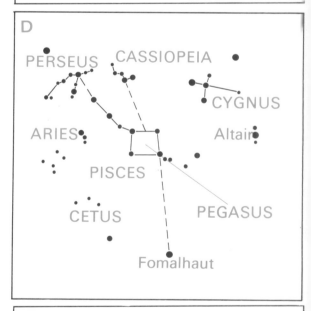

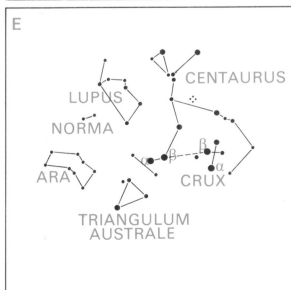

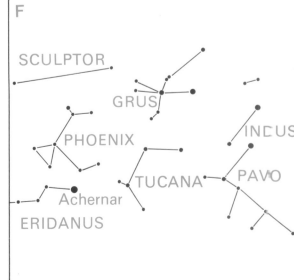

List of Constellations

Constellation with abbreviation and English name			R.A.		Declination		Area (sq. degs.)	Map
			h m	h m				
Andromeda[1]	And	Andromeda	22 56 to	02 36	+21°.4 to	+52°.9	722·28	2
Antlia	Ant	The Air-pump	09 25	11 03	−24·3	−40·1	238·90	3
Apus	Aps	The Bee	13 45	18 17	−67·5	−82·9	206·32	3
Aquarius	Aqr	The Water-bearer	20 36	23 54	+03·1	−25·2	979·85	2,4
Aquila[2]	Aql	The Eagle	18 38	20 36	−11·9	+18·6	652·47	2,4
Ara	Ara	The Altar	16 31	18 06	−45·5	−67·6	237·06	3
Aries[3]	Ari	The Ram	01 44	03 27	+10·2	+30·9	441·39	2
Auriga[4]	Aur	The Charioteer	04 35	07 27	+27·9	+56·1	657·44	1,2
Boötes[5]	Boö	The Herdsman	13 33	15 47	+07·6	+55·2	906·83	1
Cælum	Cae	The Sculptor's Tools	04 18	05 03	−27·1	−48·8	124·86	4
Camelopardalis	Cam	The Giraffe	03 11	14 25	+52·8	+85·1	756·83	1
Cancer	Cnc	The Crab	07 53	09 19	+06·8	+33·3	505·87	1
Canes Venatici[6]	CVn	The Hunting Dogs	12 04	14 05	+28·0	+52·7	465·19	1
Canis Major[7]	CMa	The Great Dog	06 09	07 26	−11·0	−33·2	380·11	3
Canis Minor[8]	CMi	The Little Dog	07 04	08 09	−00·1	+13·2	183·37	1
Capricornus	Cap	The Sea-goat	20 04	21 57	−08·7	−27·8	413·95	4
Carina[9]	Car	The Keel	06 02	11 18	−50·9	−75·2	494·18	3
Cassiopeia[10]	Cas	Cassiopeia	22 56	03 06	+46·5	+77·5	598·41	2
Centaurus[11]	Cen	The Centaur	11 03	14 59	−29·9	−64·5	1060·42	3
Cepheus[12]	Cep	Cepheus	20 01	08 30	+53·1	+88·5	587·79	2
Cetus[13]	Cet	The Whale	23 55	03 21	−25·2	+10·2	1231·41	2,4
Chamæleon	Cha	The Chameleon	07 32	13 48	−75·2	−82·8	131·59	3
Circinus	Cir	The Compasses	13 35	15 26	−54·3	−70·4	93·35	3
Columba	Col	The Dove	05 03	06 28	−27·2	−43·0	270·18	3,4
Coma Berenices	Com	Berenice's Hair	11 57	13 33	+13·8	+33·7	386·47	1
Corona Australis	CrA	The Southern Crown	17 55	19 15	−37·0	−45·6	127·69	4
Corona Borealis[14]	CrB	The Northern Crown	15 14	16 22	+25·8	+39·8	178·71	1
Corvus	Crv	The Crow	11 54	12 54	−11·3	−24·9	183·80	3
Crater	Crt	The Cup	10 48	11 54	−06·5	−24·9	282·40	3
Crux Australis[15]	Cru	The Southern Cross	11 53	12 55	−55·5	−64·5	68·45	3
Cygnus[16]	Cyg	The Swan	19 07	22 01	+27·7	+61·2	803·98	2
Delphinus	Del	The Dolphin	20 13	21 06	+02·2	+20·8	188·54	2
Dorado	Dor	The Swordfish	03 52	06 36	−48·8	−70·1	179·17	3,4
Draco[17]	Dra	The Dragon	09 18	21 00	+47·7	+86·0	1082·95	1
Equuleus	Eql	The Foal	20 54	21 23	+02·2	+12·9	71·64	2
Eridanus[18]	Eri	The River	01 22	05 09	+00·1	−58·1	1137·92	2,4
Fornax	For	The Furnace	01 44	03 48	−24·0	−39·8	397·50	4
Gemini[19]	Gem	The Twins	05 57	08 06	+10·0	+35·4	513·76	1
Grus[20]	Gru	The Crane	21 25	23 25	−36·6	−56·6	365·51	4
Hercules	Her	Hercules	15 47	18 56	+03·9	+51·3	1225·15	1
Horologium	Hor	The Clock	02 12	04 18	−39·8	−67·2	248·88	4
Hydra[21]	Hya	The Watersnake	08 08	14 58	+06·8	−35·3	1302·84	1,3
Hydrus	Hyi	The Little Snake	00 02	04 33	−58·1	−82·1	243·04	4
Indus	Ind	The Indian	20 25	23 25	−45·4	−74·7	294·01	4
Lacerta	Lac	The Lizard	21 55	22 56	+34·9	+56·8	200·69	2

Brightest stars (magnitudes in brackets)

[1]Alpheratz or Sirrah (2) Mirach (2), Almaak or Alamak (2)
[2]Altair (1)
[3]Hamal (2)
[4]Capella (0), Menkarlina (2),
[5]Arcturus (0)
[6]Cor Caroli (2¾)
[7]Sirius (−1½), Adhara (1½), Wezea, Mirzam (2)
[8]Procyon (0½)
[9]Avoir, Canopus (−0¾), Miaplacidus, Tureis (2)
[10]Shedir (2), Caph (2), Gamma Cas (2) var.
[11]Alpha Centauri (0), Agena (1)
[12]Alderamin (2¾)
[13]Diphda (2), Mira (variable)
[14]Alphekka or Gamma (2)
[15]Acrux (1), Beta Crucis (1½), Gamma Crucis (1½)
[16]Deneb (1), Sadr (2), Gienah (2)
[17]Etamin (2¼)
[18]Achernar (1)
[19]Pollux (1), Castor (1½), Alhena (2)
[20]Alnair (2)
[21]Alphard (2)

Constellation with abbreviation and English name			R.A.		Declination		Area (sq. degs.)	Map
			h m	h m				
Leo[22]	Leo	The Lion	09 18 to	11 56	−06·4 to	+33·3	946·96	1,3
Leo Minor	LMi	The Little Lion	09 19	11 04	+23·1	+41·7	231·96	1
Lepus	Lep	The Hare	04 54	06 09	−11·0	−27·1	290·29	3,4
Libra	Lib	The Balance	14 18	15 59	−00·3	−29·9	538·05	3
Lupus	Lup	The Wolf	14 13	16 05	−29·8	−55·3	333·68	3
Lynx	Lyn	The Lynx	06 13	09 14	+33·4	+62·0	545·39	1
Lyra[23]	Lyr	The Lyre	18 12	19 26	+25·6	+47·7	286·48	2
Mensa	Men	The Table	03 20	07 37	−69·9	−85·0	153·48	3,4
Microscopium	Mic	The Microscope	20 25	21 25	−27·7	−45·4	209·51	4
Monoceros	Mon	The Unicorn	05 54	08 08	−11·0	+11·9	481·57	1,3
Musca	Mus	The Fly	11 17	13 46	−64·5	−75·2	138·36	3
Norma	Nor	The Rule	15 25	16 31	−42·2	−60·2	165·29	3
Octans	Oct	The Octant	00 00	24 00	−74·7	−90·0	291·05	3,4
Ophiuchus[24]	Oph	The Serpent-bearer	15 58	18 42	+14·3	−30·1	948·34	1,3
Orion[25]	Ori	Orion	04 41	06 23	−11·0	+23·0	594·12	1,2,3,4
Pavo[26]	Pav	The Peacock	17 37	21 30	−56·8	−75·0	377·67	4
Pegasus[27]	Peg	Pegasus	21 06	00 13	+02·2	+36·3	1120·79	2
Perseus[28]	Per	Perseus	01 26	04 46	+30·9	+58·9	615·00	2
Phœnix[29]	Phe	The Phœnix	23 24	02 24	−39·8	−58·2	469·32	4
Pictor	Pic	The Painter	04 32	06 51	−43·1	−64·1	246·73	3,4
Pisces	Psc	The Fishes	22 49	02 04	−06·6	+33·4	889·42	2,4
Piscis Austrinus[30]	Psa	The Southern Fish	21 25	23 04	−25·2	−36·7	245·37	4
Puppis[31]	Pup	The Poop	06 02	08 26	−11·0	−50·8	673·43	3
Pyxis	Pyx	The Mariner's Compass	08 26	09 26	−17·3	−37·0	220·83	3
Reticulum	Ret	The Net	03 14	04 35	−53·0	−67·3	113·94	4
Sagitta	Sge	The Arrow	18 56	20 18	+16·0	+21·4	79·93	2
Sagittarius[32]	Sgr	The Archer	17 41	20 25	−11·8	−45·4	867·43	4
Scorpius[33]	Sco	The Scorpion	15 44	17 55	−08·1	−45·6	496·78	3
Sculptor	Scl	The Sculptor	23 04	01 44	−25·2	−39·8	474·76	4
Scutum	Sct	The Shield	18 18	18 56	−04·0	−16·0	109·11	4
Serpens[34]	Ser	The Serpent						
Caput (the Head)			15 08	16 20	−03·4	+25·7	428·48	1,3
Cauda (the Body)			17 14	18 56	+06·3	−16·0	208·44	1,3
Sextans	Sxt	The Sextant	09 39	10 49	+06·6	−11·3	313·51	1,3
Taurus[35]	Tau	The Bull	03 20	05 58	+01·1	+30·9	797·25	2
Telescopium	Tel	The Telescope	18 06	20 26	−45·4	−56·9	251·51	4
Triangulum	Tri	The Triangle	01 29	02 48	+25·4	+37·0	131·85	2
Triangulum Australe[36]	Tra	The Southern Triangle	14 50	17 09	−60·3	−70·3	109·98	3
Tucana	Tuc	The Toucan	22 05	01 22	−56·7	−75·7	294·56	4
Ursa Major[37]	UMa	The Great Bear	08 05	14 27	+28·8	+73·3	1279·66	1
Ursa Minor[38]	UMi	The Little Bear	00 00	24 00	+65·6	+90·0	255·86	1,2
Vela[39]	Vel	The Sails	08 02	11 24	−37·0	−57·0	499·65	3
Virgo[40]	Vir	The Virgin	11 35	15 08	+14·6	−22·2	1294·43	1,3
Volans	Vol	The Flying Fish	06 35	09 02	−64·2	−75·0	141·35	3
Vulpecula	Vul	The Fox	18 56	21 28	+19·5	+29·4	268·17	2

Brightest stars (magnitudes in brackets)

[22]Regulus(1), Denebola (2), Algeiba (2)
[23]Vega (0)
[24]Ras Alhague (2)
[25]Rigel (0), Betelgeux (1), Bellatrix (2), Alnilam (2), Alnitak (2), Saiph (2), Mintaka (2¼)
[26]Alpha Pavonis (2)
[27]Markab (2½), Scheat (2½ var.)
[28]Mirphak (2), Algol (2 var.)
[29]Ankaa
[30]Fomalhaut (1)
[31]Suhail Hadar (2)
[32]Kaus Australis (2), Nunki (2)
[33]Antares (1), Shaula (1¾), Sargas (2)
[34]Unukalhai (2¾)
[35]Aldebaran (1), Al Nath (2)
[36]Atria (2)
[37]Alioth (2), Alkaid (2), Dubhe (2)
[38]Polaris (2), Kocab (2)
[39]Gamma Vel (2), Delta Vel (2), Lambda Vel (2)
[40]Spica (1)

A Note on Stellar Nomenclature

Many stars have proper names, most of which are Arabic, but these are not now in general use except for 1st-magnitude stars (such as Sirius and Rigel), and a few stars of lesser brightness but special importance (such as Polaris and Mizar). In 1603 J. Bayer drew up a star catalogue in which he allotted each star a Greek letter; thus the brightest star in a constellation became Alpha, the second Beta, and so on, though the strict sequence of brightness was not always followed. The Greek alphabet is as follows:

α Alpha	η Eta	ν Nu	τ Tau
β Beta	θ Theta	ξ Xi or Si	υ Upsilon
γ Gamma	ι Iota	ο Omicron	φ Phi
δ Delta	κ Kappa	π Pi	χ Chi
ε Epsilon	λ Lambda	ρ Rho	ψ Psi
ζ Zeta	μ Mu	σ Sigma	Ω Omega

The brightest star in Crux (The Cross) is Alpha Crucis; the second brightest, Beta Crucis; and so on. However, there are only 24 Greek letters, and the scheme was limited. In the catalogue drawn up at Greenwich by Flamsteed, from 1675, each star was numbered; the 5th-magnitude Alcor became 80 Ursæ Majoris (80th not in order of brilliancy but in order of right ascension). The Flamsteed numbers are still in use.

Variable stars are usually given Roman letters; as R Andromedæ. For nebulæ and clusters, the catalogue published in 1781 by Messier is still used; the Andromeda galaxy is M.31. and the Orion nebula M.42. Officially, the Messier numbers have been superseded by the NGC numbers (New General Catalogue of Clusters and Nebulæ) compiled by J. L. E. Dreyer, so that, for instance, M.31 is also known as NGC 224.

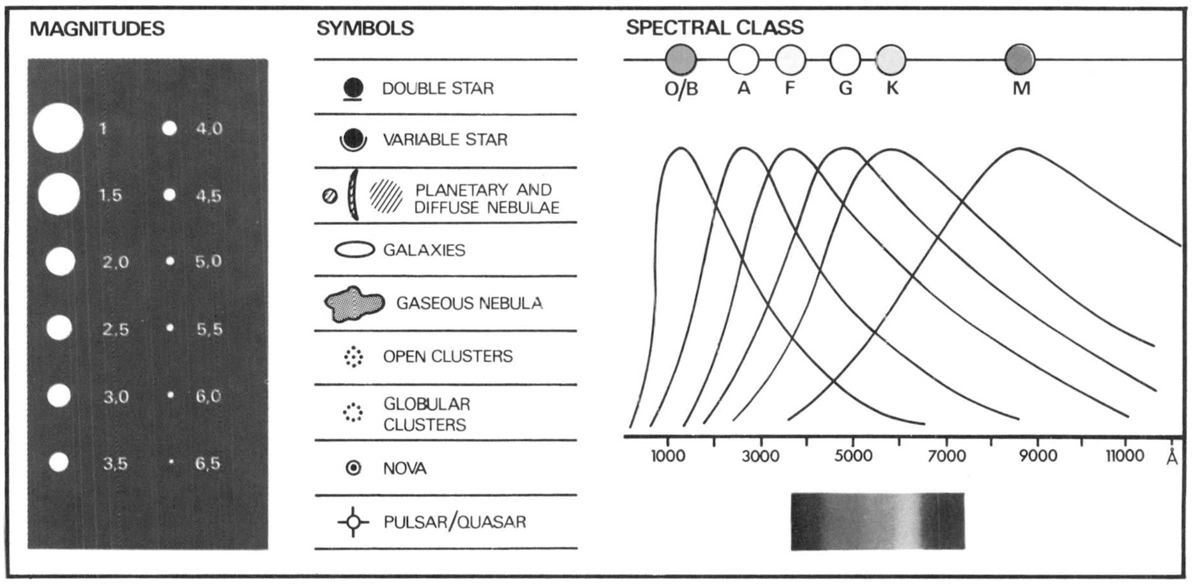

MAGNITUDES
1
1.5
2.0
2.5
3.0
3.5
4.0
4.5
5.0
5.5
6.0
6.5
SYMBOLS
DOUBLE STAR
VARIABLE STAR
PLANETARY AND DIFFUSE NEBULAE
GALAXIES
GASEOUS NEBULA
OPEN CLUSTERS
GLOBULAR CLUSTERS
NOVA
PULSAR/QUASAR
SPECTRAL CLASS
O/B
A
F
G
K
M
1000
3000
5000
7000
9000
11000
Å

The Northern Sky 1

The Northern Sky 1: Stars

The Northern Sky I: Stars

Ursa Major, the Great Bear, is probably the most famous constellation in the sky. It is also one of the easiest to recognize, even though it has no star as bright as the 1st magnitude; the three most brilliant (Alioth, Alkaid and Dubhe) are only slightly above magnitude 2. The familiar pattern which is known in England as the Plough and in the United States as the Big Dipper cannot be mistaken, and makes an excellent "signpost" which is of help in finding other groups. Because Ursa Major is so far north it never sets over northern European latitudes or much of the North American continent.

Two of its stars are of special interest. Mizar or Zeta Ursæ Majoris (magnitude 2) has a fainter star, Alcor, close beside it; telescopically Mizar is itself a fine binary. Megrez or Delta Ursæ Majoris is more than a magnitude fainter than Mizar, but ancient astronomers rated it as being of the 2nd magnitude, and it may have faded over the past 2000 years.

Merak and Dubhe, in the Great Bear, point to Polaris, the Pole Star, in Ursa Minor (the Little Bear), which is within 1° of the polar point. Ursa Minor itself looks not unlike a faint and distorted version of Ursa Major; its other bright star, Kocab (Beta Ursæ Minoris), is of the 2nd magnitude, and is decidedly orange in hue. Between the Bears sprawls the faint, extensive constellation of Draco (the Dragon); Thuban or Alpha Draconis, between Kocab and Alkaid, used to be the pole star in ancient times.

Hercules and Ophiuchus

Adjoining Draco is a very large but rather faint constellation, Hercules. Though named after the great hero of ancient legend, the group is by no means conspicuous. The two brightest stars in it, Beta and Zeta, are only slightly above the 3rd magnitude, and even the shape of Hercules is not very well-marked. There are, however, two interesting features. Between Zeta and Eta there lies the great globular cluster M.13, just visible to the naked eye on a clear night, and on the boundary of Hercules there is Rasalgethi (Alpha Herculis), a red giant star with a faint greenish companion visible in small telescopes. Rasalgethi is a variable star, but its magnitude is always between 3 and 4. Close behind it is the 2nd magnitude Rasalhague, the brightest star in another large, rather dim group; Ophiuchus (the Serpent-Bearer).

The three stars in the "tail" of the Great Bear point to Arcturus in Boötes (the Herdsman), which is the brightest star in the northern hemisphere of the sky. Its magnitude is —0·06, and it is light orange in colour, with a K-type spectrum. Arcturus is 41 light-years away, and 100 times as luminous as the Sun. The rest of Boötes is not very notable, though Epsilon (Izar) is a beautiful double. Close to Boötes is the prominent semi-circle of stars marking Corona Borealis (the Northern Crown). The leading star is of the 2nd magnitude.

Leo

Using the Pointers in the Great Bear "the wrong way" (i.e. away from Polaris) leads to Leo (the Lion), with its prominent curve of stars making up the Sickle; Regulus is of the 1st magnitude, while Algieba (Gamma Leonis) is a fine binary. Part of Virgo (the Virgin) is also shown on this map, though the leading star, Spica, is in the southern hemisphere. This region is particularly rich in faint galaxies, and between Virgo and Ursa Major lies the constellation of Coma Berenices (Berenice's Hair), which looks almost like a large, dim cluster. Canes Venatici (the Hunting Dogs) has only one star as bright as the 2nd magnitude.

Two more Zodiacal constellations lie on this map. Gemini (the Twins) is very rich in both bright stars and interesting telescopic objects; of its two chief stars, Pollux is of the 1st magnitude and Castor $1\frac{1}{2}$. Between Gemini and Leo lies Cancer (the Crab), which has no bright stars, but is distinguished by the presence of the famous open cluster Præsepe. Little need be said about the large, very dull Lynx (the Lynx) and Camelopardus (the Giraffe), but the one brilliant star of Canis Minor (the Little Dog) – Procyon – is the eighth brightest star in the sky.

Polar Star Trails *left*
This photograph shows star-trails in the north polar region of the sky photographed by Peter Gill. The short curved trail near the centre is that of Polaris, which has a declination of +89 °01′44″. When using it for fixing positions, navigators have to make a correction to compensate for the fact that Polaris is almost one degree away from the polar point.

The stars in Ursa Minor (the Little Bear) are well shown here. Kocab or Beta Ursæ Minoris, the "end star" in the Ursa Minor pattern, is of spectral type K, and is clearly orange. The colour is quite easily detectable with the naked eye, and will be found to be even more pronounced if the star is viewed through binoculars.

Sickle of Leo *below and right*
The Sickle of Leo, photographed by T. J. C. A. Moseley (Armagh Observatory). Regulus, the brightest star, is of the 1st magnitude; Gamma Leonis is a well-known binary. The open cluster Præsepe, in Cancer, can be seen toward the right-hand side of the photograph. The long trail showing up on the plate is that of the U.S. balloon satellite Echo II.

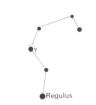

Double Stars in Hercules *right*
Hercules, showing the binary star Zeta, the red variable Alpha (Rasalgethi), and the magnificent globular cluster M.13. Hercules is not a conspicuous constellation but it is not difficult to find, since it lies between Vega on the one side and Corona Borealis on the other.

Zeta (R.A. 16h.39m.4, decl + 31 °41′) is a fine binary, with a period of 34 years; the components are of magnitudes 3 and 6·5. The separation and position angle change relatively rapidly. The separation was at its greatest (1″.6) in 1954 and will be so again in 1988.

Alpha (R.A. 17h.12m.4, decl. + 14°27′) is also a double. The primary is of spectral type M. and varies irregularly in magnitude between 3 and 4. The companion, of magnitude 6·1, is strongly greenish in hue, though the colour is accentuated by contrast. The separation is 4″.4, so that the pair may be well seen in small telescopes. The primary is an exceptionally large red giant. It is sometimes still known by its old proper name of Rasalgethi; though lettered Alpha, it is not the brightest star in Hercules.

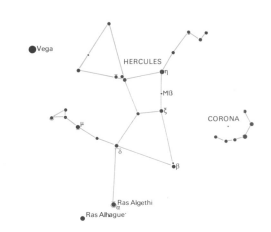

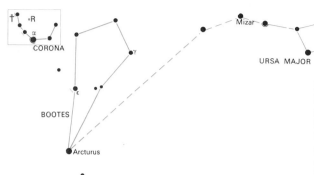

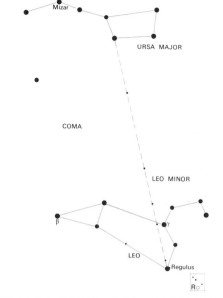

R Coronæ Borealis *left*

R Coronæ Borealis is easy to find, since it lies in the "bowl" of the Northern Crown (R.A. 15h.46m.5 decl. +28°19').

Arcturus may be located by using the stars of the Great Bear as pointers, as shown in the upper diagram; Corona lies on the boundary of Boötes, and is quite prominent, particularly since the brightest star in it (Alpha Coronæ known by its proper names of Gemma or Alphekka) is of the 2nd magnitude.

R Coronæ is usually of magnitude 6, and is on the fringe of naked-eye visibility. Binoculars show it very clearly, together with the most useful comparison star (lettered M in the star field *left*) with a magnitude of 6·6. At irregular intervals, R Coronæ fades abruptly, and may become as faint as magnitude 15, so that large telescopes have to be used in order to see it. (See light-curve, left.)

Also in the constellation, and shown in the star-field, is T Coronæ, a recurrent nova, normally of the 10th magnitude, but which flared up to the 2nd in 1866 and again in 1946.

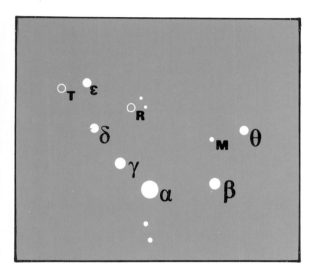

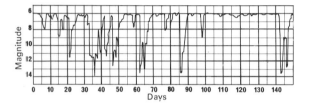

R Leonis *left*

The variable star R Leonis (R.A. 9h.44m.9, decl. +11°40'), close to Regulus, is a typical long-period variable of the Mira Ceti type. Its period is 312·6 days, and the magnitude range is from 5·4 to 10·5, but – as with all members of its class – both range and period are subject to fluctuation. The spectrum, as is the case with so many long-period variables, is of type M, and the red colour is very pronounced.

When it is at maximum, R Leonis is easily seen with the aid of binoculars, and it may even be possible to observe it with the naked eye; it never becomes so faint that it would be lost to view with a small telescope.

The telescopic star field and mean light-curve are both given to the left; the most convenient comparison stars for use when R Leonis is near maximum are of magnitude 5·4 and 5·9 respectively. To find R Leonis, sweep from Regulus in the direction of the 4th-magnitude star Omicron Leonis.

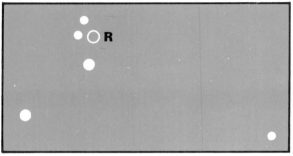

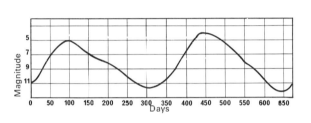

X Leonis *left*

The faint variable star X Leonis close to R, belongs to the "dwarf nova" group. It is normally of magnitude 15, beyond the reach of any but large telescopes, but every 22 days or so – the intervals are irregular – it flares up to magnitude 12, and comes within the range of smaller instruments.

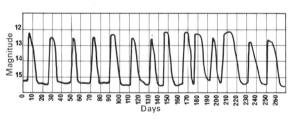

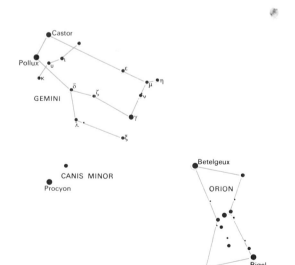

Auriga *above*

This is one of the brightest northern groups. Capella (Auriga) is a yellow star of magnitude 0.2 and is the fifth brightest star in the sky. The 2nd-magnitude star at the base of the constellation, formerly known as Gamma Auriga, is now called Beta Tauri, and is one of the few examples of a star which has been included in two constellations. Epsilon is an eclipsing binary (magnitude 3.3 to 4.0), the giant component of which is one of the largest known stars. Theta is double (magnitude 2.7, 7.1, separation 2".8).

There are three open clusters, M.36, M.37 and M.38, which lie in the area.

above

The constellation Auriga (the Charioteer), photographed in 1970 by Commander H. R. Hatfield from his private observatory in Kent.

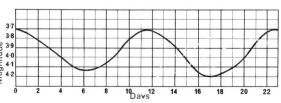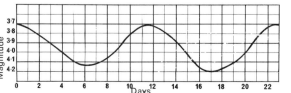

Gemini *left*

Gemini, the Twins, a constellation particularly rich in interesting objects, for observation, is easy enough to find in the sky. Zeta Geminorum (R.A. 7h.01m.1, decl. +20°39') *below left* a typical Cepheid variable, has a period of about 11 days; mag. 3.7 to 4.2. Useful comparison stars are Delta (3.5), Lambda (3.65), Iota (3.9) and Upsilon (4.2).

Another naked-eye variable is Eta (R.A. 6h.11m.95, decl. +22°31'), mag. range 3.1 to 3.9, period approximately 233 days; the light-curve is subject to marked irregularities, and the variations are slow. Comparison stars are Mu (2.9), Delta (3.5), Xi (3.4) and Nu (4.1) Eta Geminorum, and the nearby comparison star Mu, are of spectral type M.

The Northern Sky 1: Clusters and Nebula

Præsepe *below*
Photograph by H. R. Hatfield with the 12-in reflector at his observatory, Sevenoaks, Kent (exposure 15 minutes). Præsepe, M.44 (R.A. 8h.37m.4, decl. + 20°00') is 525 light-years away. Unlike the Pleiades, it contains no nebulosity, and Baade has estimated its age at about 400 million years. The apparent diameter is 70'.

M.13 in Hercules *right*
The finest globular cluster in the northern sky, discovered by Edmond Halley in 1714, and just visible to the naked eye when the sky is clear and dark. (U.S. Naval Observatory photograph). M.13 (R.A. 16h.39m.9, decl. + 36°33') has an apparent diameter of 23'; the real diameter is 100 light-years (Bečvár). The number of stars contained in this cluster has been estimated as half a million (O. Struve); the distance is 22,500 light-years. To find M.13, look between Zeta and Eta Herculis, about two thirds of the way toward Eta.

Finding Præsepe *below*
Præsepe lies midway between Pollux and Regulus Close to it are Delta Cancri (magnitude 4·2) and Gamma Cancri (4·7). Cancer is an obscure constellation shaped rather like Orion, Præsepe is easily visible with the naked eye, but strong moonlight will conceal it.

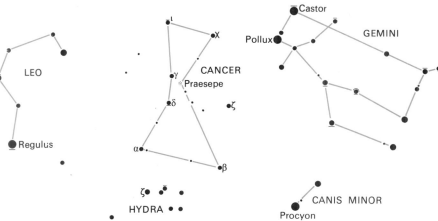

Owl Nebula *above*
Photograph of the Owl Nebula, M.97 in Ursa Major (R.A. 11h.12m.0, decl. + 55°18'), taken with the 60-in. reflector at the Mount Wilson Observatory. This is a planetary nebula which was discovered by the French astronomer Méchain in 1781.

It is 10,000 light-years away; and since its integrated magnitude is only 12, it is a faint object except when seen with large telescopes. Its mass has been estimated at 15 per cent that of the Sun.

Drawing of Owl Nebula *above*
Sir William Herschel described M.97 as "a globular body of equal light throughout", and thought that it might lie outside the Galaxy. In 1848, the Earl of Rosse drew it with his 72-in. reflector, and described "two stars considerably apart in the central region; dark penumbra around each spiral arrangement". In 1853 he concluded that the nebula had no spiral form. Rosse's drawing *above* shows M.97 in a recogniz-able form indicating why it was given its name. The nebula lies 1½ °S and 2 °E following Merak or Beta Ursæ Majoris, in the Great Bear. In the same low-power field is a spiral galaxy, NGC 3556, which has an integrated magnitude of 10·7, and is thus brighter than the Owl Nebula.

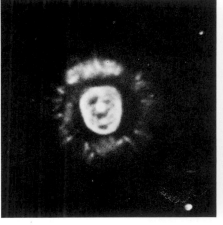

Planetary Nebula NGC 2392
Four photographs taken in red, yellow, violet and ultra-violet light with the 200-in. Palomar reflector. The position of the planetary is R.A. 7h.26m.2, decl. + 21 °01', in Gemini. The integrated magnitude is 8·3, making it one of the brightest of its class.

Rosette Nebula *right*
The Rosette Nebula, NGC 2237, photographed with
the 48-in. Schmidt telescope at Palomar. The position
is R.A. 6h.29m.6, decl. + 04°40', in the rather faint and
obscure constellation of Monoceros. The apparent
dimensions are 64' x 61', and the distance is 3400
light-years.
   This is a typical emission nebula; the brightest star
has a spectral type 09. The photograph is very
spectacular, but it must be emphasized that the colour
cannot be seen by an observer at the eye-end of a
telescope, and is revealed only with a time-exposure
taken with a large telescope.

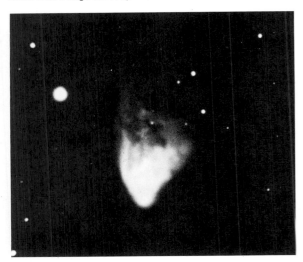

Hubble's Variable Nebula *above*
Photograph of Hubble's variable nebula, NGC 2261.
The position is R.A. 6h.36m.4, decl. + 08°46', in
Monoceros; the distance is 6500 light-years.
   The star which lights up the nebulosity, R Monocerotis,
is variable, and its changing brightness affects both
the apparent outline and the magnitude of the nebula.

Gaseous Nebula IC 443 *below*
Photograph taken in red light (48-in. Schmidt telescope,
Palomar). The position of the nebula is R.A. 6h. 13m.9,
decl. + 22° 58' (in Gemini); its dimensions are 27' x 5'.
The object to the extreme right of the photograph is the
over-exposed image of a brighter star.
   It has been suggested, very plausibly, that IC 443 is a
supernova remnant, and that the gases are still expanding
from the old explosion-centre.

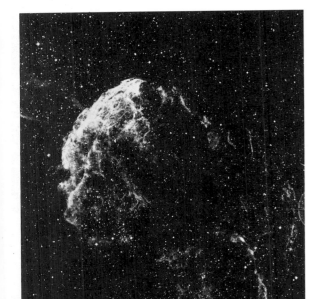

Globular Cluster M.3 *right*
   M.3 (R.A. 13h.39m.9, decl. + 28°38'; in Canes
Venatici) was discovered by Messier in 1764. Its
magnitude is 6·4. It lies ½ °N and 6 °E of the 4th-magnitude
star Beta Comæ. Its distance is 48,500 light-years.
   A. Sandage has estimated that there are 44,500 stars of
magnitude 22½ or brighter within 8' radius of the
centre, and that the total mass of the cluster is 245,000
times that of the Sun. The photograph was taken with the
200-in. Palomar reflector.

Northern Sky 1: Galaxies

Galaxies
Of all the millions of galaxies, only those which are relatively close to us on the scale of the universe can be studied in detail. Many areas of the sky are rich in them.

A Field of Galaxies *below*
The photograph shows four galaxies : NGC 3185 (upper centre), a barred spiral ; NGC 3187 (upper right), another barred spiral ; NGC 3193 (upper left), elliptical ; and NGC 3190 (lower right), a normal spiral. The stars seen in the foreground belong to our own Galaxy.

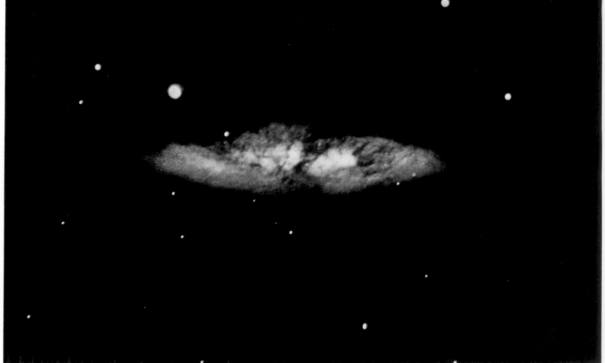

An Irregular Galaxy *above right*
M.82 (NGC 3034) in Ursa Major, a strong radio source. It is 10 million light-years away ; the structure indicates that a tremendous explosion took place in its nucleus 1½ million years before our present view of it.

A Spiral Galaxy *right*
The Whirlpool Galaxy, M.51 in Canes Venatici. It was discovered by Messier in 1773, and is of magnitude 8. It is 37 million light-years away, and was the first galaxy to be seen as a spiral (by Lord Rosse in 1845).

The Most Distant Known Galaxy *below*
This photograph, taken in 1960 with the 200-in. Palomar reflector, shows the radio galaxy 3C-295, in Boötes, indicated by the white arrow. It looks not unlike a star, and its true nature was discovered only because it is a strong source of radio emission. According to the red shift in its spectrum, it is 5000 million light-years away, and is receding at nearly half the velocity of light. It is the most distant radio galaxy whose distance has been measured, though it is not so remote as some of the quasars.

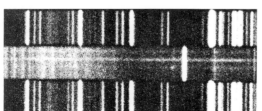

Spectrum of 3C-295 *above*
The spectrum of the galaxy is shown in the centre, with laboratory spectra above and below. The bright line is due to light from the night sky. 3C/295 is represented by the small light-spot to the left, which is strongly red-shifted.

Spiral Galaxies

These six photographs show spiral galaxies of different types, in order of increasing "tightness" of the arms.

A Loose Spiral *right*

The galaxy NGC 2403, in Camelopardalis : a very open spiral, with poorly defined nucleus and irregular arms ; there is no real symmetry. (Palomar 200-in. photograph.) Dimensions : 16'·8 x 10'·0 ; mag. 8·9 ; type Sc.

Sc-Type Spiral *far right*

M.101 (NGC 5457) in Ursa Major, also type Sc, but its nucleus and arms are better defined. Discovered by Méchain in 1781, it is of mag. 9·6. At 11·5 million light-years away, it is a close spiral. Though considerably smaller than our Galaxy, it is face-on and well displayed. Apparent dimensions : 10' x 8'.

Galaxy of Type Sb *right*

NGC 2903, in Leo, is a good example of an Sb spiral, with arms which are much better-defined than with the very loose systems. The magnitude is 9·1, and the apparent dimensions 11' x 4'.6. The galaxy is not face-on, as with M.101, but lies at a fairly sharp angle to us, so that its form is not seen to the best advantage. The spiral arms are extensive.

Spiral with Large Nucleus *far right*

NGC 2841 in Ursa Major (Palomar 200-in. photograph). This is an Sb-type system, with a magnitude of 9·3 ; dimensions 6'.4 x 2'.4. The nucleus is larger and brighter than with NGC 2903, and the spiral arms more tightly wound. As in the other photographs, the isolated foreground stars belong to our own Galaxy.

Face-On Spiral Galaxy *right*

M.94 (NGC 4736) ; sometimes classed as type Sa, sometimes as type Sb. There is a large nucleus, with tightly wound, relatively inconspicuous arms. Mag. 8, dimensions 5' x 3'·5. It is 2¾ °N and 1 °W of Alpha Canum Venaticorum, and 32·6 million light-years away, and according to Holmberg its mass is equivalent to 250,000 million Suns.

The "Black-Eye" Galaxy *far right*

M.64 (NGC 4826) in Coma is a tight spiral, 44 million light-years away, and of magnitude 6·6. M.64 is a large, massive galaxy ; according to Holmberg its mass is equal to 790,000 million Suns. (60-in. Mount Wilson photograph.)

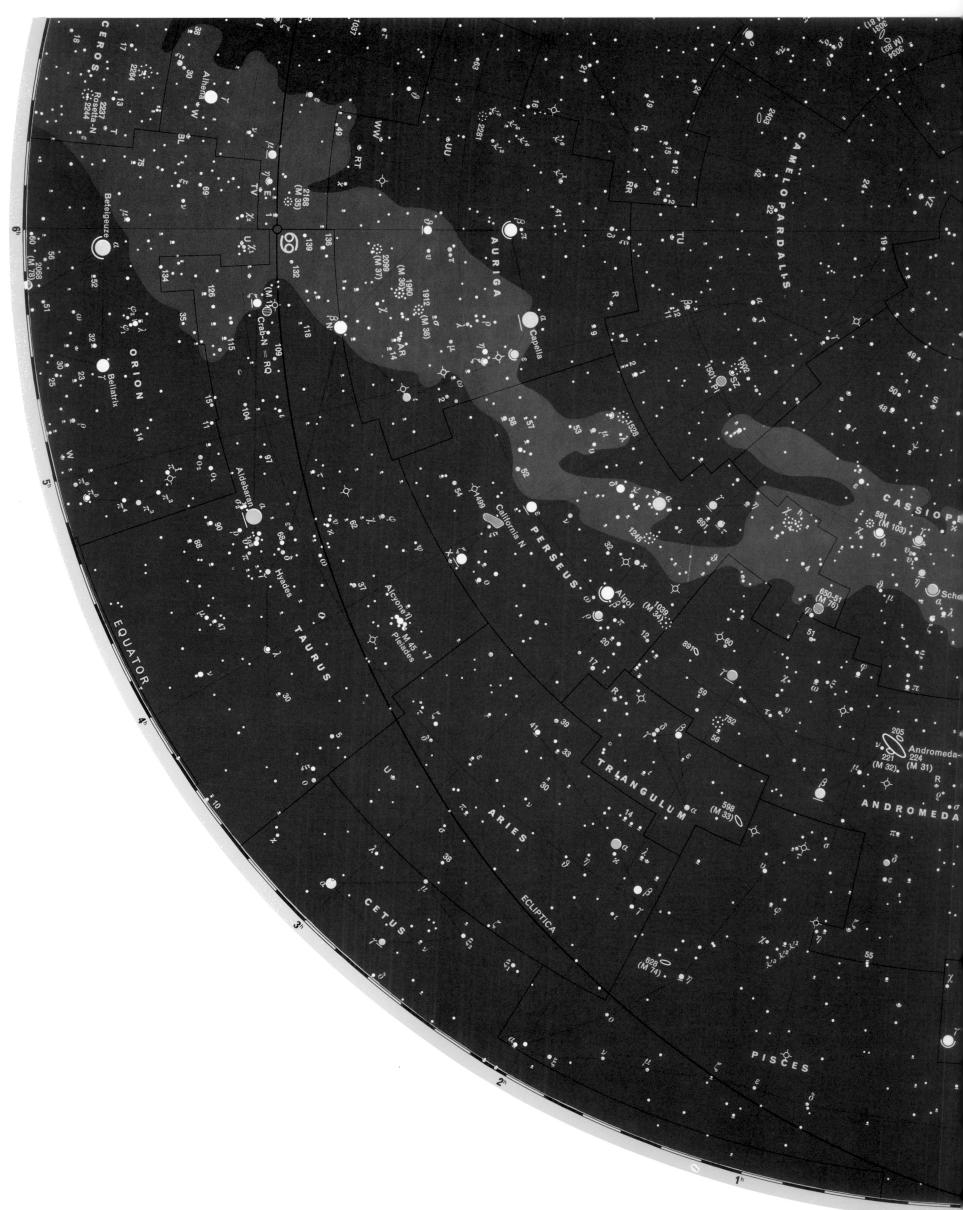

This map includes some of the brightest and most interesting constellations in the sky. There are two particularly brilliant stars, Vega and Capella, which lie on opposite sides of the celestial pole, and at roughly equal distances from it. From Europe and the northern United States, Capella is almost overhead during winter evenings, while Vega occupies a similar position during evenings in summer.

Capella and Vega

The two stars are not alike. Capella, in the prominent constellation of Auriga (the Charioteer), is a yellow giant, 150 times as luminous as the Sun and 47 light-years away. Vega, in Lyra (the Lyre), is only 50 times as powerful as the Sun, and its distance from us is little over half that of Capella, but it has a hotter surface (spectral type A) and its lovely blue colour is noticeable even with the naked eye. There are no other bright stars in Lyra, but there are many notable objects, including the eclipsing binary Beta Lyræ, the double-double star Epsilon, and the famous planetary nebula M.57, always nicknamed "the Ring".

Vega and Capella are of magnitude 0; in the whole sky only Sirius, Canopus, Alpha Centauri, and Arcturus are brighter. Vega makes up a prominent triangle with two other 1st magnitude stars, Deneb in Cygnus (the Swan) and Altair in Aquila (the Eagle), and this is often nicknamed the Summer Triangle, as it dominates the night sky during summer in Europe.

Cygnus

Cygnus is a large and important constellation, often known as the Northern Cross for reasons which are obvious to anyone who has looked at it. Deneb, of magnitude 1·3, is exceptionally luminous. It is at least 10,000 times as powerful as the Sun, and even this may be an underestimate. On the other hand Altair, in Aquila, is a mere 16 light-years away and 9 times as luminous as the Sun, so that it is one of our closest stellar neighbours; of the 1st magnitude stars only Sirius and Alpha Centauri are closer. Altair is easily recognized, because it has a fainter star to either side of it. In the general area between Lyra, Cygnus and Aquila there are several small but interesting constellations, of which the most prominent is Delphinus (the Dolphin). Albireo (Beta Cygni) is the most beautiful coloured double in the sky.

Almost equally conspicuous is Cassiopeia, the proud queen of the Perseus legend, marked by a very prominent W of stars; the brightest are of the second magnitude. The best way to find Cassiopeia is to run an imaginary line from Mizar in the Great Bear through Polaris, and extend it for an equal distance to the far side of the pole. Adjoining Cassiopeia is the much fainter and less distinctive figure of her husband, Cepheus. The brightest star in Cepheus, Alderamin, is of the 3rd magnitude.

The Square of Pegasus

Two of the stars in Cassiopeia can be used as "pointers" to the Square of Pegasus. In mythology, Pegasus was a flying horse; in the sky he is very large, but not quite as bright as might be thought from his aspect on a map. Moreover, the brightest star, Alpheratz or Sirrah, has been transferred to the adjacent constellation of Andromeda – yet another character in the Perseus legend – where we find the famous Great Spiral, M.31, and the magnificent binary Almaak or Alamak (Gamma Andromedæ), with a 2nd magnitude yellow primary and a 5th magnitude blue companion. Perseus himself has one star, Mirphak, of above the 2nd magnitude, and between Perseus and Cassiopeia is the double cluster in the Sword-Handle, dimly visible to the naked eye and easily seen with binoculars. Also in Perseus is Algol, the prototype eclipsing binary, which is usually about as bright as the Pole Star, though at minimum the magnitude drops to 3½.

The three Zodiacal constellations on this map are of different types. Taurus (the Bull) has the brilliant orange Aldebaran and the two most famous open clusters, the Hyades and the Pleiades; Aries has one 2nd magnitude star, Hamal, and a wide, easy binary, Gamma; Pisces (the Fishes) has neither bright stars nor interesting objects, and occupies a large, dull area of the sky adjoining Pegasus.

Cygnus *right*
The constellation Cygnus, photographed by Commander H. R. Hatfield at his observatory in Kent. Deneb is an A2-type supergiant. Beta Cygni or Albireo (R.A. 19h.28m.42s., decl. +27°52') is a giant of type K1, absolute magnitude −2·2; it has a companion magnitude 5·4, distance 34".6. Albireo is probably the most beautiful double star in the sky; the primary is yellow, the companion green.

The smaller illustration *below right* shows the long-period variable Chi Cygni (R.A. 19h.48m.6., decl. +32°47'), which was approaching maximum when the photograph was taken; it may be compared with Eta Cygni, magnitude 4. Near Gamma is P Cygni, a nova-like variable whose magnitude has remained consistently between 4·5 and 5·5 over the course of the past 100 years.

Epsilon Lyræ *below right*
Epsilon Lyræ, the double-double or quadruple star near Vega (R.A. 18h.42m.41s., decl. +39°37'). The separation between Epsilon^1 and Epsilon^2 is 207", and the two may be seen separately with the naked eye. A 3-in. telescope is powerful enough to show that each component is again double. The separation of Epsilon^1 is 2".8 and for Epsilon^2 it is 2".3. All four of the stars in the system are of type A. The revolution period of the two main pairs round their common centre of gravity is too long to be measured with any accuracy.

Telescopically, several stars may be seen between the two main pairs; these, however, are not associated with the system.

Aries *below*
Aries, showing the position of the double star Gamma Arietis (R.A. 1h.50m.47s., decl. +19°03'). The magnitude of the components are 4·7 and 4·8, giving a combined naked-eye magnitude of 4·0; both are of type A, and are typical Main Sequence stars, considerably more luminous than the Sun. As the separation is 8".2, Gamma Arietis ranks as a wide, easy double. It is a binary system, but the revolution period is extremely long.

The star is easy to identify. Adjoining Andromeda and Triangulum is Hamal or Alpha Arietis, which is of the 2nd magnitude. It makes up a group with Beta Arietis (magnitude 2·7) and Gamma. There are no other bright stars in Aries, but the group is quite distinctive.

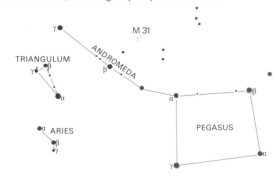

Variable Stars in Perseus *right*
Algol (Beta Persei) is an eclipsing binary, with a range from magnitude 2·2 to 3·5; good comparison stars are Alpha Persei (1·9), Epsilon (3·0) and Zeta (2·9). Rho is a red M-type irregular variable with a range of 3·3 to 4·2, though for the greater part of the time the magnitude remains between 3·6 and 4·1; the variations which it undergoes are always slow. Good comparisons with Rho are Kappa (4·0) and 16 Persei (4·3).

Variable Stars in Cepheus *below*
Cepheus, showing the variable stars Delta and Mu. Delta (R.A. 22h.27m.3., decl. +58°10) is the prototype Cepheid, with a period of 5·4 days; it makes up a triangle with Zeta (magnitude 3·3) and Epsilon (4·2) which form useful comparison stars, since Delta varies between 4·1 and 5·2. Mu (R.A. 21h.42m.0., decl. +58°33') has a range of between 3·6 and 5·1, but is usually of about magnitude 4½, so that it may be compared with Mu Cephei (4·5). Mu Cephei is Herschel's "Garnet Star".

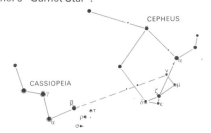

Square of Pegasus *above*
The Square is made up of four stars: Alpha Andromedæ (upper left), magnitude 2·1; Beta Pegasi (upper right), variable; Alpha Pegasi (lower right), 2·5; and Gamma Pegasi (lower left), 2·8. Beta Pegasi is an M-type red supergiant. Its extreme magnitude range is between 2·1 and 2·9, though its usual range lies between magnitudes 2·3 and 2·8.

Beta Pegasi is a star suitable for naked-eye observation, partly because of its brightness and partly because the comparison stars are so convenient. However, the effects of extinction must be allowed for. If, for instance, an observer in Britain or the northern United States compares Beta with Alpha, he will have to allow for the fact that Alpha will be considerably lower over the horizon, so that its light will be reduced. This reduction, or extinction, amounts to 3 magnitudes for an object of altitude 1° above the horizon; at altitude 2°, 2·5 magnitudes; at 10°, 1 magnitude; at 25°, only 0·3 magnitude. Above 43°, extinction may be neglected for naked-eye observations. This photograph was taken by H. R. Hatfield.

Cassiopeia *right*
near right Close to Beta
Cassiopeiæ (magnitude
2·3), the right-hand star of
the W, are three stars in
rough alignment. The
central member is Rho
Cassiopeiæ, a famous
irregular variable with a
range from 4 to 6·2.
Photograph by H. R.
Hatfield.
far right Star-trails
photographed by Peter
Gill. Cassiopeia is shown,
and the different colours
of the stars are evident ;
Alpha or Shedir (spectral
type K) shows up clearly
as orange. The trail
across the picture is of
the U.S. balloon satellite
Echo 2.

Pleiades Cluster *above,*
right and far right
These three photographs
of the Pleiades cluster, M.45
(R.A. 3h.44m.1., decl.
+23°58') show the effects
of different exposures.
near right The whole area
of Taurus, showing the
Pleiades to the top right
and also the more open
cluster of the Hyades,
round Aldebaran.
far right Another photograph of the Pleiades ; the
exposure is 20 minutes and
many more stars are seen.
above The Pleiades as
photographed with the
48-in. Schmidt telescope
at Palomar, showing the
cluster as well as the
associated nebulosity. The

brightest star in the Pleiades
cluster is Alcyone or Eta
Tauri, magnitude 2·9.
Other stars of the group
visible with the naked eye are
Electra, Atlas, Merope,
Maia and Taygete. It
has been said that the old
records indicate the
naked-eye visibility of
7 stars, and that one may
have faded considerably.
This may or may not be so,
but certainly Pleione is
somewhat variable in
brilliancy.
The cluster is 410 light-
years away, and has an
apparent diameter of 120
minutes of arc. The leading
stars are hot and white ;
there are 500 of them in a
sphere 50 light-years across.

Double Cluster NGC 869-864 (Chi-h Persei) *right*
Known as the Sword-Handle, it is easy to locate, since it lies midway between Cassiopeia and Alpha Persei. It is visible to the naked eye, and is excellently seen with binoculars or a low magnification on a telescope. Each cluster is 36' in apparent diameter, corresponding to a real diameter of 75 light-years; their distance from us is 7000 light-years.

Both clusters are loose in form. Each contains approximately 350 stars, though NGC 869 is slightly the richer of the two in terms of numbers.

Rather surprisingly neither cluster was included in Messier's catalogue. The name of the "Sword-Handle", which it is sometimes known by, is unofficial, and there is no implied association with the great gaseous nebula M.42, the Sword of Orion.

This particular photograph was taken at the United States Naval Observatory.

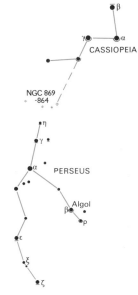

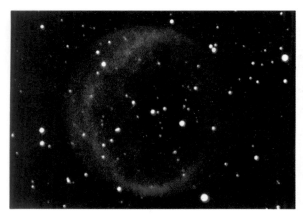

Planetary Nebula *above*
Photograph of the planetary nebula NGC 6781. Position, R.A. 19h.16m.0., decl. +06°26' (in Aquila). The apparent diameter is 106 seconds of arc, which is rather large for a planetary, but the integrated magnitude is only 12·5, and the central star is of magnitude 15·4, so that it is a faint object.

The photograph was taken with the 48-in. Schmidt telescope at Palomar.

Ring Nebula M.57 (NGC 6720)
below
R.A. 18h.15m.7., decl. +32°58'. M.57 is visible with a small telescope, midway between Beta Lyræ (the eclipsing binary) and Gamma Lyræ, mag. 3·0(L). The integrated magnitude is 9·3; the distance, 1410 light-years

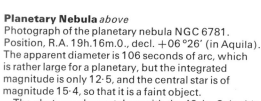

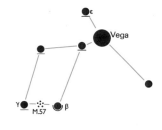

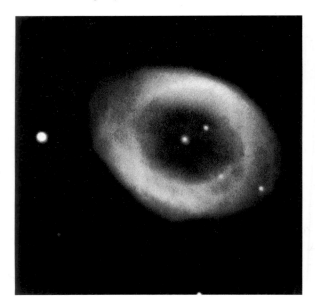

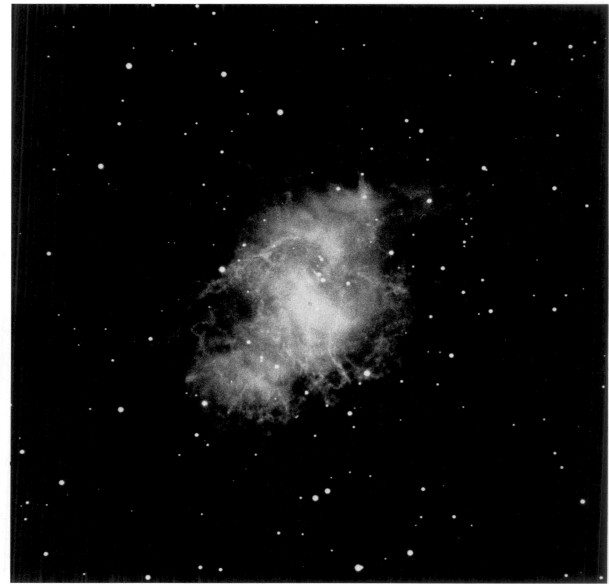

Veil Nebula *right*
Colour photograph of the
Veil or Cirrus Nebula in
Cygnus, NGC 6992.
The position is R.A.
20h.54m.0., decl. +31°30′
The photograph was taken
with the 48-in. Schmidt
telescope at Palomar. The
nebula is 2500 light-years
away, and is probably the
result of a supernova
explosion, though the
outburst took place in
prehistoric times and
there is naturally no
record of it.

It has been calculated
that the gas filaments were
ejected 50,000 years ago,
with an initial velocity of
5000 miles per second ;
the expansion velocity has
now been reduced to 75
miles per second. It has
been estimated that in
25,000 years' time the
nebula will cease to be
luminous.

North America Nebula
below
The North America Nebula
in Cygnus, photographed
with the 48-in. Schmidt
telescope at Palomar. The
nebula, NGC 7000 (R.A.
20h.57m.0., decl.
+44°08′) is 1000 light-
years away, and is
apparently associated with
the exceptionally luminous
A2-type supergiant Deneb
or Alpha Cygni. The
apparent dimensions are
120′ × 100′.

The comparatively
dark areas seen in the
photograph are due to an
intervening cloud of opaque
dust, which cuts out the light
of the nebula as well as
of the background stars.

The nickname of the
North America Nebula is
unofficial, and comes
from a somewhat rough
resemblance between the
shape of the nebula and
that of the North American
continent.

Crab Nebula *left*
M.1 (NGC 1952), the
Crab, was discovered by
John Bevis in 1731, and
independently by Messier
in 1758. It is the remnant
of the supernova of 1054,
and is a source of radio
waves and X-rays, and it
contains a pulsar. This
pulsar — certainly the
remnant of the supernova
itself — has been called
"The Power-house of the
Crab Nebula".

Though the integrated
magnitude is only 8·4, the
Crab is not hard to find.
The position is R.A.
5h.31m.5., decl. +21°59′,
1 °N and 1 °W of Zeta Tauri
(magnitude 3·0). It is
0°.4W of the 6th-magnitude
star W742, which will be
in the same field of
view in a low-power
eyepiece.

The view of the Crab
obtained visually is
disappointing ; little can be
seen apart from a blur of
light. Long-exposure
photographs taken with
large telescopes are needed
to bring out the intricate
structure.

Veil Nebula *below*
The Veil or Cirrus Nebula in Cygnus, photographed
with the Palomar 48-in. Schmidt. The whole area of
nebulosity is shown, including the portion visible in the
colour photograph above also taken with the 48-in.
Schmidt at Palomar (the most powerful instrument of its
type in the world). The approximate symmetry is evident,
supporting the theory that the nebula represents the
debris of a supernova explosion.

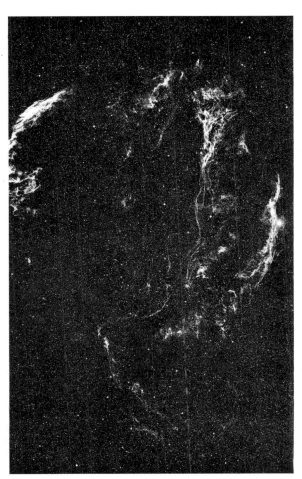

The northern Sky 2: galaxies

Great Spiral in Andromeda *right*
M.31 (NGC 224), the Great Spiral in Andromeda, photographed with the 48-in. Schmidt telescope at Palomar. Position: R.A. 0h.40m.0, decl. +41°.00'. M.31 is 2·2 million light-years away, and it is a member of the Local Group. According to Baade, the angle between our line of sight and the central-plane of M.31 is 11°.7 so that it is not far from an edge-on position. Were we seeing it face-on, we would see much more of it but the surface brightness would be less.

Andromeda *below*
The constellation Andromeda (photographed by H. R. Hatfield), showing the Great Spiral, M.31, as a hazy spot. The Spiral is clearly visible to the naked eye, 1°.2W and very slightly north of Nu Andromedæ. The whole of the constellation is shown in this illustration, together with Triangulum and part of Pegasus. Alpheratz is included in the Square of Pegasus; it used to be known as Delta Pegasi, but it has now been transferred to Andromeda and has taken the official name of Alpha Andromedæ.

Locating the Andromeda Spiral *below*
The diagram shows how to locate the Andromeda Spiral, near Nu Andromedæ (magnitude 4½). Associated with M.31 are two much smaller elliptical galaxies, M.32 and NGC 205. The Triangulum Spiral, M.33, close to the border between Triangulum and Andromeda, is much fainter than the Great Spiral, M.31, but binoculars will show it up quite clearly lying almost midway between Alpha Trianguli and Beta Andromedæ. Like M.31, M.33 is a member of the Local Group.

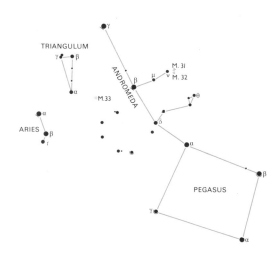

Nucleus of M.31 *right*
The nucleus of the Andromeda Spiral (U.S. Naval Observatory). In most photographs, the nucleus of a spiral galaxy tends to be over-exposed so that the detail in the arms is brought out clearly. In this view the nucleus has been correctly exposed, so the arms are not well seen.

The nucleus of the galaxy is predominantly of Population II ; the brightest stars are evolving off the Main Sequence into the giant branch ; in the arms, the brightest stars are still on the Main Sequence (Population I regions).

Radio Source in Perseus (NGC 1275) *above*
The "jet" issuing from the main source is a notable feature. Photograph taken with the 200-in. Hale reflector at Palomar.

Radio Source in Cassiopeia *above*
Red light photograph (Palomar 200-in. reflector). Only faint gas wisps are visible ; probably a supernova remnant.

M.74 *near right*
Galaxy M.74 (NGC 628) (200-in. photograph). R.A. 1h.34m.0., decl. +15°.32' (in Pisces) ; distance 26 million light-years. An open spiral ; integrated magnitude 10·2.

NGC 7741 *far right*
(200-in. photograph.) R.A. 23h.41m.4., decl. +25°48' (in Pegasus). Mag. 11·6 ; apparent diameter 2'.0. A fine barred spiral.

NGC 891 *near right*
(60-in. photograph.) R.A. 2h.19m.3., decl. +42°07' (in Andromeda). Mag. 11. The longest apparent diameter, 11'.8. A spiral galaxy, edge-on.

NGC 147 *far right*
(200-in. photograph, red light.) R.A. 0h.30m.4., decl. +48°14' (in Cassiopeia). E4-type galaxy , mag. 12 ; dimensions 4'.5 x 2'.5.

The Triangulum Spiral M.33 *above*
R.A. 1h.31m.0., decl. +30°24'. The integrated magnitude is 6·7, but the surface brightness low, and it is best seen with a wide-field, low-power eyepiece. Dimensions : 55' x 40'.

M.33 is a good example of a normal open-type spiral. The distance is 2·35 million light-years ; it is the most distant known member of the Local Group. The mass is equal to 7900 million Suns, or $\frac{1}{25}$ the mass of our Galaxy. The arms and the central region are less easy to distinguish than those of galaxies like M.74.

The Southern Sky 1: stars

Just as the Great Bear is the most famous constellation of the northern sky, so the Southern Cross – Crux Australis – is the best-known group in the south. It is not visible from any part of Europe or the United States, and it is not, therefore, one of the ancient constellations; it was named only in the 17th century.

The Southern Cross

Crux is the smallest constellation in the entire sky, but it contains three stars above the 2nd magnitude and one of the 3rd. The brightest of them, Acrux or Alpha Crucis, is a superb binary; the magnitudes of the components are 1·6 and 2·1, giving a combined naked-eye magnitude of 0·9, and the separation is almost 5 seconds of arc, so that the pair may be seen well with a small telescope. Beta Crucis (magnitude 1·3) is a very luminous B-type star; Gamma Crucis (magnitude 1·6) is a red giant, whose colour contrasts sharply with the whiteness of Acrux and Beta. Also in Crux is the open cluster, Kappa Crucis (NGC 4755), known popularly as "the Jewel Box". Crux does not look very much like a cross. Its shape is more like that of a kite. Two of its stars indicate the direction of the south celestial pole, in Octans.

Centaurus

Almost surrounding Crux is the imposing constellation of Centaurus (the Centaur), which again is too far south to be seen in Europe or most of the United States. There are two 1st magnitude stars, Alpha and Beta, which point to the Southern Cross. Of these, Alpha Centauri is the nearest bright star in the sky; and the closest of all our stellar neighbours, Proxima (4·2 light-years), is actually a faint member of the Alpha Centauri system. Alpha itself shines as a star of magnitude −0·3, inferior only to Sirius and Canopus, but it is not a single star; it is a binary, with components which are of magnitudes 0·4 and 1·7 and a revolution period of 80 years. Alpha Centauri is a magnificent telescopic object. Air navigators call it Rigel Kent.

Adjoining it is the very remote Beta Centauri, 3000 times as luminous as the Sun, and bluish-white. There are several other bright stars in the constellation as well as the globular cluster Omega Centauri (NGC 5139), much the finest in the sky; it is easily visible to the naked eye.

Argo Navis

Adjoining Centaurus is the vast constellation formerly named Argo Navis, the Ship. Argo is so large that it has been officially divided into several parts, as shown on the map. Its most important components are Carina (the Keel), Puppis (the Poop), and Vela (the Sails). Canopus, in Carina, is the brightest star in the sky apart from Sirius; it is an extremely luminous supergiant. It is an F-type star, and dominates its whole region by its brilliance.

Though all these groups are in the far south, Europeans and North Americans can at least see Sirius, the brightest star in the sky; it lies in line with the three stars in the Belt of Orion, and is the leader of Canis Major (the Great Dog). The magnitude of Sirius is −1·4, but it owes its eminence to its closeness – only 8·6 light-years – rather than its luminosity; it is "only" 26 times as powerful as the Sun.

The Scorpion

Of the Zodiacal groups in this map, much the most imposing is Scorpius or Scorpio (the Scorpion), with its 1st magnitude leader, the red Antares. Like Altair, Antares has a fainter star to either side of it. The Scorpion is one of the few constellations which slightly resembles its namesake, as it consists of a long line of stars, many of them bright. Antares itself rises well above the horizon in most of Europe, but the rest of the constellation is not really well seen except from more southerly countries. The adjacent Zodiacal group, Libra (the Balance), is dim and dull; Virgo (the Virgin), partly in the northern sky, has one 1st magnitude star, Spica, which can be found by following the curve from the Great Bear through and past Arcturus.

On this map, too, lies much of Hydra, which is the largest of all constellations (just as Crux is the smallest). Yet in spite of its size, there is only one 2nd magnitude star: Alphard, the Solitary One, to which Castor and Pollux act as pointers. Alphard is decidedly reddish in colour.

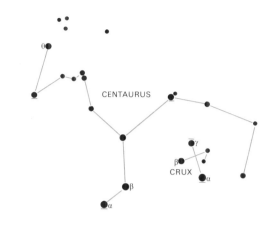

Centaurus and Crux *left*

The chart shows Centaurus and Crux, two of the most prominent constellations of the far south; both are invisible from British or North American latitudes, although a few stars in Centaurus do rise above the British horizon. Alpha Centauri (R.A. 14h.36m.11s.2., decl. −60°27'49'') is a magnificent binary, with a period of 80·1 years. The spectra of the two components are G4 and K5. The position angle naturally alters quite rapidly, but the average separation of the components is 4'', so that Alpha Centauri is a splendid object when seen in a small telescope. At its distance of 4·3 light-years, it is also the nearest of the bright stars. The closest of all the stars, Proxima (4·2 light-years), is a faint member of the Alpha Centauri system.

Alpha Crucis (R.A. 12h.23m.8, decl. −62°49') is also a fine object; magnitudes 1·6, 2·1 (combined magnitude 0·9); spectra B1,B1; position angle 114°, separation 4''.7. Unlike Alpha Centauri, Alpha Crucis is very remote (distance 230 light-years).

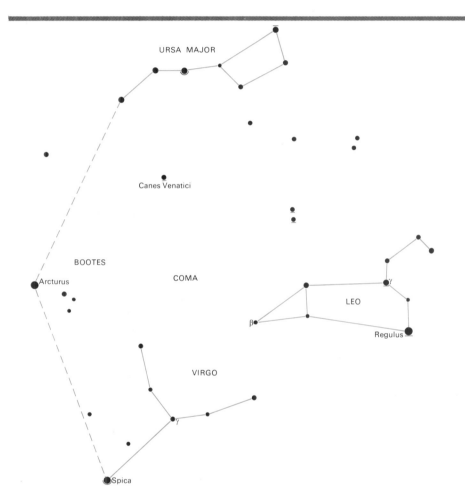

Gamma Virginis *left*

A well-known binary, found by using the Great Bear as a direction-indicator; Arcturus and Spica are located, and Gamma Virginis is then found at the root of the "Y" of Virgo (R.A. 12h.39m.07s. decl. −01°32''). The components, both of spectrum F0, are equal (mag. 3·6), and together they give a naked-eye combined magnitude of 2·9.

The period of the Gamma Virginis binary is 180 years, and its orbit is of high eccentricity. In 1780 the separation was 5''.7. By 1836 it had decreased to only 0''.4, and then increased again, reaching 6''.2 in 1921.

At present the separation is decreasing, but the star is still a wide, easy pair (5'') and it is possible to see it excellently with the aid of a small telescope. By the end of the century it will have become close and difficult, and by 2010 it will appear single except in very large telescopes.

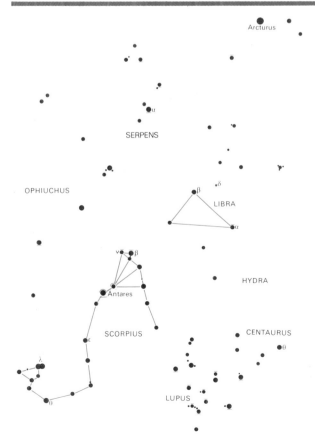

Scorpius *left*

Scorpius is one of the most distinctive of the Zodiacal constellations. Its brightest star, Antares (R.A. 16h.26m.20s.2, decl. −26°19'22''), is a vast red giant, with a diameter of 250million miles; it is 360 light-years away, and its luminosity is 3400 times that of the Sun. The apparent magnitude is 1.1. Antares has a companion of magnitude 6.5 at position angle 274° and separation 2''.9; the spectrum of the companion is B4 but the contrast with the redness of the primary makes the companion look green. Antares is a fine object in a telescope of more than 4-in. aperture. Beta Scorpii and Nu Scorpii also shown on the chart and listed in the catalogue of double stars on page 178 can also be separated with small telescopes.

Adjoining Scorpius is the obscure Zodiacal constellation of Libra. The brightest star, Beta Libræ (magnitude 2·7), has a spectrum of type B8; it has been alleged that the colour is greenish. Delta Libræ is an eclipsing binary, with a magnitude range of from 4·8 to 5·9 and a period of 2·3 days. It is of the Algol type, and is bright enough to be followed throughout its changes with a pair of binoculars. (The position is R.A. 14h.58m.3., decl. −08°19'.)

Crux Australis *below*
This photograph of Crux Australis, the Southern Cross, shows the concentration of bright stars, as well as the dark nebulosity known as the Coal-Sack. Crux was added to the map of the sky by Royer in 1679.

The False Cross *below*
The False Cross lies in the old constellation of Argo Navis ; since Argo has now been divided up, two False Cross stars, Iota (mag. 2·2) and Epsilon (1·7), are in Carina, while Kappa (2·6) and Delta (2·0) are in Vela.
    Canopus (F-type supergiant ; mag. —0·86 ; R.A. 6h.22m.50s.5, decl. —52°40'44") is the brightest star in Carina. Also in Carina is the erratic variable Eta (R.A. 10h.43m.1s., decl. —59°25'16"). In 1843 it reached mag. —0·8, but has now faded below naked-eye visibility (mag. 7·7). Associated with extensive nebulosity.

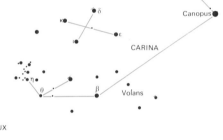

Star-Fields *above*
Star-fields in the Milky Way, photographed at the U.S. Naval Observatory, Flagstaff, Arizona. The trail of the U.S. balloon satellite Echo I can be seen on the plate.

Trifid Nebula in Sagittarius *below*
M.20 (NGC 6514) (Palomar 200-in. photograph). The position is R.A. 17h.58m.9., decl. —23°02'. The integrated magnitude is 9, so that the nebula is visible in small telescopes even though photographs taken with large instruments are needed to show its delicate details adequately.

The Trifid, which forms a typical emission nebula, contains hot early-type stars, the effect of which is to excite the gas to luminosity. The distance is 2300 light-years, and the estimated real diameter of the nebula is 30 light-years. The "trifid" appearance arises from the presence of dark matter associated with the nebula. Apparent dimensions are 29' x 27', so that the shape is close to being symmetrical. M.20 is also a radio source at 9·4 centimetres.
    The nebula lies 2°S and 2½°W of the 4th-magnitude star Mu Sagittarii. One degree NE of M.20 lies the compact galactic cluster M.21.

The Galaxy *below*
The Galaxy, seen edge-on. The Sun is 30,000 light-years from the centre of the system. The direction of the centre is toward the rich star-clouds in the constellation of Sagittarius, but it cannot be seen through the obscuring intervening matter.

Star-Clouds in Sagittarius *below*
A photograph taken with the 48-in. Schmidt telescope at Palomar in red light, showing the richest part of the whole Milky Way. There is also considerable dark nebulosity, which blocks out the light of the background stars. We cannot see through to the centre of the Galaxy ; there is too much obscuring interstellar material in the way.

Open Cluster NGC 5897 *left*
The open cluster NGC 5897, in Libra, photographed with the 200-in. reflector at Palomar. This is a very loose cluster in comparison with the Pleiades, and there is no associated nebulosity. It is typical of the less condensed galactic clusters which will not retain their identity indefinitely.

Star-Clouds in Sagittarius *below*
Star-clouds in Sagittarius, taken with the 18-in. Schmidt telescope at Palomar. In addition to the star-clouds themselves the photograph shows two famous gaseous nebulæ M.8 (the Lagoon Nebula) in the centre, and M.20 (the Trifid Nebula) above (also shown on page 167).

Planetary Nebula in Hydra *below*
The position of this nebula (NGC 3242) is R.A. 10h.22m.4., decl. —18°23' (in Hydra). The distance is 1800 light-years. The integrated magnitude is 9·0, and the central star is of magnitude 11·4. The dimensions are 40'' x 35''. The general aspect is clearly quite different from that of the brighter and better-known Ring Nebula M.57 Lyræ.
   This photograph was taken in red light with the Palomar 200-in. reflector.

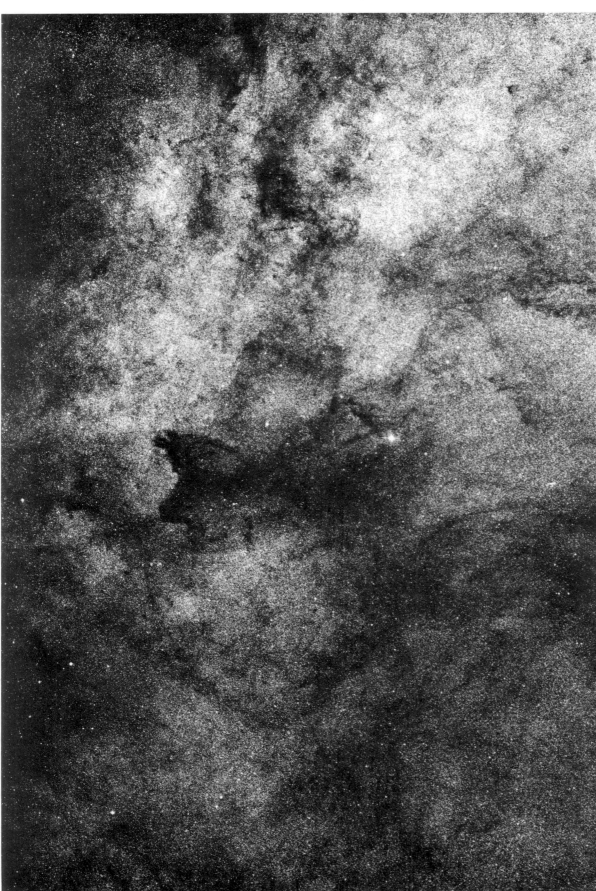

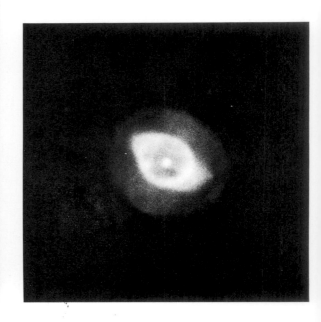

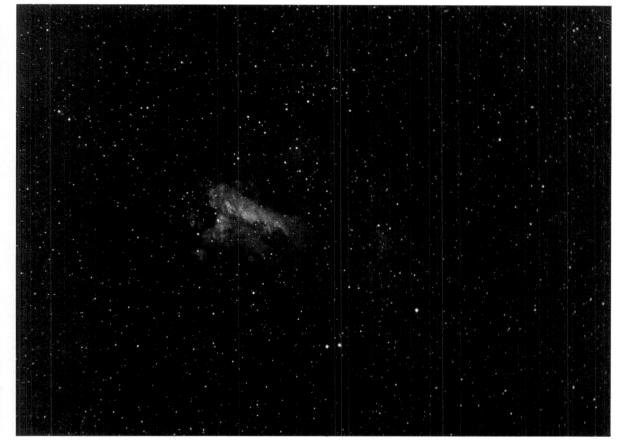

Lagoon Nebula *above*
200-in. Palomar photograph of the Lagoon Nebula, M.8
(NGC 6523). The position is R.A. 18h.00m.1.,
decl. —24 °23', in Sagittarius. M.8 is an emission nebula,
and is a radio source at 9·4 centimetres.

The integrated magnitude is 6·0, and so the nebula is
easily visible in binoculars. It is on the fringe of
naked-eye visibility, 5 °WNW of the group of stars made
up of Zeta, Phi and Lambda Sagittarii. The distance of M.8
is 4850 light-years ; the apparent size is 90' x 40'.

Associated with the Lagoon Nebula is the galactic
cluster NGC 6530, which has an integrated magnitude of
6·3 and a diameter of 10'. This is a cluster which
contains a total complement of 25 stars.

Omega Nebula *left*
The Omega Nebula, M.17 (NGC 6618) first discovered by
the French astronomer de Chéseaux in 1746, is easy
to pick out with the aid of binoculars ; the integrated
magnitude is 7·0. The position is R.A. 18h.17m.9.,
decl. —16 °12'.

The nebula is 5870 light-years away and its
apparent size is 46' x 37' According to recent estimates,
M.17 is more massive than the Orion Nebula, and
has a mass 800 times that of the Sun, though it
is much more remote than the Orion Nebula and so
appears as a less conspicuous object in our skies.
As is the case with many diffuse nebulæ, there are
bright areas as well as indications of dark obscuring
matter.

The Nebula is on the borders of Sagittarius and
Scutum. The nearest naked-eye star, 1½ °N and 2 °E of
the nebula, is Gamma Scuti, which is of magnitude 4·7.
This photograph was taken with the 48-in. Schmidt
telescope at Palomar.

"Sombrero Hat" Galaxy *above*
M.104, the so-called "Sombrero Hat" Galaxy, photographed at the U.S. Naval Observatory.

M.104 (NGC 4594) is an Sb-type spiral ; position, R.A. 12h.37m.4, decl. 11 °21'. It is a member of the Virgo cluster of galaxies, and lies at a distance of 41 million light-years. It was discovered by Méchain in 1781.

The integrated magnitude of M.104 is 8·7 ; the galaxy is not difficult to find in its position in the sky. It forms an isosceles triangle with Spica and Gamma Virginis, toward Corvus.

The band of dark obscuring material along the main plane is very distinctive. According to G. de Vaucouleurs, the main plane of the system is inclined by only 6 °

to our line of sight.

The spiral is a massive one ; Holmberg gives its mass as 1·3 million million solar masses, more than 10 times that of our own Galaxy. It is, however, only a weak source of radio radiation. According to Pease, unlike most other galaxies M.104 revolves as though its arms were "unwinding".

Cluster of Galaxies in Hydra *above*
Photographed with the 200-in. Palomar reflector. The distance of the group is 1800 million light-years. There are, of course, many foreground stars which belong to our own Galaxy ; on a photograph of this kind it is not easy to tell, at a glance, which objects are stars and which are external systems. The shapes of the individual galaxies are not easy to make out, and no individual stars in them are visible, so that our distance-measurements depend mainly on the red shifts in their spectra, interpreted as Doppler effects.

NGC 4038 and 4039 in Corvus *right*
Peculiar Sc galaxies in Corvus (48-in. Schmidt, Palomar) ; each is a radio source. There can be no doubt that the two systems are genuinely associated.

Clouds of Magellan *below*

The diagram below shows the positions of the two Clouds of Magellan. Both are prominent objects with the naked eye, but are too far south to be seen from any part of Europe or the United States.

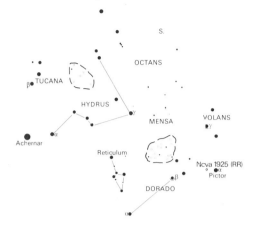

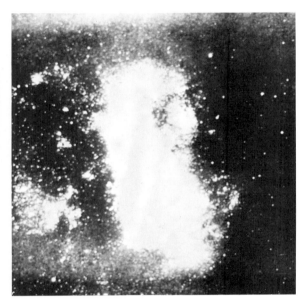

The Large Cloud of Magellan *above*

An independent galaxy associated with our own, officially known as Nubecula Major. It lies mainly in Mensa, partly in Dorado ; it is classed as an irregular system, though some authorities claim that it shows signs of spiral structure. It is visible with the naked eye even in moonlight.

Centaurus A *above*

A radio source now thought to be a single system rather than a collision of galaxies, though the cause of the intense radio emission has not yet been satisfactorily explained. The radio source consists of two principal components and is very much larger than the optical source.

M.83 (NGC 5236) *below*

Sc-type spiral galaxy in Hydra ; R.A. 13h.34m.2., decl. —29 °37′. A massive galaxy, larger than our own, M.83 was discovered by Lacaille in 1752. It lies near the boundary with Corvus ; the nearest bright naked-eye star visible from northern latitudes is Beta Corvi (3rd magnitude). The distance of M.83 is 8·5 million light-years.

30 Doradûs *above*

Outer portion of the great looped nebula 30 Doradûs, in the Large Cloud of Magellan, photographed in red light, with a 30-minute exposure, at the Radcliffe Observatory in South Africa. The intricate structure is well shown. The Large Cloud contains many gaseous nebulæ, some of which are much larger and more massive than the Orion Nebula, M.42. Were these gaseous nebulæ contained in our own Galaxy, they would be most spectacular objects.

Until recently the Clouds had not been satisfactorily studied, as they are too far south to be available for observation by any of the great telescopes which are situated in the northern hemisphere ; now, with the completion of large southern telescopes, this deficiency is at last being remedied.

The brightest star on this map is Rigel, in Orion, which is only slightly below zero magnitude, and is particularly luminous and remote. Orion itself – probably the most distinctive constellation in the sky – is divided by the celestial equator, and so not all of it is shown on this chart; Betelgeux, in particular, lies well in the northern hemisphere. However, it is best to describe Orion as a whole, since it cannot possibly be misidentified.

Orion

In mythology, Orion was a famous hunter, and it is therefore appropriate that in the sky he should be surrounded by animals; his dogs (Canis Major and Canis Minor), the Hare (Lepus) and others. The outline of the Hunter himself takes the form of a quadrilateral, inside which are the three 2nd magnitude stars of the Belt; the celestial equator passes very close to the northernmost of these, Mintaka or Delta Orionis. Below the Belt is the Sword, marked by the great gaseous nebula M.42, which can easily be seen with the naked eye.

Adjoining Orion is Eridanus (the River), a very long, sprawling constellation which winds its way southward, ending in the 1st magnitude Achernar, which never rises above the horizon in Europe or North America. Today, Achernar is the only really brilliant star in Eridanus, but if ancient observations are to be believed there used to be another. It was recorded that Acamar or Theta Eridani – "the Last in the River" – was also of the 1st magnitude. It is now only of the 3rd, and it may have faded, though the old records should not be trusted too far. In any case, Acamar is a splendid double star, with components of magnitudes 3·4 and 4·4 separated by 8 seconds of arc. A small telescope will show it excellently.

The South Polar Region

The south polar region of the sky is divided into various faint, undistinguished constellations which have been formed in relatively modern times, and named appropriately. (For obvious reasons they were unknown to the Greeks and the other civilizations of the ancient world, since they never rise above the horizon in the Mediterranean area.) Yet the far south of the sky is of immense interest to astronomers, since it includes the two Clouds of Magellan.

The Large Cloud, visible with the naked eye even under conditions of bright moonlight, looks like a detached part of the Milky Way, though it is in fact a separate galaxy; it lies partly in Dorado (the Swordfish) but extends into Mensa (the Table). The Small Cloud is situated in Tucana (the Toucan), but spreads into Hydrus (the Watersnake – not to be confused with the much larger Hydra). Astronomers who work in the northern hemisphere never cease to regret the fact that the Clouds of Magellan are inaccessible to them, and that it is not possible for them to be studied from great observatories such as Palomar and Mount Wilson.

The Southern Birds

The other areas of this map are of less interest, though Sagittarius (the Archer), partly visible in Europe, contains several bright stars, and the rich regions in it indicate the direction of the centre of the Galaxy. Grus (the Crane) is a distinctive group, and its leader, Alnair, is above the 2nd magnitude; the other Birds, Tucana (the Toucan), Pavo (the Peacock) and Phœnix (the Phœnix) are less prominent, though Tucana contains part of the Small Cloud as well as a splendid globular cluster, and Pavo includes one of the brightest of the Cepheid variables, Kappa Pavonis (magnitude range 4 to 5·5, period 9·1 days). Much farther north is Cetus (the Whale), with the famous long-period variable star, Mira.

The Southern Fish

Lastly there is the Southern Fish, Piscis Austrinus, which contains one 1st magnitude star, Fomalhaut, but nothing else of note. Fomalhaut is always low down as seen from Europe or the northern United States, but it is in fact almost as brilliant as Spica, and brighter than Deneb. It is relatively close, with a distance of 24 light-years, and in luminosity it is equal to 13 Suns.

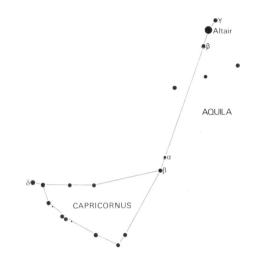

Alpha Capricorni *above*
The position of the brighter member of this naked-eye pair, known as Alpha Capricorni, is R.A. 20h.14m.9, decl. —12°40'. The line of stars that is made up of Gamma Aquilæ, Altair and Beta Aquilæ points to it. The separation between the two members of the pair is 376''; the magnitudes are 3·7 and 4·3. The fainter component is again double (magnitudes 3·7, 11; separation 7''; position angle 158°), and the smaller component of this pair is also double.

Beta Capricorni is another wide pair; R.A. 20h.18.m 12s.2 decl. —14°56'27''; magnitudes 3·3 and 6; separation 205'', position angle 290°. The fainter component has a 10·6 magnitude companion at 1''.3, position angle 103°.

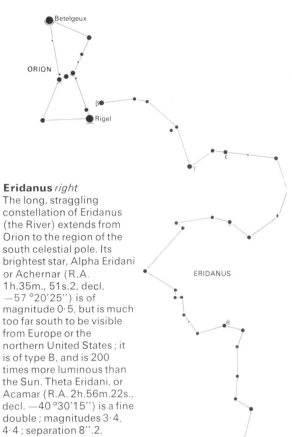

Eridanus *right*
The long, straggling constellation of Eridanus (the River) extends from Orion to the region of the south celestial pole. Its brightest star, Alpha Eridani or Achernar (R.A. 1h.35m., 51s.2, decl. —57°20'25'') is of magnitude 0·5, but is much too far south to be visible from Europe or the northern United States; it is of type B, and is 200 times more luminous than the Sun. Theta Eridani, or Acamar (R.A. 2h.56m.22s., decl. —40°30'15'') is a fine double; magnitudes 3·4, 4·4; separation 8''.2, position angle 088°.

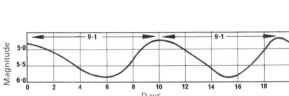

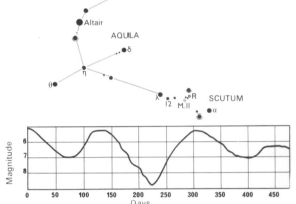

Kappa Pavonis *above*
The Cepheid variable Kappa Pavonis (R.A. 18h.51m.8, decl. —67°18') is one of the brightest members of the class; the magnitude range is from 4.8 to 5.7, and the period is 9·1 days (see light-curve).

The spectral type of Kappa Pavonis changes between F5 and G5. It is always visible with the naked eye.

R Scuti *above*
The red variable star R Scuti (R.A. 18h.44m.8, decl. —05°46') lies in the same binocular field with the open cluster M.11, nicknamed the "Wild Duck".

R Scuti has a range of from magnitude 5.0 to 8.6; it is an RV Tauri type star, and is the brightest member of its class.

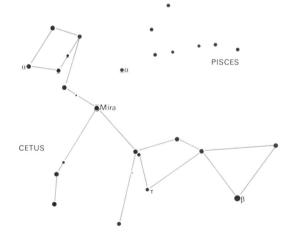

Mira *above*
Mira (Omicron Ceti) is the prototype long-period variable (R.A. 02h.16m.8s., decl. —03°12') (light-curve *right*). It was probably the first variable star to be recognized as such, being known in antiquity, and rediscovered by Fabricius in 1596. It is of spectrum M, and has a period of 331 days.

The range is, on average, from 3 to 10; but at some maxima Mira may attain or even exceed the 2nd magnitude. In 1969 it rose to 2·1, brighter than it had been for a quarter of a century. It remains visible to the naked eye for about 6 months at a time. Mira has a 10th-magnitude companion at a distance of 0s.8, position angle 131°; this is either a white dwarf or a sub-dwarf.

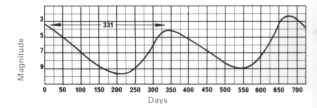

Sword of Orion *right*
Cf all the gaseous nebulæ, M.42 (NGC 1976), the Sword of Orion, is the most spectacular. Bayer, in his famous catalogue of 1603, lettered it Theta Orionis, though this designation is now restricted to the famous multiple star, the Trapezium, which lies within the nebula and is responsible for its luminosity.

The nebula is easily visible with the naked eye, south of the three prominent stars making up the Hunter's Belt. This photograph, taken by H. R. Hatfield, shows the whole area; M.42 is near the middle of the picture, with the 3rd-magnitude star Iota Orionis below.

The discovery of the nebula itself seems to stand to the credit of N. Peiresc in 1610. A small telescope will show it excellently, together with the four stars of Theta, the Trapezium. The brightest component of the Trapezium lies at R.A. 5h.32m.55s., decl. —05°26'51''. All four members of the group are of spectral type 09. The nebulosity in M.42 is of the emission type; this is an H-II region, where the hydrogen atoms are ionized by the intense radiation from the hot stars contained in the nebulosity.

Great Nebula in Orion *below*
Colour photograph of M.42, the Great Nebula in Orion, taken in red light at the U.S. Naval Observatory. In the photograph, the detailed structure of the gas-cloud is excellently displayed.

The Orion Nebula is one of the nearest to us of the gaseous nebulæ, and is therefore a particularly prominent object in the southern sky. It can be picked out easily with the naked eye.

Great Nebula in Orion *below*
Another photograph of M.42, taken this time with the 200-in. reflector at the Palomar observatory. The distance of the nebula is 1500 light-years; the integrated magnitude is 4.

There can be little doubt that stars are at present in the process of being formed inside the nebula. There are also some irregular variable stars which seem still to be contracting out of the interstellar material from which they have been formed. In fact, between 1947 and 1954 a group of 5 stars in M.42 appeared to increase to 7.

Clusters and Nebulae

Planetary Nebula NGC 7293 *below*
Photograph of the planetary nebula NGC 7293 ;
R.A. 22h.27m.0., decl. —21°06', in Aquarius. This is the
brightest of all planetaries. The integrated magnitude is
5·5, with a central star of magnitude 13·3 ; the
dimensions are 900'' x 720'', and the nebula is an easy
object to detect with the aid of a small telescope. This
colour photograph was taken with the 48-in.
Schmidt telescope at Palomar.

Gaseous Nebula M.16 *above*
The gaseous nebula M.16 (NGC 6611), near the boundary
between Scutum and Sagittarius. It is easy to find ;
R.A. 18h.16m.0., decl. —13 °48'. Though Scutum
contains no bright stars, it is a rich region, and adjoins the
"tail" of Aquila (star-finder *right*). The photograph was
taken in red light with the 200-in. Palomar reflector.
Dark nebulosity is much in evidence. The integrated
magnitude is 6·4, so that it is possible to see the nebula
perfectly clearly in binoculars ; the distance from us
is 5870 light-years.

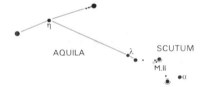

The Horse's Head Nebula *below*
NGC 2024, the Horse's Head Nebula in Orion, taken
with the 48-in. Schmidt telescope at Palomar. The
R.A. is 5h.39m.4., decl. —01 °52', so that the nebula
lies close to Zeta Orionis, the southernmost star in the
Hunter's Belt.
    The nickname is easily appreciated, since the main
dark mass bears a distinct resemblance to the head of
a knight in chess. NGC 2024 is part of the general
nebulosity which covers much of Orion, but it is some
distance away from the Sword, M.42.

Galaxies

Spiral Galaxy NGC 253 *above*

The edge-on spiral galaxy NGC 253, in the obscure constellation of Sculptor, photographed with the 48-in. Schmidt telescope at Palomar. Position: R.A. 0h.45m.1., decl. —25°34'. This is a relatively bright galaxy, with a visual magnitude of 8·9 (the photographic magnitude is 7·0). The apparent dimensions are 24'.6 x 4'.5.

In this photograph it is possible to see the structure of the galaxy to a considerable extent, and the general aspect of NGC 253 is not unlike that of the much closer and brighter Andromeda Spiral, though the arms appear to be "looser". In Hubble's classification the Andromeda Spiral is of type Sb, while NGC 253 is of type Sc

South Galactic Pole *right*

Position of the south galactic pole. This is, as would be expected, an area of the sky which is to some extent barren of bright stars.

In the region that has been included in the diagram reproduced here, Fomalhaut in the small constellation of Piscis Austrinus (the Southern Fish) is of the 1st magnitude, while Diphda (Beta Ceti) and Alpha Phœnicis are of the 2nd.

There is relatively little obscuration from interstellar material occurring in this part of the sky, and the whole area of the galactic pole is in fact rich in faint galaxies; a similar situation occurs in the case of the north galactic pole in Coma Berenices (Berenice's Hair), which is not far from Ursa Major.

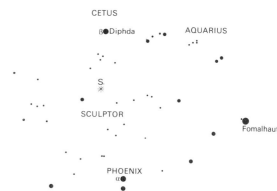

Spiral Galaxy in Fornax *above*

The photograph, taken with the Palomar 200-in. reflector, shows a barred spiral galaxy in the inconspicuous constellation of Fornax (the Furnace). In this example the arms are tightly wound and emerge from an inner ring, and the overall effect is symmetrical; the bar through the nucleus of the galaxy stands out with exceptional clarity.

The stars scattered around on the area of the plane are in the foreground, and belong to our own Galaxy.

Spiral NGC 1300 *above*

Another photograph taken with the 200-in. reflector at Palomar, showing the barred spiral NGC 1300 in Eridanus (R.A. 3h.17m.5., decl. —19°35'). This is a faint galaxy, and it has photographic magnitude of 11·1.

The structure is clearly different from that of the Fornax galaxy. The nucleus is relatively smaller, and the bar is more extended, with arms which are much more loosely wound.

Dwarf Galaxy in Sextans *above*

This photograph (200-in. reflector) shows a dwarf galaxy in Sextans. It is a member of our Local Group, and contains no detectable luminosity. The whole galaxy is composed of Population II objects, and its structure is so "loose" that it is not easy to recognize it immediately as an external system.

The bright star slightly below the centre of the photograph, to the right, is actually in the foreground, and belongs to our Galaxy.

Catalogue of stellar objects

Catalogue of Double Stars

Star	R.A. h m	Decl. ° ′	Mag.	P.A. °	Dist. ″	Map
Beta Tucanæ	00 29·3	−63 14	4·5, 4·5	170	27·1	4
Eta Cassiopeiæ	00 46·1	+57 33	3·6, 7·5	293	10·1	2
Lambda Tucanæ	00 53·1	−69 48	5·3, 7·3	080	20·8	4
Beta Phœnicis	01 03·9	−46 59	4·1, 4·1	352	1·3	4
Zeta Piscium	01 11·1	+07 19	5·6, 6·4	063	23·6	2
Omega Andromedæ	01 24·6	+45 09	5·0, 12·0	103	2·4	2
Epsilon Sculptoris	01 43·3	−25 18	5·4, 9·4	048	4·7	4
Alpha Ursæ Minoris	01 48·8	+89 07	2·0, 9·0	217	18·3	1–2
Gamma Arietis	01 50·8	+19 03	4·8, 4·8	359	8·2	2
Lambda Arietis	01 55·1	+23 21	4·8, 7·4	046	37·4	2
Alpha Piscium	01 59·4	+02 31	4·3, 5·2	297	2·1	2
Gamma$_1$ Andromedæ	02 00·8	+42 06	2·3, 5·1	063	10·0	2[1]
Gamma$_2$ Andromedæ			5·5, 6·3	var	var	
Iota Trianguli	02 09·5	+30 04	5·4, 7·0	071	3·6	2
Omicron Ceti	02 16·8	−03 12	var 10·0	131	0·8	4
Iota Cassiopeiæ	02 24·9	+67 11	4·7, 7·0	240	2·3	2
Omega Fornacis	02 31·6	−28 27	5·0, 8·7	245	10·8	4
Nu Ceti	02 33·2	+05 23	5·0, 9·8	083	7·8	2
Gamma Ceti	02 40·7	+03 02	3·7, 6·4	293	3·0	2
Theta Persei	02 40·8	+49 01	4·2, 10·0	301	18·2	2
Epsilon Arietis	02 26·4	+21 08	5·3, 5·6	208	1·5	2
Theta Eridani	02 56·3	−40 30	3·4, 4·4	088	8·5	4
Zeta Persei	03 51·0	+31 44	2·9, 9·4	208	13·0	2
Epsilon Persei	03 54·5	+39 52	3·0, 8·1	009	9·0	2
40 Eridani	04 13·0	−07 44	4·5, 9·4	105	82·8	4
Phi Tauri	04 17·3	+27 14	5·1, 8·7	250	52·1	2
Chi Tauri	04 19 5	+25 31	5·4, 8·2	025	20·0	2
Alpha Tauri	04 33·0	+16 25	1·1, 11·0	112	31·4	2
Iota Pictoris	04 49·8	−53 33	5·6, 6·4	058	12·0	3–4
Omega Aurigæ	04 55·9	+37 49	5·0, 8·0	355	5·8	2
Gamma Cæli	05 02·6	−35 33	4·6, 8·5	311	3·1	4
Rho Orionis	05 10·7	+02 48	4·6, 8·6	063	7·0	1–2
Kappa Leporis	05 10·9	−13 00	4·5, 7·5	258	2·6	4
Beta Orionis	05 12·1	−08 15	0·1, 7·0	206	9·2	4
Eta Orionis	05 22·0	−02 26	3·7, 5·1	083	1·5	4
Theta Pictoris	05 23·6	−52 22	6·3, 6·8	287	38·2	3–4
Delta Orionis	05 29·4	−00 20	2·5v, 6·9	000	52·8	4
Lambda Orionis	05 32·4	+09 54	3·7, 5·6	042	4·4	2
Theta Orionis	05 32·8	−05 25	5·9, 6·8 6·8, 6·8			4[2]
Iota Orionis	05 33·0	−05 56	2·9, 7·4	142	11·4	4
Sigma Orionis	05 36·2	−02 38	3·7, 3·8, 3·9	Multiple		4
Alpha Columbæ	05 37·8	−34 06	2·8, 11·5	000	12·6	3–4
Zeta Orionis	05 38·2	−01 58	2·1, 4·2	164	2·4	3–4
Gamma Leporis	05 42·4	−22 28	3·8, 6·4	351	95·0	3–4
Theta Aurigæ	05 56·3	+37 13	2·7, 7·5	320	3·9	1
Eta Geminorum	06 11·9	+22 31	var, 8·8	278	1·3	1
Epsilon Monocerotis	06 21·1	+04 37	4·5, 6·5	027	13·2	1
Beta Monocerotis	06 26·4	−07 00	3·9, 4·6	132	7·4	3[3]
Alpha Canis Majoris	06 42·9	−16 39	−1·4, 8·7	var	var	3[4]
Pi Canis Majoris	06 53·5	−20 04	4·6, 9·5	018	12·0	3
Mu Canis Majoris	06 55·5	−24 34	5·7, 7·2	265	1·0	3
Epsilon Canis Majoris	06 56·7	−28 54	1·6, 8·1	160	7·4	3
Gamma Volantis	07 09·1	−70 25	3·9, 5·8	299	13·8	3
Lambda Geminorum	07 15·2	+16 38	3·7, 10·0	033	10·0	1
Tau Canis Majoris	07 16·6	−24 52	4·4, 10·5	090	8·3	3
Sigma Puppis	07 27·6	−43 12	3·3, 8·5	074	22·4	3
Alpha Geminorum	07 31·4	+32 00	2·0, 2·9	171	2·2	1
Kappa Puppis	07 36·8	−26 41	4·5, 4·6	318	9·8	3
Kappa Geminorum	07 41·4	+24 31	3·7, 9·5	236	6·8	1
Epsilon Volantis	08 07·8	−68 28	4·5, 8·0	022	6·1	3
Gamma Velorum	08 07·9	−47 12	2·2, 4·8	220	41·0	3
Zeta Cancri	08 09·3	+17 48	5·1, 6·0	089	5·9	1[5]
Theta Volantis	08 38·9	−70 13	5·3, 9·8	108	45·0	3
Delta Velorum	08 43·3	−54 31	2·0, 6·5	164	2·0	3
Iota Cancri	08 43·7	+28 57	4·2, 6·6	307	30·7	1
Iota Ursæ Majoris	08 55·8	+48 15	3·1, 10·8	014	5·0	1
Upsilon Carinæ	09 45·9	−64 50	3·2, 6·0	126	4·6	3
Gamma Leonis	10 17·2	+20 06	2·6, 3·8	122	4·3	1
Xi Ursæ Majoris	11 15·5	+31 49	4·4, 4·8	var	var	1[6]
Nu Ursæ Majoris	11 15·8	+33 22	3·7, 9·7	147	7·2	1
Gamma Crateris	11 22·4	−17 25	4·1, 9·5	097	5·2	3
Alpha Crucis	12 23·8	−62 49	1·6, 2·1	114	4·7	3
Gamma Crucis	12 28·4	−56 50	1·6, 6·7	031	110·6	3
Gamma Centauri	12 38·7	−48 41	3·1, 3·2			3[7]
Gamma Virginis	12 39·1	−01 11	3·6, 3·6	309	5·2	3[8]
Iota Crucis	12 42·7	−60 42	4·7, 7·8	027	26·4	3
Beta Muscæ	12 43·2	−67 50	3·9, 4·2	007	1·6	3
Alpha Canum Venaticorum	12 53·7	+38 35	2·9, 5·4	228	19·7	1
Zeta Ursæ Majoris	13 21·9	+55 11	2·4, 4·0	150	14·5	1[9]
Iota Boötis	13 44·9	+17 42	4·5, 10·6	007	5·7	1
Alpha Centauri	14 36·2	−60 38	0·3, 1·7	var	var	3[10]
Zeta Boötis	14 38·8	+13 57	4·6, 4·6	313	1·2	1
Epsilon Boötis	14 42·8	+27 17	2·7, 5·1	338	2·9	1
Xi Boötis	14 49·1	+19 18	4·8, 6·9	350	6·7	1
Pi Lupi	15 01·7	−46 51	4·7, 4·8	076	1·7′	3
Kappa Lupi	15 68·5	−08 33	4·1, 6·0	144	27·0	3
Gamma Circini	15 19·4	−59 09	5·2, 5·3	051	1·3	3
Eta Coronæ Borealis	15 21·1	+30 28	5·7, 6·0	var	var	1[11]
Delta Serpentis	15 32·4	+10 42	4·2, 5·2	179	3·9	1
Zeta Coronæ Borealis	15 37·5	+36 48	5·1, 6·0	305	6·3	1
Gamma Coronæ Borealis	15 40·6	+26 27	4·2, 5·6	var	var	1[12]
Eta Lupi	15 56·8	−38 15	3·6, 7·7	021	15·3	3
Beta Scorpii	16 02·5	−19 40	2·9, 5·1	023	13·7	3[13]
Kappa Herculis	16 05·8	+17 11	5·3, 6·5	012	29·4	1
Nu Scorpii	16 09·1	−19 20	4·3, 6·5	337	41·4	3
Eta Draconis	16 23·3	+61 38	2·9, 8·8	142	6·1	1
Alpha Scorpii	16 26·3	−26 19	1·1, 6·5	274	2·9	3
Zeta Herculis	16 39·4	+31 42	3·1, 5·6	var	var	1[14]
Alpha Herculis	17 12·4	+14 27	var, 5·4	109	4·6	1
Rho Herculis	17 22·0	+37 11	4·5, 5·5	317	4·0	1
Nu Draconis	17 31·2	+55 13	5·0, 5·0	312	62·0	1
Epsilon Lyræ	18 42·7	+39 37	4·7, 4·5	172	207·8	2[15]
Epsilon$_1$			5·1, 6·0	002	2·8	
Epsilon$_2$			5·1, 5·4	101	2·3	
Zeta Lyræ	18 43·0	+37 32	4·3, 5·9	150	43·7	2
Beta Lyræ	18 48·2	+33 18	var, 7·8	149	46·6	2
Theta Serpentis	18 53·8	+04 08	4·5, 4·5	103	22·6	2
Zeta Sagittarii	18 59·4	−29 57	3·4, 3·5			4[16]
Gamma Coronæ Australis	19 03·0	−37 08	5·0, 5·1	054	2·7	4
Beta Cygni	19 28·7	+27 52	3·0, 5·4	055	34·6	2
Delta Cygni	19 43·4	+45 00	3·0, 6·5	246	2·1	2
Alpha$_1$ Capricorni	20 14·9	−12 40	4·6, 9·0	221	45·5	4[17]
Alpha$_2$ Capricorni	20 15·3	−12 42	3·8, 10·6	158	7·1	4
Beta Delphini	20 35·2	+14 25	4·1, 5·1			2[18]
Gamma Delphini	20 44·3	+15 57	4·5, 5·5	269	10·4	2
61 Cygni	21 04·7	+38 30	5·5, 6·4	140	27·0	2
Zeta Aquarii	22 26·3	−00 17	4·4, 4·6	266	2·0	4
Delta Cephei	22 27·3	+58 10	var, 7·5	192	41·0	2
Sigma Cassiopeiæ	23 56·5	+55 29	5·1, 7·2	332	2·1	2

[1] Binary 61 years
[2] Multiple: The Trapezium
[3] Both components again double. [4] Period 50 years
[5] Both components again double: periods 59·6 years, 17·6 years
[6] Period 59·9 years
[7] Period 84·5 years [8] Binary, period 180 years
[9] Mizar. Naked-eye pair with Alcor [10] Period 80·1 years
[11] Period 41·6 years [12] Period 91 years [13] Beta again double
[14] Period 34·4 years
[15] Quadruple [16] Period 20·8 years [17] Wide naked eye pair
[18] Period 26·6 years

Catalogue of Variable Stars

(Period: in days. Type: M=Mira type.
EA=Eclipsing binary, Algol type.
EB=Eclipsing binary, Beta Lyræ type.
C=Cepheid. RR=RR Lyræ type.
SR=Semi-regular. I=Irregular.
RV=RV Tauri type. RCrB=R Coronæ Borealis type.
RN=Recurrent nova. N=Nova.)

Variable	R.A. h m	Decl. ° ′	Mag. max	Mag. min	Period	Type	Map
T Andromedæ	00 19·8	+26 43	7·7	14·3	280·4	M	2
T Cassiopeiæ	00 20·5	+55 31	7·3	12·4	445	M	2
R Andromedæ	00 21·4	+38 18	6·0	14·9	409·2	M	2
Alpha Cas.	00 37·7	+56 16	2·2	2·5		I	2
Gamma Cas.	00 53·7	+60 27	1·6	3·3		I	2
R Piscium	01 28·1	+02 37	7·1	14·8	344·1	M	2
UV Ceti	01 36·4	−18 13	6·8	12·9	Flare star		4
R Arietis	02 13·3	+24 50	7·5	13·7	186·7	M	2
Omicron Ceti	02 16·8	−03 12	1·8	10·1	331·6	M	4
S Persei	02 19·3	+58 22	7·9	11·1		SR	2
R Ceti	02 23·5	−00 24	7·2	14	166·2	M	4
U Ceti	02 31·3	−13 22	6·8	13·4	234·5	M	4
R Trianguli	02 34·0	+34 03	5·4	12·0	266·4	M	2
T Arietis	02 45·5	+17 18	7·5	11·3	319·6	SR	2
R Horologii	02 52·3	−50 06	4·7	14·3	402·7	M	4
T Horologii	02 59·5	−50 50	7·2	13·7	217·2	M	4
Rho Persei	03 02·0	+38 39	3·3	4·2		I	2
Beta Persei	03 04·9	+40 46	2·2	3·5	2·8	EA	2
R Persei	03 26·9	+35 30	8·1	14·8	210·0	M	2
T Eridani	03 53·1	−24 11	7·4	13·2	252·0	M	4
Lambda Tauri	03 57·9	+12 21	3·5	4·0	4	EA	2
R Reticuli	04 33·0	−63 08	6·8	14·0	278·3	M	4
R Pictoris	04 44·8	−49 20	6·7	10·0	171·0	SR	3–4
R Leporis	04 57·3	−14 53	5·9	10·5	432·5	M	4
Epsilon Aurigæ	04 58·4	+43 45	3·3	4·6	989·8	Eclips.	2
Zeta Aurigæ	04 59·0	+41 00	5·0	5·7	972·2	Eclips.	2
T Leporis	05 02·7	−21 58	7·4	13·5	368·1	M	4
W Orionis	05 02·8	+01 07	5·9	7·7	212	SR	2
R Aurigæ	05 30·3	+53 32	6·7	13·7	458·9	M	1–2
T Columbæ	05 17·5	−33 45	6·6	12·7	225·3	M	3
S Orionis	05 26·5	−04 44	7·5	13·5	416·3	M	4
Beta Doradûs	05 33·2	−62 31	4·5	5·7	9·8	C	3–4
U Aurigæ	05 38·9	+32 01	7·5	15·5	407·3	M	1–2
R Octantis	05 41·1	−86 26	6·4	13·2	405·1	M	3
Alpha Orionis	05 52·5	+07 24	0·1	1·0	±2070	SR	1
U Orionis	05 52·9	+20 10	5·3	12·6	272·2	M	1–2
Eta Geminorum	06 11·9	+22 31	3·1	3·9	233·4	SR	1
V Monocerotis	06 20·2	−02 10	6·0	13·7	334·7	M	3
T Monocerotis	06 22·5	+07 07	6·3	7·8	27·0	C	1
X Geminorum	06 43·9	+30 20	7·6	13·6	263·5	M	1
R Lyncis	06 57·2	+55 24	7·2	14·0	378·6	M	1
Zeta Geminorum	07 01·1	+20 39	3·7	4·4	10·2	C	1
R Geminorum	07 04·3	+22 47	6·0	14·0	369·9	M	1
R Canis Min.	07 06·0	+10 06	7·4	11·6	337·9	M	1
VZ Camelopardalis	07 20·7	+82 31	4·8	5·2	23·7	SR	1
Z Puppis	07 30·5	−20 33	7·2	14·6	509·9	M	3
T Geminorum	07 46·3	+23 52	8·0	15·0	287·6	M	1
V Puppis	07 56·8	−49 07	4·5	5·1	1·5	EB	3
R Cancri	08 13·8	+11 53	6·2	11·8	362·1	M	1
V Cancri	08 18·9	+17 27	7·5	13·9	272·1	M	1
V Carinæ	08 27·7	−59 57	7·4	8·1	6·7	C	3
T Velorum	08 36·1	−47 11	8·3	9·1	4·6	C	3
S Hydræ	08 51·0	+03 16	7·4	13·3	256·7	M	2
X Cancri	08 52·6	+17 25	5·9	7·3	170?	SR	1
T Pyxidis	09 02·6	−32 11	7·0	14·0		RN	3[1]
W Cancri	09 06·9	+25 27	7·4	14·4	393·3	M	1
R Carinæ	09 31·0	−62 34	3·9	10·0	308·6	M	3
R Leonis Mi.	09 42·6	+34 45	6·3	13·2	372·3	M	1
X Carinæ	09 43·9	−62 17	4·2	5·5	35·5	C	3
R Leonis	09 44·9	+11 40	5·4	10·5	312·6	M	1
R Ursæ Maj.	10 41·1	+69 02	6·7	13·4	301·7	M	1
Eta Carinæ	10 43·1	−59 25	−0·8	7·9		I	3
V Hydræ	10 49·2	−20 59	6·0	12·5	533·0	M	3
VW Ursæ Maj.	10 55·5	+70 16	7·2	7·8	125	SR	1
X Centauri	11 46·7	−41 28	7·0	13·9	314·6	M	3
Z Ursæ Maj.	11 53·9	+58 09	6·6	9·1	198	SR	1
R Corvi	12 17·0	−18 59	6·7	14·4	316·7	M	3
RY Ursæ Maj.	12 18·1	+61 36	7·0	8·0	311·2	SR	1
R Crucis	12 20·9	−61 21	6·8	8·0	5·8	C	3
U Centauri	12 30·7	−54 23	7·2	14·0	220·2	M	3
T Ursæ Maj.	12 34·1	+59 46	6·6	13·4	256·9	M	1
R Virginis	12 36·0	+07 16	6·2	12·1	145·6	M	1
S Ursæ Maj.	12 41·8	+61 22	7·4	12·3	226·1	M	1
RY Draconis	12 54·5	+66 16	6·5	8·0	172·5	SR	1
S Virginis	13 30·4	−06 56	6·3	13·2	378·0	M	3
T Centauri	13 38·9	−33 21	5·5	9·0	90·6	SR	3
Theta Apodis	14 00·4	−76 33	6·4	8·6	119	SR	3
R Centauri	14 12·9	−59 41	5·4	11·8	546·6	M	3
S Boötis	14 21·2	+54 02	8·0	13·8	270·7	M	1
R Camelopardalis	14 21·3	+84 04	7·9	14·4	269·7	M	1
R Boötis	14 35·0	+26 57	6·7	12·8	223·3	M	1
Delta Libræ	14 58·3	−08 19	4·8	5·9	2·3	EA	3
S Coronæ Bor.	15 19·4	+31 33	6·6	14·0	360·7	M	1
R Normæ	15 32·3	−49 21	6·5	13·9	490·2	M	3
R Coronæ Bor.	15 46·5	+28 19	5·8	15		RCrB	1
R Serpentis	15 48·4	+15 17	5·7	14·4	356·8	M	1
T Coronæ Bor.	15 57·4	+26 04	2·0	10·8		RN	1
R Herculis	16 04·0	+18 30	8·2	15·0	318·5	M	1
U Serpentis	16 04·9	+10 04	7·8	14·0	238·2	M	1
W Coronæ Bor.	16 13·6	+37 54	7·8	14·3	238·0	M	1
U Herculis	16 23·6	+19 00	7·0	13·4	406·0	M	1
R Draconis	16 32·5	+66 51	6·9	13·0	245·6	M	1
W Herculis	16 33·4	+37 27	7·7	14·4	279·8	M	1
S Herculis	16 49·6	+15 01	7·0	13·8	307·4	M	1
RR Scorpii	16 53·4	−30 30	5·0	12·4	278·7	M	3
R Ophiuchi	17 04·9	−16 02	7·0	13·6	302·5	M	3
Alpha Herculis	17 12·1	+14 27	3·0	4·0		SR	1
S Octantis	17 46·0	−86 48	7·4	14·0	258·8	M	3
T Draconis	17 55·7	+58 14	7·2	13·5	421·7	M	1
R Pavonis	18 08·1	−63 38	7·5	13·8	230·3	M	4
W Lyræ	18 13·2	+36 39	7·5	13·0	196·4	M	1–2
T Lyræ	18 30·7	+36 58	7·8	9·6		I	1–2
R Scuti	18 44·8	−05 46	6·3	8·6	144	RV	4
Beta Lyræ	18 48·2	+33 18	3·4	4·3	12·9	EB	2
Kappa Pavonis	18 51·8	−67 18	4·0	5·5	9·1	C	4
R Lyræ	18 53·8	+43 53	4·0	5·0	±47	SR	1–2
V Aquilæ	19 01·7	−05 46	6·7	8·0	353	SR	4
R Sagittarii	19 13·8	−19 24	6·7	12·8	268·6	M	4
U Sagittæ	19 16·6	+19 31	6·4	9·0	3·4	EA	2
CH Cygni	19 23·2	+50 09	6·6	7·8	97	SR	2
RR Lyræ	19 23·9	+42 41	6·9	8·0	0·6	RR	2
U Aquilæ	19 26·7	−07 09	6·8	8·0	7·0	C	4
R Cygni	19 35·5	+50 05	6·5	14·2	426·3	M	2
T Pavonis	19 45·1	−71 54	7·0	14·0	244·1	M	4
Chi Cygni	19 48·6	+32 47	3·3	14·2	407·C	M	2
Eta Aquilæ	19 49·9	+00 53	4·1	5·3	7·2	C	2
Z Cygni	20 00·0	+49 54	7·6	14·7	263·8	M	2
R Delphini	20 12·5	+08 56	7·6	13·7	284·5	M	2
P Cygni	20 15·9	+37 53	3	6		I	2
U Cygni	20 18·1	+47 44	6·7	11·4	464·7	M	2
T Microscopii	20 24·9	−28 26	7·7	9·6	347	SR	4
EU Delphini	20 35·6	+18 06	6·0	6·9	59	SR	2
HR Delphini Nova 1967	20 40·0	+18 59	3·6	12?		N	2
U Delphini	20 43·2	+17 54	5·6	7·5		I	2
W Aquarii	20 43·9	−04 16	8·7	14·9	381·0	M	4
T Aquarii	20 47·3	−05 20	7·2	14·2	202·1	M	4
R Vulpeculæ	21 02·2	+23 37	7·4	13·4	136·8	M	2
T Cephei	21 08·9	+68 17	5·4	11·0	389·3	M	2
T Indi	21 16·9	−45 14	7·7	9·4	320	SR	4
W Cygni	21 34·1	+45 09	5·0	7·6	130·9	SR	2
S Cephei	21 35·9	+78 24	7·4	12·9	487·5	M	2
SS Cygni	21 41·0	+43 21	8·4	12·2		I	2
Mu Cephei	21 42·0	+58 33	3·6	5·1		I	2
R Gruis	21 45·3	−47 09	7·4	14·9	332·5	M	4
Delta Cephei	22 27·3	+58 10	3·5	4·4	5·4	C	2
W Cephei	22 34·5	+58 10	6·9	8·6	1100?	SR	2
AR Cephei	22 52·6	+84 47	7·1	7·8	116	SR	2
S Aquarii	22 54·6	−20 37	7·6	15·0	379·2	M	4
Beta Pegasi	23 01·3	+17 49	2·1	2·9		I	2
R Pegasi	23 04·1	+10 16	7·1	13·8	377·5	M	2
V Cassiopeiæ	23 09·5	+59 25	7·3	12·8	227·9	M	2
R Aquarii	23 41·2	−15 34	5·8	11·5	386·9	M	4
Rho Cassiopeiæ	23 51·9	+57 13	4·1	6·2		I	2
R Phœnicis	23 53·9	−50 04	7·5	14·4	268·0	M	4
R Cassiopeiæ	23 55·9	+51 07	5·5	13·0	431·2	M	2
W Ceti	23 59·6	−14 58	7·1	14·6	350·9	M	4

[1] Outbursts 1920, 1944 [2] Outbursts 1866, 1946

Planetary Nebulæ not in Messier's Catalogue

With Integrated Magnitude above 8·5

NGC	R.A. h m	Decl. °	Dia. (secs. of arc)	Magnitude	Constellation	Map
246	00 44·6	−12 09	240 x 210	8·5	Cetus	4
2392	07 26·2	+21 01	47 x 43	8·3	Gemini	1
3132	10 04·9	−40 11	84 x 53	8·2	Antlia	3
3918	11 47·8	−56 54	32 x 13	8·4	Centaurus	3
7009	21 01·4	−11 34	44 x 26	8·4	Aquarius	4
7293	22 27·0	−21 06	900 x 720	6·5	Aquarius	4
I.1470	23 03·2	+59 59	70 x 45	8·1	Cepheus	2
7635	23 18·5	+60 54	205 x 180	8·5	Cassiopeia	2

Galaxies not in Messier's Catalogue

With Integrated Magnitude above 10.0

NGC	R.A. h m	Decl. °	Phot. Mag.	Diameter	Type	Constellation	Map
55	00 12·5	−39 30	7·8	25·0 x 3·0	S	Sculptor	4
205	00 37·6	+41 25	8·9	10·0 x 4·5	E6	Andromeda	2
247	00 44·6	−21 01	9·5	18·2 x 4·5	S	Cetus	4
253	00 45·1	−25 34	7·0	24·6 x 4·5	S	Sculptor	4
—	00 50	−73	1·5	216 x 216	I	Small Magellanic Cloud	3—4
—	05 26	−69	0·5	432 x 432	I	Large Magellanic Cloud	3—4
2403	07 32·0	+65 43	8·8	16·8 x 10·0	Sc	Camelopardalis	1
2903	09 29·3	+21 44	9·5	11·0 x 4·6	Sb	Leo	1
4449	12 25·9	+44 22	9·9	4·1 x 3·4	I	Canes Venatici	1
4632	12 39·8	+32 49	9·7	12·6 x 1·4	Sc	Canes Venatici	1
4945	13 02·4	−49 01	9·2	11·5 x 2·0	S	Centaurus	3
5128	13 22·4	−42 45	7·2	10·0 x 8·0	I	Centaurus	3
6822	19 42·1	−14 53	9·2	16·2 x 11·2	I	Sagittarius	4
6946	21 33·9	+59 58	9·7	9·0 x 7·5	Sc	Cygnus	2
7793	23 55·3	−32 51	9·7	6·0 x 4·0	S	Sculptor	4

Globular Clusters not in Messier's Catalogue

With Integrated Magnitude above 7·5

NGC	R.A. h m	Decl. °	Dia.	Magnitude (v—visual p—photographic)	Constellation	Map
104	00 21·9	−72 21	23·0	3·0p	Tucana	4
288	00 50·2	−26 52	10·0	7·2p	Sculptor	4
362	01 00·6	−71 07	5·3	6·8p	Tucana	4
1851	05 12·4	−40 05	5·3	8·1v	Columba	3
2808	09 10·9	−64 39	6·3	5·7v	Carina	3
3201	10 15·5	−46 09	7·7	7·4v	Vela	3
4833	12 56·0	−70 36	4·7	6·8v	Musca	3
5139	13 23·8	−47 03	23·0	3·7v	Omega Centauri	3
5897	15 14·5	−20 50	7·3	7·3v	Libra	3
6362	17 26·6	−67 01	6·7	7·1p	Ara	3
6388	17 32·6	−44 43	3·4	7·1p	Scorpius	3
6397	17 36·8	−53 39	19·0	4·7p	Ara	3
6541	18 04·4	−43 44	6·3	5·8p	Corona Australis	4
6723	18 56·2	−36 42	5·8	6·0p	Sagittarius	4
6752	19 06·4	−60 04	13·3	4·6p	Pavo	4

The Messier Objects *Identification uncertain

M	NGC	R.A. (1950.00) h m	Decl. o	Magnitude (vis.)	Size (Minutes of arc)	Distance	Constellation	Description
1	1952	5 31·5	+21 59	8·4	6 x 4	1050 pc	Tau	Crab nebula
2	7089	21 30·9	− 1 03	6·3	12	16 kpc	Aqr	Globular cluster
3	5272	13 39·9	−28 38	6·4	19	14 kpc	CVn	Globular cluster
4	6121	16 20·6	−26 24	6·4	23	2·3 kpc	Sco	Globular cluster
5	5904	15 16·0	+ 2 16	6·2	20	8·3 kpc	Ser	Globular cluster
6	6405	17 36·8	−32 11	5·3	26	630 pc	Sco	Galactic cluster
7	6475	17 50·6	−34 48	4:	50	250 pc	Sco	Galactic cluster
8	6523	18 00·1	−24 23	6:	90 x 40	1·5 kpc	Sgr	Lagoon nebula
9	6333	17 16·2	−18 28	7·3	6	7·9 kpc	Oph	Globular cluster
10	6254	16 54·5	− 4 02	6·7	12	5·0 kpc	Oph	Globular cluster
11	6705	18 48·4	− 6 20	6·3	12	1·7 kpc	Sct	Galactic cluster
12	6218	16 44·6	− 1 52	6·6	12	5·8 kpc	Oph	Globular cluster
13	6205	16 39·9	+36 33	5·7	23	6.9 kpc	Her	Globular cluster
14	6402	17 35·0	− 3 15	7·7	7	7·2 kpc	Oph	Globular cluster
15	7078	21 27·6	+11 57	6·0	12	15 kpc	Peg	Globular cluster
16	6611	18 16·0	−13 48	6·4	8	1·8 kpc	Ser	Gaseous nebula
17	6618	18 17·9	−16 12	7·0	46 x 37	1·8 kpc	Sgr	Omega nebula
18	6613	18 17·0	−17 09	7·5	7	1·5 kpc	Sgr	Galactic cluster
19	6273	16 59·5	−26 11	6·6	5	6·9 kpc	Oph	Globular cluster
20	6514	17 58·9	−23 02	9:	29 x 27	1·6 kpc	Sgr	Trifid nebula
21	6531	18 01·7	−22 30	6·5	12	1·3 kpc	Sgr	Galactic cluster
22	6656	18 33·3	−23 58	5·9	17	3·0 kpc	Sgr	Globular cluster
23	6494	17 54·0	−19 01	6·9	27	660 pc	Sgr	Galactic cluster
24	6603	18 15·5	−18 26	4·6	4	5·0 kpc	Sgr	Galactic cluster
25	IC4725	18 28·8	−19 17	6·5	35	600 pc	Sgr	Galactic cluster
26	6694	18 42·6	− 9 27	9·3	9	1·5 kpc	Sct	Galactic cluster
27	6853	19 57·4	+22 35	7·6	8 x 4	200 pc	Vul	Dumbbell
28	6626	18 21·5	−24 54	7·3	15	4·6 kpc	Sgr	Globular cluster
29	6913	20 22·1	+38 22	7·1	7	1·2 kpc	Cyg	Galactic cluster
30	7099	21 37·5	−23 25	8·4	9	13 kpc	Cap	Globular cluster
31	224	0 40·0	+41 00	4·8	160 x 40	700 kpc	And	Galaxy
32	221	0 40·0	+40 36	8·7	3 x 2	700 kpc	And	Galaxy
33	598	1 31·0	+30 24	6·7	60 x 40	700 kpc	Tri	Galaxy
34	1039	2 38·8	+42 34	5·5	30	440 pc	Per	Galactic cluster
35	2168	6 05·8	+24 21	5·3	29	870 pc	Gem	Galactic cluster
36	1960	5 32·0	+34 07	6·3	16	1·3 kpc	Aur	Galactic cluster
37	2099	5 49·1	+32 32	6·2	24	1·3 kpc	Aur	Galactic cluster
38	1912	5 25·3	+35 48	7·4	18	1·3 kpc	Aur	Galactic cluster
39	7092	21 30·4	+48 13	5·2	32	250 pc	Cyg	Galactic cluster
40*	—	12 21·0	+58 20	9·0, 9·3	—	—	UMa	Double Star
41	2287	6 44·9	−20 41	4·6	32	670 pc	CMa	Galactic cluster
42	1976	5 32·9	− 5 25	4:	66 x 60	460 pc	Ori	Orion nebula
43	1982	5 33·1	− 5 18	9:		460 pc	Ori	Orion nebula
44	2632	8 37·4	+20 00	3·7	90	158 pc	Cnc	Præsepe
45	—	3 44·1	+23 58	1·6	120	126 pc	Tau	Pleiades
46	2437	7 39·5	−14 42	6·0	27	1·8 kpc	Pup	Galactic cluster
47*	2422	7 34·3	−14 22	5·2	25	548 pc	Pup	Galactic cluster
48*	2548	8 11·2	− 5 38	5·5	35	480 pc	Hya	Galactic cluster
49	4472	12 27·2	+ 8 16	8·6	4 x 4	11 Mpc	Vir	Galaxy
50	2323	7 00·6	− 8 16	6·3	16	910 pc	Mor	Galactic cluster
51	5194	13 27·8	+47 27	8·1	12 x 6	2 Mpc	CVn	Whirlpool
52	7654	23 22·0	+61 19	7·3	13	2·1 kpc	Cas	Galactic cluster
53	5024	13 10·5	+18 26	7·6	14	20 kpc	Com	Globular cluster
54	6715	18 52·0	−30 32	7·3	6	15 kpc	Sgr	Globular cluster
55	6809	19 36·9	−31 03	7·6	15	5·8 kpc	Sgr	Globular cluster
56	6779	19 14·6	+30 05	8·2	5	14 kpc	Lyr	Globular cluster
57	6720	18 51·7	+32 58	9·3	1 x 1	550 pc	Lyr	Planetary nebula
58	4579	12 35·0	+12 05	8·2	4 x 3	11 Mpc	Vir	Galaxy
59	4621	12 39·5	+11 55	9·3	3 x 2	11 Mpc	Vir	Galaxy
60	4649	12 41·1	+11 49	9·2	4 x 3	11 Mpc	Vir	Galaxy
61	4303	12 19·4	+ 4 45	9·6	6	11 Mpc	Vir	Galaxy
62	6266	16 58·1	−30 03	8·9	6	6·9 kpc	Oph	Globular cluster
63	5055	13 13·6	+42 18	10·1	8 x 3	4 Mpc	CVn	Galaxy
64	4826	12 54·3	+21 57	6·6	8 x 4	6 Mpc	Com	Galaxy
65	3623	11 16·3	+13 22	9·5	8 x 2		Leo	Galaxy
66	3627	11 17·6	+13 16	8·8	8 x 2		Leo	Galaxy
67	2682	8 47·8	+12 00	6·1	18	830 pc	Cnc	Galactic cluster
68	4590	12 36·8	−26 29	9:	9	12 kpc	Hya	Globular cluster
69	6637	18 28·1	−32 23	8·9	4	7·2 kpc	Sgr	Globular cluster
70	6681	18 40·0	−32 21	9·6	4	20 kpc	Sgr	Globular cluster
71	6838	19 51·5	+18 39	9:	6	5·5 kpc	Sge	Globular cluster
72	6981	20 50·7	−12 44	9·8	5	18 kpc	Aqr	Globular cluster
73	6994	20 56·4	−12 50	9·0	3		Aqr	Star group
74	628	1 34·0	+15 32	10·2	8	8 Mpc	Psc	Galaxy
75	6864	20 03·2	−22 04	8·0	5	24 kpc	Sgr	Globular cluster
76	650	1 38·8	+51 19	12·2	2 x 1	2·5 kpc	Per	Planetary nebula
77	1068	1 40·3	− 0 13	8·9	2	16 Mpc	Cet	Galaxy
78	2068	5 44·2	+ 0 02	8·3	8 x 6	500 pc	Ori	Gaseous nebula
79	1904	5 22·2	−24 34	7·9	8	13 kpc	Lep	Globular cluster
80	6093	16 14·1	−22 52	7·7	5	11 kpc	Sco	Globular cluster
81	3031	9 51·5	+69 18	7·9	16 x 10	3·0 Mpc	UMa	Galaxy
82	3034	9 51·9	+69 56	8·8	7 x 2	3 Mpc	UMa	Galaxy
83	5236	13 34·2	−29 37	10·1	10 x 8	4 Mpc	Hya	Galaxy
84	4374	12 22·5	+13 10	9·3	3	11 Mpc	Vir	Galaxy
85	4382	12 22·9	+18 28	9·3	4 x 2	11 Mpc	Com	Galaxy
86	4406	12 23·7	+13 13	9·7	4 x 3	11 Mpc	Vir	Galaxy
87	4486	12 28·3	+12 40	9·2	3	11 Mpc	Vir	Galaxy
88	4501	12 29·5	+14 42	10·2	6 x 3	11 Mpc	Com	Galaxy
89	4552	12 33·1	+12 50	9·5	2	11 Mpc	Vir	Galaxy
90	4569	12 34·3	+13 26	10·0	6 x 3	11 Mpc	Vir	Galaxy
91*	4571	12 35·0	+14 02	0·2	—	—	Vir	Planetary nebula
92	6341	17 15·6	+43 12	6·1	12	11 kpc	Her	Globular cluster
93	2447	7 42·4	−23 45	6·0	18	1·1 kpc	Pup	Galactic cluster
94	4736	12 48·5	+41 24	7·9	5 x 4	6 Mpc	CVn	Galaxy
95	3351	10 41·3	+11 58	10·4	3	9 Mpc	Leo	Galaxy
96	3368	10 44·1	+12 05	9·1	7 x 4	9 Mpc	Leo	Galaxy
97	3587	11 12·0	+55 18	12·0	3	800 pc	UMa	Owl nebula
98	4192	12 11·3	+15 11	10·7	8 x 2	11 Mpc	Com	Galaxy
99	4254	12 16·3	+14 42	10·1	4	11 Mpc	Com	Galaxy
100	4321	12 20·4	+16 06	10·6	5	11 Mpc	Com	Galaxy
101	5457	14 01·4	+54 35	9·6	22	3 Mpc	UMa	Galaxy
102*	—	—	—	—	—	—	—	—
103	581	1 29·9	+60 27	7·4	6	2·6 kpc	Cas	Galactic cluster
104	4594	12 37·4	−11 21	8·7	7 x 2	4·4 Mpc	Vir	Sombrero nebula
105	3379	10 45·2	+12 51	9·2	2 x 2		Leo	Galaxy
106	4258	12 16·5	+47 35	8·6	20 x 6		UMa	Galaxy
107	6171	16 29·7	−12 57	9·2	8		Oph	Globular cluster
108	3556	11 08·7	+55 57	10·7	8 x 2		UMa	Galaxy
109	3992	11 55·0	+53 39	10·8	7		UMa	Galaxy

Bright Diffuse Nebulæ not in Messier's Catalogue

	R.A. h m	Decl. ° ′	Dia. (secs. of arc)	Magnitude	Constellation	Map	Description
NGC 1499	04 00·1	+36 17	145 x 40	4·0	Perseus	1	California Nebula
NGC 1554	04 19·9	+19 25	variable	var	Taurus	2	Hind's Nebula (T Tauri neb)
NGC 2070	05 39·9	−69 04	20 x 20		Dorado	3	Tarantula Nebula
2237/9	06 29·6	+04 40	64 x 61		Monoceros	1	Rosette Nebula
2261	06 26·4	+08 46	variable	var	Monoceros	1	Hubble's Nebula (R Monoc.)
2264	06 38·2	+09 57	60 x 30	4·7	Monoceros	1	Cone Nebula (S Monocerotis)
6960	20 43·6	+30 32	70 x 6		52 Cygnus	2	Cirrus Nebula
—	20 46·0	+31 30	40 x 20		Cygnus	2	Cirrus Nebula
1.5067/0	20 46·9	+44 11	85 x 75	1·3	Cygnus	2	Pelican Nebula (nr Deneb)
6992/5	20 44·3	+31 30	78 x 8		Cygnus	2	Cirrus Nebula
7000	20 57·0	+44 08	120 x 100	1·3	Cygnus	2	N. America Nebula

The Brightest Stars List includes all the stars above magnitude 2·00 (Epoch 1970·0)

Star		Magnitude (apparent)	Magnitude (absolute)	R.A. h m s	Declination ° ′ ″	Spectrum
Sirius	Alpha Canis Majoris	−1·43	+1·5	06 43 49·6	−16 40 25	A0
Canopus	Alpha Carinæ	−0·73	−8?	06 23 17·1	−52 40 44	F0
	Alpha Centauri	−0·27	+4·1	14 37 32·9	−60 42 46	G0+K5
Arcturus	Alpha Boötis	−0·06	−0·3	14 14 17·5	+19 20 16	K0
Vega	Alpha Lyræ	0·04	+0·5	18 35 55·3	+38 45 17	A0
Capella	Alpha Aurigæ	0·09	−0·6	05 14 28·2	+45 58 10	G0
Rigel	Beta Orionis	0·15	−8·2	05 13 05·7	− 8 14 06	B8p
Procyon	Alpha Canis Minoris	0·37	+2·7	07 37 43·9	+ 5 18 11	F5
Achernar	Alpha Eridani	0·58	−1·3	01 36 35·9	−57 23 20	B5
Betelgeux	Alpha Orionis	var.	var.	05 53 32·8	+ 7 24 10	M0
Agena	Beta Centauri	0·66	−4·3	14 01 41·4	−60 13 45	B1
Altair	Alpha Aquilæ	0·80	+2·4	19 49 19·1	+ 8 47 16	A5
Aldebaran	Alpha Tauri	0·85	−0·6	04 34 11·8	+16 27 01	K5
Acrux	Alpha Crucis	0·87	−3·4, −2·9	12 24 54·9	−62 55 59	B1+B1
Antares	Alpha Scorpii	0·98	−5·0	16 27 33·9	−26 22 01	M0+A3
Spica	Alpha Virginis	1·00	−2·9	13 23 36·6	−11 00 19	B2
Fomalhaut	Alpha Piscis Austrini	1·16	+2·0	22 55 59·7	−29 46 54	A3
Pollux	Beta Geminorum	1·16	+1·0	07 43 29·0	+28 06 00	K0
Deneb	Alpha Cygni	1·26	−6·2	20 40 24·4	+45 10 21	A2p
—	Beta Crucis	1·31	−4·5	12 45 57·3	−59 31 30	B1
Regulus	Alpha Leonis	1·36	−0·2	10 06 46·5	+12 06 52	B8
Castor	Alpha Geminorum	1·58	+0·9	07 32 41·2	+31 57 19	A0+A2
—	Gamma Crucis	1·61	−2·4	12 29 29·4	−56 56 44	M3
Adhara	Epsilon Canis Majoris	1·63	−4·4	06 57 26·8	−28 55 49	B1
Alioth	Epsilon Ursæ Majoris	1·68	−0·2	12 52 42·8	+56 07 20	A0p
Bellatrix	Gamma Orionis	1·70	−3·4	05 23 31·2	+ 6 19 26	B2
Shaula	Lambda Scorpii	1·71	−3·0	17 31 34·1	−37 05 01	B2
Avoir	Epsilon Carinæ	1·74	−3·0	08 21 54·0	−59 24 45	K0+B
Alnilam	Epsilon Orionis	1·75	−7·0	05 34 41·4	− 1 13 11	B0
Al Nath	Beta Tauri	1·78	−1·5	05 24 23·6	+28 35 01	B8
Miaplacidus	Beta Carinæ	1·80	−0·9	09 12 52·8	−69 35 37	A0
Atria	Alpha Trianguli Australe	1·88	−0·5	16 45 28·3	−68 58 31	K2
Mirphak	Alpha Persei	1·90	−3·9	3 22 10·3	+49 45 21	F5
Alkaid	Eta Ursæ Majoris	1·91	−1·6	13 46 21·6	+49 27 45	B3
Alnitak	Zeta Orionis	1·91	−6·4	05 39 14·6	− 1 57 26	B0
—	Gamma Velorum	1·92	−4·5	08 08 36·4	−47 14 51	Oap+W
Alhena	Gamma Geminorum	1·93	−0·1	06 35 58·8	+16 25 35	A0
Dubhe	Alpha Ursæ Majoris	1·95	−0·5	11 01 53·4	+61 54 48	K0
Kaus Australis	Epsilon Sagittarii	1·95	−1·3	18 22 10·8	−34 24 02	A0
Wezea	Delta Canis Majoris	1·98	−5·9	07 07 10·3	−26 20 40	F8p
Mirzam	Beta Canis Majoris	1·99	−4·5	06 21 22·7	−17 56 24	B1
Polaris	Alpha Ursæ Minoris	1·99v	−3·7	02 03 18·9	+89 07 34	F8

Galactic Clusters not in Messier's Catalogue

With Integrated Magnitude above 7·0

(Class: c=v. loose d=loose, poor e=moderately rich f=fairly rich g=rich and condensed)

NGC	R.A. h m	Declination ° ′	Diameter	Magnitude	Number of Stars	Class	Constellation	Map
752	01 54·7	+37 25	45	7·0	70	d	Andromeda	2
869	02 15·5	+56 55	36	4·4	350	f	Perseus (h Persei)	2
884	02 18·9	+56 53	36	4·7	300	e	Perseus (Chi)	2
1245	03 11·2	+47 03	30	6·9	40	e	Perseus	2
1444	03 45·6	+52 31	4	6·4	15		Perseus	2
1502	04 03·0	+62 11	7	5·3	15	e	Camelopardalis	1
1528	04 11·4	+61 07	25	6·2	80	e	Perseus	2
Mel 25	04 16·7	+15 31	330	0·8	40	c	Taurus (Hyades)	2
1647	04 43·2	+18 59	40	6·3	30	c	Taurus	2
1746	05 00·6	+23 44	45	6·0	60	e	Taurus	2
2169	06 05·7	+13 58	5	6·4	18	d	Orion	1
2175	06 06·8	+20 20	18	6·7	15		Orion	1
2244	06 29·7	+04 54	40	6·2	16	c	Monoceros	1
2264	06 38·4	+09 56	30	4·7	23		Monoceros	1
2281	06 45·8	+41 07	17	6·7	30	e	Auriga	1
2301	06 49·2	+00 31	15	5·8	60	d	Monoceros	1
2353	07 12·3	−10 12	20	5·3	25	d	Monoceros	3
2422	07 34·3	−14 22	25	4·5	50	d	Puppis	3
2423	07 34·8	−13 45	20	6·9	60	d	Puppis	3
2451	07 43·6	−37 51	45	3·6	50		Puppis	3
2477	07 50·5	−38 25	25	5·7	300	g	Puppis (Globular?)	3
2516	07 59·7	−60 44	60	3·0	80	g	Carina	3
3547	08 08·9	−49 07	15	5·1	50	d	Vela	3
2546	08 10·6	−37 29	40	4·6	50		Puppis	3
I.2931	08 38·8	−52 53	40	2·6	20	c	O Velorum	3
I.2395	08 43·4	−48 00	10	4·6	16	e	Vela	3
H.3	08 44·6	−52 36	7	6·2	35	e	Vela	3
3114	10 01·1	−59 53	30	4·4	100	e	Carina	3
3228	10 19·7	−51 28	30	6·5	12	f	Vela	3
I.2581	10 25·4	−57 23	5	5·2	35	f	Carina	3
I.2602	10 41·0	−64 03	70	1·6	32	c	Carina	3
3532	11 03·4	−58 24	60	3·3	130	f	Carina	3
3766	11 34·2	−61 13	10	5·1	60	g	Centaurus	3
Mel.111	12 22·6	+26 24	275	2·7	30	c	Coma Berenices	1
4755	12 50·6	−60 05	10	5·2	50	g	Crux	3
5460	14 04·5	−48 05	30	6·3	25	d	Centaurus	3
6025	15 59·4	−60 22	10	5·8	30	d	Triangulum Australe	3
6067	16 09·3	−54 05	15	6·7	120	f	Norma	3
6087	16 14·7	−57 47	20	6·0	35	d	Norma	3
6124	16 22·2	−40 35	25	6·3	120	e	Scorpius	3
6167	16 30·6	−49 30	18	6·4	110		Ara	3
6193	16 37·6	−48 40	20	5·0	30	e	Ara	3
6383	17 31·4	−32 33	6	5·5	12	e	Scorpius	3
6530	18 01·6	−24 20	10	6·3	25	e	Sagittarius	3, 4
6633	18 25·1	+06 32	20	4·9	65	d	Ophiuchus	1
I.4756	18 36·6	+05 26	70	5·1	80	d	Serpens	1
6716	18 51·6	−19 57	7	6·9	20		Sagittarius	4
6871	20 04·0	+35 38	37	5·6	60		27 Cygnus	2
6910	20 21·3	+40 37	8	6·7	40	d	Cygnus	2
I.1396	21 37·5	+57 16	50	5·1	30		Cepheus	2
7160	21 52·3	+62 22	7	6·6	25		Cepheus	2

Glossary

A

Aberration of starlight: The apparent displacement of a star from its true position in the sky, due to the fact that light has a definite velocity (186,000 miles per second). The Earth is moving round the Sun, and thus the starlight seems to reach it "at an angle". The apparent positions of stars may be affected by up to 20·5 seconds of arc.

Absolute magnitude: The apparent magnitude that a star would have if it were observed from a standard distance of 10 **parsecs**, or 32·6 **light-years**. The absolute magnitude of the Sun is +4·8.

Absolute zero: The lowest limit of temperature: —273·16 °C. This value is used as the starting-point for the Kelvin scale of temperature, so that absolute zero = 0° Kelvin.

Absorption of light in space: Space is not completely empty, as used to be thought. There is appreciable material spread between the planets, and there is also material between the stars; the light from remote objects is therefore absorbed and reddened. This effect has to be taken into account in all investigations of very distant objects.

Absorption spectrum: A spectrum made up of dark lines against a bright continuous background. The Sun has an absorption spectrum; the bright background or continuous spectrum is due to the Sun's brilliant surface (**photosphere**), while the dark absorption lines are produced by the solar atmosphere. These dark lines occur because the atoms in the solar atmosphere absorb certain characteristic wave-lengths from the continuous spectrum of the photosphere.

Acceleration: Rate of change of velocity. Conventionally, increase of velocity is termed acceleration; decrease of velocity is termed deceleration, or negative acceleration.

Aerolite: A **meteorite** whose composition is stony.

Aeropause: A term used to denote that region of the atmosphere where the air-density has become so slight as to be disregarded for all practical purposes. It has no sharp boundary, and is merely the transition zone between "atmosphere" and "space".

Air resistance: Resistance to a moving body caused by the presence of **atmosphere**. An artificial satellite will continue in orbit indefinitely only if its entire orbit is such that the satellite never enters regions where air resistance is appreciable.

Airglow: The faint natural luminosity of the night sky, due to reactions going on in the Earth's upper atmosphere.

Airy disk: The apparent size of a star's disk produced even by a perfect optical system. Since the star can never be focused perfectly, 84 per cent of the light will concentrate into a single disk, and 16 per cent into a system of surrounding rings.

Albedo: The reflecting power of a planet or other non-luminous body. A perfect reflector would have an albedo of 100 per cent.

Altazimuth mounting: A type of **telescope mounting** in which the telescope may move independently in elevation (i.e. about a horizontal axis) and in **azimuth** (i.e. about a vertical axis).

Altitude: The angular distance of a celestial body above the horizon, ranging from 0° at the horizon to 90° at the **zenith**.

Angström unit: The unit for measuring the wavelength of light and other electromagnetic vibrations. It is equal to one hundred-millionth part of a centimetre. Visible light ranges between about 7500Å (red) down to about 3900Å (violet).

Antenna: A conductor, or system of conductors, for radiating or receiving radio waves. Systems of antennæ coupled together to increase sensitivity, or to obtain directional effects, are known as antenna arrays and as radio telescopes when used in **radio astronomy**.

Aphelion: The orbital position of a planet or other body when farthest from the Sun.

Arc, degree of: One three hundred and sixtieth part of a full circle (360°).

Arc, minute of: One sixtieth part of a degree of arc. One minute of arc (1') is in turn divided into 60 seconds of arc (60").

Ashen light: The faint luminosity of the night side of the planet Venus, seen when Venus is in the crescent stage. It is probably a genuine phenomenon rather than a mere contrast effect, but its cause is not certainly known.

Asteroids: The minor planets, most of which move round the Sun between the orbits of Mars and Jupiter. Several thousands of asteroids are known; much the largest is Ceres (diameter 429 miles). Only one asteroid (Vesta) is ever visible with the naked eye.

Astrology: A pseudo-science which claims to link the positions of the planets with human destinies. It has no scientific foundation.

Astronomical unit: The distance between the Earth and the Sun. It is equal to 92,957,000 miles, usually rounded off to 93,000,000 miles.

Astrophysics: The application of the laws and principles of physics to all branches of astronomy. It has often been defined as "the physics and chemistry of the stars".

Atmosphere: The gaseous mantle surrounding a planet or other body. It can have no definite boundary, but merely thins out until the density is no greater than that of surrounding space.

Atom: The smallest unit of a chemical element which retains its own particular character. (Of the 92 elements known to occur naturally, hydrogen is the lightest and uranium is the heaviest.)

Aurora: Auroræ, or polar lights, occur at a height of about 80 to 100 km. and are the result of the ionization (see **Ion**) of the Earth's atmosphere in that region. Some years ago this was believed to be caused by charged particles sent out from the Sun, which produced the ionization by collision with the atoms of the upper atmosphere. The process is not now believed to be so straightforward, and is probably connected with the zones of trapped particles in the **Van Allen Zones** surrounding the Earth. It seems likely that the interaction of the solar particles releases particles from these Van Allen belts, and that it is the dumping of these particles into the atmosphere which causes the ionization.

Azimuth: The horizontal direction or bearing of a celestial body, reckoned from the north point of the observer's horizon. Because of the Earth's rotation, the azimuth of a body is changing all the time.

B

Baily's Beads: Brilliant points seen along the edge of the Moon's disk at a total solar **eclipse**, just before totality and again just after totality has ended. They are due to the Sun's light shining through valleys between mountainous regions on the limb of the Moon.

Barycentre: The centre of gravity of the Earth—Moon system. Because the Earth is 81 times more massive than the Moon, the barycentre lies within the terrestrial globe.

Binary star: A star made up of two components which are genuinely associated, and are moving round their common centre of gravity. They are very common; sometimes the components are more or less equal, sometimes very unequal. Widely separated components may have revolution periods of thousands of years; with other binaries, the separations are so small that the components are almost touching each other, and cannot be seen separately, though they can be detected by means of spectroscopy (see **Spectroscope**). See also **Eclipsing binary; Spectroscopic binary.**

Bode's Law: An empirical relationship between the distances of the planets from the Sun, discovered by J. D. Titius in 1772 and made famous by J. E. Bode. The law seems to be fortuitous, and without any real scientific basis.

Bolide: A brilliant **meteor**, which may explode during its descent through the Earth's atmosphere.

Bolometer: A very sensitive radiation detector, used to measure slight quantities of radiation over a very wide range of wavelengths.

C

Carbon-nitrogen cycle: The stars are not "burning" in the usual sense of the word; they are producing their energy by converting hydrogen into helium, with release of radiation and loss of mass. One way in which this conversion takes place is by a whole series of reactions, involving carbon and nitrogen as catalysts. It used to be thought that the Sun shone because of this process, but modern work has shown that another cycle, the so-called proton-proton reaction, is more important in stars of solar type. The only stars which do not shine because of the hydrogen-into-helium process are those which are at a very early or relatively late stage in their evolution.

Cassegrain reflector: A type of reflecting telescope (see **Reflector**) in which the light from the object under study is reflected from the main mirror to a convex secondary, and thence back to the eyepiece through a hole in the main mirror.

Celestial sphere: An imaginary sphere surrounding the Earth, concentric with the Earth's centre. The Earth's axis indicates the positions of the celestial poles; the projection of the Earth's equator on to the celestial sphere marks the celestial equator.

Centrifuge: A motor-driven apparatus with a long arm, at the end of which is a cage. When men (or animals) are put into the cage, and revolved and rotated at high speeds, it is possible to study effects comparable with the accelerations experienced in space-craft. All astronauts are given tests in a centrifuge during their training.

Cepheid: An important type of variable star. Cepheids have short periods of from a few days to a few weeks, and are regular in their behaviour. It has been found that the period of a Cepheid, is linked with its real luminosity: the longer the period, the more luminous the star. From this, it follows that once a Cepheid's period has been measured its distance can be worked out. Cepheids are luminous stars, and may be seen over great distances; they are found not only in our Galaxy, but also in external galaxies. The name comes from Delta Cephei, the brightest and most famous member of the class.

Chromatic aberration: A defect found in all lenses, resulting in the production of "false colour". It is due to the fact that light of all wavelengths is not bent or refracted equally; for instance, blue light is refracted more strongly than red, and so is brought to focus nearer the lens. With an astronomical telescope, the object-glass is compound – i.e. made up of several lenses composed of different kinds of glass. In this way chromatic aberration may be reduced, though it can never be entirely cured.

Chromosphere: The part of the Sun's atmosphere lying above the bright surface or **photosphere**, and below the outer **corona**. It is visible with the naked eye only during a total solar **eclipse**, when the Moon hides the photosphere; but by means of special instruments, it may be studied at any time.

Circular velocity: The velocity with which an object must move, in the absence of air resistance, in order to describe a circular **orbit** round its primary.

Circumpolar star: A star which never sets, but merely circles the celestial pole and remains above the horizon.

Clusters, stellar: A collection of stars which are genuinely associated. An *open cluster* may contain several hundred stars usually together with gas and dust; there is no particular shape to the cluster. *Globular clusters* contain thousands of stars, and are regular in shape; they are very remote, and lie near the edge of the Galaxy. Both open and globular clusters are also known in external galaxies. *Moving clusters* are made up of widely separated stars moving through space in the same direction and at the same velocity. (For instance, five of the seven bright stars in the Great Bear are members of the same moving cluster.)

Colour index: A measure of a star's colour, and hence of its surface temperature. The ordinary or visual **magnitude** of a star is a measure of the apparent brightness as seen with the naked eye; the photographic magnitude is obtained by measuring the apparent size of a star's image on a photographic plate. The two magnitudes will not generally be the same, because in the old standard plates red stars will seem less prominent than they appear to the eye. The difference between visual and photographic magnitude is known as the colour index. The scale is adjusted so that for a white star, such as Sirius, colour index = 0. A blue star will have negative colour index; a yellow or red star will have positive colour index.

Colures: Great circles on the celestial sphere. The equinoctial colure, for instance, is the great circle which passes through both celestial poles and also the **First Point of Aries** (vernal equinox), i.e. the point where the ecliptic intersects the celestial equator.

Coma: (1) The hazy looking patch surrounding the nucleus of a **comet**. (2) The blurred haze surrounding the images of stars on a photographic plate, due to optical defects in the equipment.

Collimator: An optical arrangement for collecting light from a source into a parallel beam.

Comet: A member of the **Solar System**, moving round the Sun in an orbit which is generally highly eccentric. It is made up of relatively small particles (mainly ices) together with tenuous gas; the most substantial part of the comet is the nucleus, which may be several kilometres in diameter. A comet's tail always points more or less away from the Sun, due to the effects of **solar wind**. There are many comets with short periods, all of which are relatively faint; the only bright comet with a period of less than a century is Halley's. The most brilliant comets have periods so long that they cannot be predicted. See also **Sun-grazers.**

Conjunction: (1) The apparent close approach of a planet to a star or to another planet; it is purely a line-of-sight effect, since the planet is very much closer to us than the star. (2) *Inferior conjunction* for Mercury and Venus; the position when the planet has the same right ascension as the Sun. See **Inferior planets.** (3) *Superior conjunction:* the position of a planet when it is on the far side of the Sun with respect to the Earth.

Constellation: A group of stars named after a living or a mythological character, or an animal or an inanimate object. The names are highly imaginative, and have no real significance. Neither is a constellation made up of stars which are genuinely associated with each other; the individual stars lie at very different distances from the Earth, and merely happen to be in roughly the same direction in space. The International Astronomical Union recognizes 88 separate constellations.

Corona: The outermost part of the Sun's atmosphere; it is made up of very tenuous gas at a very high temperature, and is of great extent. It is visible to the naked eye only during a total solar **eclipse**.

Coronagraph: A type of **telescope** designed to view the solar **corona** in ordinary daylight; ordinary telescopes are unable to do this, partly because of the sunlight scattered across the sky by the Earth's atmosphere, and partly because of light which is scattered inside the telescope — mainly by particles of dust. The coronagraph was invented by the French astronomer B. Lyot.

Cosmic rays: High-velocity particles reaching the Earth from outer space. The heavy cosmic-ray primaries are broken up when they enter the top part of the Earth's atmosphere, and only the secondary particles reach ground-level. There is still considerable doubt as to whether cosmic radiation will prove to be a major hazard in long-term space-flights.

Cosmology: The study of the universe as a whole; its nature, origin, evolution, and the relations between its various parts.

Counterglow: See **Gegenschein.**

Crab Nebula: The remnant of a **supernova** observed in 1054; an expanding cloud of gas, approximately 6000 **light-years** away according to recent measurements. It is important because it emits not only visible light, but also radio waves and X-rays. Much of the radio radiation is due to synchrotron emission (i.e. the acceleration of charged particles in a strong magnetic field). The Crab Nebula contains a **pulsar**, the first to be identified with an optical object.

Culmination: The time when a star or other celestial body reaches the observer's **meridian,** so that it is at its highest point (*upper culmination*). If the body is circumpolar, it may be observed to cross the meridian again twelve hours later (*lower culmination*). With a non-circumpolar object, lower culmination cannot be observed, as the object is then below the horizon.

Cybernetics: The study of methods of communication and control which are common to machines and to living organisms.

D

Day: In everyday language, the time taken for the Earth to spin once on its axis. (There are, however, several definitions of the term.) A *sidereal day* (see **Sidereal time**) is the rotation period measured with reference to the stars (23h.56m.4s.091). A *solar day* is the time interval between two successive noons; the length of the mean solar day is 24h.3m.56s.555 – rather longer than the sidereal day, since the Sun is moving eastward along the *ecliptic.* The *civil day* is, of course, taken to be 24 hours.

Declination: The angular distance of a celestial body north or south of the celestial equator. It may be said to correspond to latitude on the surface of the Earth.

Density: The mass of a given substance per unit volume. Taking water as unity, the density of the Earth is 5½.

Dichotomy: The exact half-phase of Mercury, Venus or the Moon.

Diffraction rings: Concentric rings surrounding the image of a star as seen in a **telescope.** They cannot be eliminated, since they are due to the wave-motion of light and the construction of a telescope. They are most evident in small instruments.

Direct motion: Bodies which move round the Sun in the same sense as the Earth are said to have *direct motion*; those which move in the opposite sense have *retrograde motion*. The term may also be applied to satellites of the planets. No planet or asteroid with retrograde motion is known, but there are various retrograde satellites and comets. The terms are also used with regard to the apparent movements of the planets in the sky. When moving eastward against the stars, the planet has direct motion; when moving westward, it is retrograding.

Diurnal motion: The apparent daily rotation of the sky from east to west. It is due to the real rotation of the Earth from west to east.

Doppler effect: The apparent change in the wavelength of light caused by the motion of the observer. When a light-emitting body is approaching the Earth, more light-waves per second enter the observer's eye than would be the case if the object were stationary; therefore, the apparent wavelength is shortened, and the light seems "too blue". If the object is receding, the wavelength is apparently lengthened, and the light is "too red". For ordinary velocities the actual colour changes are very slight, but the effect shows up in the spectrum of the object concerned. If the dark lines are shifted toward the red or long-wave, the object must be receding; and the amount of the shift is a key to the velocity of recession. Apart from the galaxies in our **Local Group,** all external systems show red shifts, and this is the observational proof that the universe is expanding. The Doppler principle also applies to radiations at radio wavelengths.

Double star: A star which is made up of two components. Some doubles are *optical*; that is to say, the components are not truly associated, and simply happen to lie in much the same direction as seen from Earth. Most double stars, however, are physically associated or binary systems (see **Binary stars**).

E

Earthshine: The faint luminosity of the night hemisphere of the Moon, due to light reflected on to the Moon from the Earth.

Eclipses: These are of two kinds: solar and lunar.
(1) A *solar eclipse* is caused by the Moon passing in front of the Sun. By coincidence, the two bodies appear almost equal in size. When the alignment is exact, the Moon covers up the Sun's bright disk for a brief period, either totally or partially (never more than about 8 minutes; usually much less). When the eclipse is total the Sun's surroundings – the **chromosphere, corona** and **prominences** – may be seen with the naked eye. If the Sun is not fully covered, the eclipse is *partial,* and the spectacular phenomena of totality are not seen. If the Moon is near its greatest distance from the Earth (apogee) it appears slightly smaller than the Sun, and at central alignment a ring of the Sun's disk is left showing round the body of the Moon; this is an *annular* eclipse, and again the phenomena of totality are not seen.
(2) A *lunar eclipse* is caused when the Moon passes into the shadow cast by the Earth; it may be either total or partial. Generally, the Moon does not vanish, as some sunlight is refracted on to it by way of the ring of atmosphere surrounding the Earth.

Eclipsing binary (or **Eclipsing variable**): A **binary star**, made up of two components moving round their common centre of gravity at an angle such that, as seen from the Earth, the components mutually eclipse each other. In the case of the eclipsing binary Algol, one component is much brighter than the other; every 2½ days the fainter star covers up the brighter, and the star seems to fade by more than a magnitude. Many eclipsing binaries are now known.

Ecliptic: The projection of the Earth's orbit on to the celestial sphere. It may also be defined as "the apparent yearly path of the Sun against the stars", passing through the constellations of the **Zodiac.** Since the plane of the Earth's orbit is inclined to the equator by 23½°, the angle between the ecliptic and the celestial equator must also be 23½°.

Ecosphere: The region round the Sun in which the temperatures are neither too hot nor too cold for life to exist under suitable conditions. Venus lies near the inner edge of the ecosphere; Mars is near the outer edge. The ecospheres of other stars will depend upon the luminosities of the stars concerned.

Electromagnetic spectrum: The full range of what is termed electromagnetic radiation: **gamma-rays, X-rays, ultra-violet,** visible light, infra-red and radio waves. Visible light makes up only a very small part of the whole electromagnetic spectrum. Of all the radiations, only visible light and some of the radio waves can pass through the Earth's atmosphere and reach ground-level.

Electron: A fundamental particle carrying a unit negative charge of electricity.

Electron density: The number of free electrons in unit volume of space. A free electron is not attached to any particular atom, but is moving independently.

Element: A substance which cannot be chemically split up into simpler substances; 92 elements are known to exist naturally on the Earth and all other substances are made up from these fundamental 92. Various extra elements have been made artificially, all of which are heavier than uranium (No. 92 in the natural sequence) and most of which are very unstable.

Elongation: The apparent angular distance of a planet from the Sun, or of a satellite from its primary planet.

Emission spectrum: A spectrum consisting of bright lines or bands. Incandescent gases at low density yield emission spectra.

Ephemeris: A table giving the predicted positions of a moving celestial body, such as a planet or a comet.

Epoch: A date chosen for reference purposes in quoting astronomical data. For instance, some star catalogues are given for "epoch 1950"; by the year 2000 the given positions will have changed slightly, because of the effects of **precession.**

Equation of time: The Sun does not move among the stars at a constant rate, because the Earth's orbit is not circular.

Astronomers therefore make use of a *mean* sun, which travels among the stars at a speed equal to the average speed of the real Sun. The interval by which the real Sun is ahead of or behind the mean sun is termed the *equation of time.* It can never exceed 17 minutes; four times every year it becomes zero.

Equator, celestial: The projection of the Earth's equator on to the **celestial sphere.** It divides the sky into two equal hemispheres.

Equatorial mount: A **telescope** mounting in which the instrument is set upon an axis which is parallel to the axis of the Earth; the angle of the axis must be equal to the observer's latitude. This means that to keep an object in view, the telescope need be turned only in **right ascension.** Most modern large telescopes have been equatorially mounted and clock-driven, but the advent of computers has created new interest in the **altazimuth mounting** for large optical telescopes. The new Russian 236-in. reflector is on an altazimuth mount.

Equinox: Twice a year the Sun crosses the celestial equator, once when moving from south to north (about 21 March) and once when moving from north to south (about 22 September). These points are known respectively as the vernal equinox, or **First Point of Aries,** and the autumnal equinox, or **First Point of Libra.** (The equinoxes are the two points at which the ecliptic cuts the celestial equator.)

Escape velocity: The minimum velocity at which an object must move in order to escape from the surface of a planet, or other body, without being given extra propulsion, and neglecting any air resistance. The escape velocity of the Earth is 7 miles per second, or about 25,000 m.p.h.; for the Moon it is only 1½ miles per second; for Jupiter, as much as 37 miles per second.

Exosphere: The outermost part of the Earth's atmosphere. It is very rarefied, and has no definite upper boundary, since it simply "thins out" into space.

F

Faculæ: Bright, temporary patches on the surface of the Sun, usually (though not always) associated with **sunspots.** Faculæ frequently appear in a position near which a spot-group is about to appear, and may persist for some time in the region of a group which has disappeared.

First Point of Aries: The vernal **equinox.** The right ascension of the vernal equinox is taken as zero, and the right ascensions of all celestial bodies are referred to it.

First Point of Libra: The autumnal equinox, described under **Equinox.**

Flares, solar: Brilliant outbreaks in the outer part of the Sun's atmosphere, usually associated with active **sunspot** groups. They send out electrified particles which may later reach the Earth, causing magnetic storms and auroræ (see **Aurora**); they are also associated with strong outbursts of solar radio emission. It has been suggested that the particles emitted by flares may present a hazard to astronauts who are in space or on the unprotected surface of the Moon.

Flare stars: Faint red dwarf stars which may brighten up by several magnitudes over a period of a few minutes, fading back to their usual brightness within an hour or so. It is thought that this must be due to intense flare activity in the star's atmosphere. Although the energies involved are much higher than for solar **flares,** it is not yet known whether the entire stellar atmosphere is involved, or only a small area, as in the case of flares on the Sun. Typical flare stars are UV Ceti and AD Leonis.

Flash spectrum: Just before the Moon completely covers the Sun at a total solar eclipse, the Sun's atmosphere is seen shining by itself, without the usual brilliant background of the **photosphere.** The dark lines in the spectrum then become bright, producing what is termed the flash spectrum. The same effect is seen just after the end of totality.

Flocculi: Patches on the Sun's surface, observed by instruments based on the

principle of the **spectroscope.** Bright flocculi are composed of calcium; dark flocculi are made up of hydrogen.

Fraunhofer lines: The dark absorption lines in the Sun's spectrum, named in honour of the German optician J. von Fraunhofer, who first studied and mapped them in 1814.

Focal length: The distance between a lens (or mirror) and the point at which the image of an object at infinity is brought to focus. The focal length divided by the aperture of the mirror or lens is termed the *focal ratio.*

Free fall: The normal state of motion of an object in space under the influence of the gravitational pull of a central body; thus the Earth is in free fall round the Sun, while an artificial satellite moving beyond the atmosphere is in free fall round the Earth. While no thrust is being applied, a lunar probe travelling between the Earth and the Moon is in free fall; the same applies for a probe in a transfer orbit between the Earth and another planet. While a vehicle is in free fall, an astronaut will have no apparent "weight", and will be experiencing zero gravity or weightlessness.

Fringe region: The upper part of the **exosphere.** Atomic particles in the fringe region have little chance of collision with each other, and to all intents and purposes they travel in free orbits, subject to the Earth's gravitation.

G

g: Symbol for the force of gravity at the Earth's surface. The acceleration due to gravity is 32 feet per second per second at sea-level.

Galaxies: Systems of stars. About 1000 million of them are within photographic range of the Palomar 200-in. reflector. They are of many kinds, and they differ widely in size and luminosity. The Galaxy which contains our Sun includes about 100,000 million stars, but is not exceptional in size; the Andromeda Galaxy (Messier 31), which is dimly visible to the naked eye, is considerably larger. Like many other galaxies – including our own – it is spiral in shape, with a well-defined nucleus and massive spiral arms. Also visible with the naked eye are the two southern Clouds of Magellan, which are irregular in outline. These clouds lie at about 180,000 light-years from us, and are among the closest of the outer systems; the distance of the Andromeda Spiral is 2,200,000 light-years. The most distant galaxies so far measured are more than 5000 million light-years away. Apart from the members of our **Local Group,** about 25 in all, the galaxies appear to be moving away from us and from each other, so that the entire universe appears to be in a state of expansion. Galaxies are of various shapes; some are spiral, some elliptical and some globular, while there are the curious "barred spirals" and also systems which are formless. **Seyfert galaxies** have condensed, bright nuclei and spiral arms. In some galaxies there is evidence of vast explosions which have taken place; and for reasons which are still obscure, some galaxies are very strong emitters of radio waves. There is also evidence of clustering. At the moment, we have to admit that very little is known about the evolutionary cycle of a typical galaxy, and neither do we know why spiral arms are produced.

Galaxy, the: The system of stars of which our Sun is a member. It contains about 100,000 million stars, arranged in a somewhat flattened form; the diameter is about 100,000 **light-years,** and the maximum breadth of the nucleus is about 20,000 light-years. The Sun lies about 25,000 light-years from the galactic centre. The Galaxy is a rather loose spiral, rotating round its centre; the Sun takes about 225 million years to complete one circuit.

Gamma-rays: Extremely short-wavelength electromagnetic radiations. Their wavelength is no more than a millionth the wavelength of visible light – even shorter than X-rays. Gamma-ray sources in the Galaxy have been detected, but not much is yet known about them. Rocket techniques have to be used, since gamma radiation coming from space is transformed and scattered in the Earth's **ionosphere** and **atmosphere** before reaching ground level.

Glossary

Gegenschein (or Counterglow): A very faint glow in the sky, exactly opposite to the Sun; it is very difficult to observe, and has never been satisfactorily photographed. Its origin is uncertain, but it is thought to be due to tenuous matter spread along the main plane of the **Solar System**, so that it may be associated with the **Zodiacal Light**.

Geocentric: Relative to the Earth as a centre: or, as measured with respect to the centre of the Earth.

Geocorona: A layer of very tenuous hydrogen surrounding the Earth near the uppermost limit of the atmosphere.

Geodesy: The science which deals with the Earth's form, dimensions, elasticity, mass, gravitation and allied topics.

Geophysics: The science dealing with the physics of the Earth and its environment. Its range extends from the interior of the Earth out to the limits of the magnetosphere. In 1957–8 an ambitious international programme, the International Geophysical Year (I.G.Y.), was organized to undertake intensive studies of geophysical phenomena at the time of a sunspot maximum. It was extended to 18 months, and was so successful that at the next sunspot minimum a more limited but still extensive programme was organized, the International Year of the Quiet Sun (I.Q.S.Y.).

Gibbous: A phase of the Moon or a planet which is more than half, but less than full.

Gravitation: The force of attraction which exists between all particles of matter in the universe. Particles attract each other with a force which is directly proportional to the product of their masses and inversely proportional to the square of the distance between them.

Great circle: A circle on the surface of a sphere (such as the Earth, or the celestial sphere) whose plane passes through the centre of the sphere. Thus a great circle will divide the sphere into two equal parts.

Green flash (or Green ray): When the Sun is setting, the last visible portion of the disk may flash brilliant green for a very brief period. This is due to effects of the Earth's atmosphere, and is best observed over a sea horizon. Venus has also been known to show a green flash when setting.

Greenwich Mean Time (G.M.T.): The time at Greenwich, reckoned according to the mean sun. It is used as the standard time throughout the world. Also known as Universal Time (U.T.).

Greenwich Meridian: The line of longitude which passes through the Airy transit circle at Greenwich Observatory. It is taken as longitude 0°, and is used as the standard throughout the world.

Gregorian reflector: A type of reflecting telescope (see **Reflector**) in which the incoming light is reflected from the main mirror on to a small concave mirror placed outside the focus of the main mirror; the light then comes back through a hole in the main mirror and is brought to focus. Gregorian reflectors are not now common.

H

Halation ring: A ring sometimes seen round a star image on a photograph. It is purely a photographic effect.

H-I and H-II regions: Clouds of hydrogen in the Galaxy. In *H-I regions* the hydrogen is neutral, and the clouds cannot be seen, but they may be studied by radio telescopes by virtue of the emission at a wavelength of 21 cm. In *H-II regions* the hydrogen is ionized (see **Ion**), generally in the presence of hot stars. The recombination of the ions and free electrons to form neutral atoms gives rise to the emission of light, by which the H-II regions can be seen.

Halo: (1) A luminous ring round the Sun or Moon, due to ice crystals in the Earth's upper atmosphere. (2) The *galactic halo:* the spherical-shaped star cloud round the main part of the **Galaxy**.

Hertzsprung-Russell Diagram (H/R Diagram): A diagram in which stars are plotted according to spectral type and luminosity. It is found that there is a well-defined band or **Main Sequence** from the upper left of the Diagram (very luminous bluish stars) down to the lower right (faint red stars); there is also a giant branch to the upper right, while the dim, hot **white dwarfs** lie to the lower left. H/R Diagrams have been of the utmost importance in studies of stellar evolution. If colour index is used instead of spectrum, the diagram is known as a colour-magnitude diagram.

Hohmann orbit: See **Transfer orbit**.

Hour angle: The time which has elapsed since a celestial body crossed the meridian of the observer.

Hour circle: A great circle on the celestial sphere which passes through both poles of the sky. The zero hour circle corresponds to the observer's meridian.

Hubble's Constant: It is known that the universe is expanding, and that the more distant galaxies are receding more quickly than those which are closer to us. Hubble's Constant relates the distance of a galaxy to its recessional velocity. The value generally accepted at present is 100 km./sec. per megaparsec.

I

Inferior planets: Mercury and Venus, whose orbits lie closer to the Sun than does that of the Earth. When their **right ascensions** are the same as that of the Sun, so that they are approximately between the Sun and the Earth, they reach *inferior conjunction.* If the declination is also the same as that of the Sun, the result will be a transit of the planet.

Infra-red radiations: Radiations with wavelength longer than that of red light (about 7500 Ångströms), which cannot be seen visually. Infra-red radiation coming from beyond the Earth is strongly absorbed by the atmosphere, so that researches have to be carried out either from high-altitude sites (such as mountain-tops) or by balloon or rocket techniques. The infra-red region extends up to the short-wave end of the radio part of the electromagnetic spectrum.

Ion: An atom which has lost or gained one or more electrons; it has a corresponding positive or negative electrical charge, since in a complete atom the positive charge of the nucleus is balanced out by the combined negative charge of the electrons. The process of producing an ion is termed *ionization.*

Ionosphere: The region above the **stratosphere**, from about 40 up to about 500 miles above sea-level. Ionization of the atoms in this region (see **Ion**) produces various layers such as the E and F layers, which reflect radio waves and make long-range communication over the Earth possible. Solar events have marked effects upon the ionosphere, and produce ionospheric storms; on occasion, radio communication is interrupted.

Irradiation: The effect which makes brightly lit or self-luminous bodies appear larger than they really are. For example, the Moon's bright crescent appears larger in diameter than the earth-lit part of the disk, though in fact this is not so.

J

Julian day: A count of the days, starting from 12 noon on 1 January 4713 B.C. The system was introduced by Scaliger in 1582. The "Julian" is in honour of Scaliger's father, and has nothing to do with Julius Cæsar or the Julian Calendar. Julian days are used by **variable star** observers, and for reckonings of phenomena which extend over very long periods of time.

K

Kepler's Laws of Planetary Motion: The three important laws announced by J. Kepler between 1609 and 1618. They are:
(1) The planets move in elliptical orbits, the Sun being situated at one focus of the ellipse, while the other focus is empty.
(2) The radius vector, or imaginary line joining the centre of the planet to the centre of the Sun, sweeps out equal areas in equal times.
(3) The squares of the sidereal periods of the planets are proportional to the cubes of their mean distances from the Sun (Harmonic Law).

Kiloparsec: 1000 **parsecs**, or 3260 **light-years**.

Kirkwood Gaps: Regions in the belt of **asteroids** between Mars and Jupiter in which almost no asteroids move. The gravitational influence of Jupiter keeps these zones "swept clear"; an asteroid which enters a Kirkwood region will be regularly perturbed by Jupiter until its orbit has been changed. They were first noted by the American mathematician Daniel Kirkwood.

L

Laser (Light Amplification by the Simulated Emission of Radiation): A device which emits a beam of light made up of rays of the same wavelength (coherent light) and in phase with each other. It can be extremely intense. Laser beams have already been reflected off the Moon.

Latitude, celestial: The angular distance of a celestial body from the nearest point on the **ecliptic**.

Librations, lunar: Though the Moon's rotation is captured with respect to the Earth, there are various effects, known as librations, which enable us to examine 59 per cent of the total surface instead of only 50 per cent, though no more than 50 per cent can be seen at any one time. There are three librations: in longitude (because the Moon's orbital velocity is not constant), in latitude (because the Moon's equator is inclined by 6° to its orbital plane), and diurnal (due to the rotation of the Earth).

Light-year: The distance travelled by light in one year. It is equal to 5,880,000,000,000 miles.

Limb: The edge of the visible disk of the Sun, Moon, a planet, or the Earth (as seen from space).

Local Group of galaxies: The group of which our **Galaxy** is a member. There are more than two dozen systems, of which the most important are the Andromeda Spiral, our Galaxy, the Triangulum Spiral, and the two Clouds of Magellan.

Longitude, celestial: The angular distance from the vernal equinox to the foot of a perpendicular drawn from a celestial body to meet the **ecliptic**. It is measured eastward along the ecliptic from 0° to 360°.

Lunation (Synodical month): The interval between successive new moons: 29 days 12 hours 44 minutes. See also **Synodic period**.

Lyot filter (monochromatic filter): A device used for observing the Sun's prominences, and other features of the solar atmosphere, without waiting for a total **eclipse**. It was invented by the French astronomer, B. Lyot.

M

Mach number: The velocity of a vehicle moving in an atmosphere divided by the velocity of sound in the same region. Near the surface of the Earth, sound travels at about 750 m.p.h.; Mach 2 would therefore be 2 x 750 = 1500 m.p.h.

Magnetic storm: A sudden disturbance of the Earth's magnetic field, shown by interference with radio communication as well as by variations in the compass needle. It is due to charged particles sent out from the Sun, often associated with solar **flares**. A *magnetic crochet* is a sudden change in the Earth's magnetic field due to changing conditions in the lower **ionosphere**. The crochet is associated with the flash phase of the flare, and commences with it; the storm is associated with the particles, which reach the Earth about 24 hours later.

Magnetosphere: The area of the Earth's magnetic field. It extends out to about 10 Earth radii, and has been studied by means of space-probes. Of the other planets, Jupiter has an extensive magnetosphere, but no magnetic fields have been detected for Mars or Venus.

Magnetohydrodynamics: The study of the interactions between a magnetic field and an electrically conducting fluid. The Swedish scientist H. Alfvén is regarded as the founder of magnetohydrodynamics.

Magnitude: This is really a term for "brightness", but there are several different types.
(1) *Apparent or visual magnitude:* the apparent brightness of a celestial body as seen with the eye. The brighter the object, the lower the magnitude. The planet Venus is of about magnitude −4½; Sirius, the brightest star, −1·4; the Pole Star, +2; stars just visible with the naked eye, +6; the faintest stars that can be recorded with the world's largest telescopes, below +20. A star's apparent magnitude is no reliable key to its luminosity.
(2) *Absolute magnitude:* the apparent magnitude that a star would have if seen from a standard distance of 10 **parsecs** (32·6 **light-years**).
(3) *Photographic magnitude:* the magnitude derived from the size of a star's image on a photographic plate.
(4) *Bolometric magnitude:* this refers to the total radiation sent out by a star, not merely to visible light.

Main sequence: The well-defined band from the upper left to lower right of a **Hertzsprung-Russell Diagram**. The Sun is a typical main sequence star.

Maser (Microwave Amplification by Simulated Emission of Radiation): The same basic principle as that of the **laser**, but applied to radio wavelengths rather than to visible light.

Mass: The quantity of matter that a body contains. It is not the same as weight, which depends upon local gravity; thus on the Moon, an Earthman has only one sixth of his normal weight, but his mass is unaltered.

Meridian, celestial: The great circle on the celestial sphere which passes through the **zenith** and both celestial poles. The meridian cuts the observer's horizon at the exact north and south points.

Messier numbers: Numbers given by the 18th-century French astronomer Messier to various nebulous objects, including open and globular clusters, gaseous nebulæ, and galaxies. Messier's catalogue contained slightly over 100 objects. His numbers are still used; thus the Andromeda Spiral is M.31, the Orion Nebula M.42, the Crab Nebula M.1 and so on.

Meteor: A small particle which enters the Earth's atmosphere (at a velocity of anything up to 45 miles per second) and becomes heated by friction, so that it destroys itself in the streak of luminosity known as a shooting-star. Few meteors penetrate below a height of 50 miles above the ground. Some meteors belong to definite showers; non-shower or sporadic meteors may appear from any direction at any moment. The origin of meteors is not definitely known, but there is a close association between comets and some of the meteor streams. It is probable that meteors represent cometary debris (see **Comet**).

Meteorite: A larger body, which is able to reach ground-level without being destroyed. There is a fundamental difference between meteorites and **meteors**; a meteorite seems to be more nearly related to an asteroid or minor planet. Meteorites may be stony (*aerolites*), iron (*siderites*) or of intermediate type. In a few cases meteorites have produced craters; the most famous example is the large crater in Arizona, which is almost a mile in diameter and which was formed in prehistoric times.

Meteoroids: the collective term for meteoritic bodies. It was once thought that they would present a serious hazard to space-craft travelling outside the Earth's atmosphere, but it now seems that the danger is very much less than was feared, even though it cannot be regarded as entirely negligible.

Micrometeorite: An extremely small particle, less than 1/250 of an inch in diameter, moving round the Sun. When a micrometeorite enters the Earth's atmosphere, it cannot produce a shooting-star effect, as its mass is too slight. Since 1957, micrometeorites have been closely studied from space-probes and artificial satellites.

Micron: A unit of length equal to one thousandth of a millimetre. There are 10,000 Ångströms to one micron. The usual symbol is u.

Midnight Sun: The Sun seen above the horizon at midnight. This can occur at some part of the year anywhere inside the Arctic or Antarctic Circles.

Milky Way: The luminous band stretching across the night sky. It is due to a line-of-sight effect; when we look along the main plane of the **Galaxy** (i.e. directly toward or away from the galactic centre) we see many stars in roughly the same direction. Despite appearances, the stars in the Milky Way are not closely crowded together. The term used to be applied to the Galaxy itself, but is now restricted to the appearance as seen in the night sky.

Millibar: The unit which is used as a measure of atmospheric pressure. It is equal to 1000 dynes per square centimetre. The standard atmospheric pressure is 1013·25 millibars (29·92 inches of mercury).

Minor planets: See **Asteroids.**

Molecule: A stable association of atoms: a group of atoms linked together. For example, a water molecule (H_2O) is made up of 2 hydrogen atoms and 1 atom of oxygen.

Month: (1) *Calendar month:* the month in everyday use.
(2) *Anomalistic month:* the time taken for the Moon to travel from one perigee to the next. (Perigee = the closest point to the Earth.)
(3) *Sidereal month:* the time taken for the Moon to complete one journey round the **barycentre,** with reference to the stars.

Multiple star: A star made up of more than two components, physically associated.

Nadir: The point on the celestial sphere immediately below the observer. It is directly opposite to the overhead point or **zenith.**

N

Nebula: A mass of tenuous gas in space, together with what is loosely termed "dust". If there are stars in or very near the nebula, the gas and dust will become visible, either because of straightforward reflection or because the stellar radiation excites the material to self-luminosity.
If there are no suitable stars, the nebula will remain dark, and will betray its presence only because it will blot out the light of stars lying beyond it. Nebulæ are regarded as regions in which fresh stars are being formed out of the interstellar material.

Neutrino: A fundamental particle which has no mass and no electric charge.

Neutron: A fundamental particle whose mass is equal to that of a **proton,** but which has no electric charge. Neutrons exist in the nuclei of all atoms apart from that of hydrogen.

Neutron star: A star made up principally or completely of **neutrons,** so that it will be of low luminosity but almost incredibly high density. Theoretically, a neutron star should represent the final stage in a star's career. It is now thought probable that the remarkable radio sources known as **pulsars** are in fact neutron stars.

Newtonian reflector: The common form of astronomical **reflector.** Incoming light is collected by a mirror, and directed on to a smaller flat mirror placed at 45°; the light is then sent to the side of the tube, where it is brought to a focus and the eyepiece is placed. Most small and many large reflectors are of Newtonian type.

Noctilucent clouds: Rare, strange clouds in the **ionosphere,** best seen at night when they continue to catch the rays of the Sun. They lie at altitudes of over 50 miles, and are quite different from normal clouds. It is possible that they are produced by meteoritic dust in the upper atmosphere.

Nodes: The points at which the orbit of a planet, a comet or the Moon cuts the plane of the **ecliptic,** either as the body is moving from south to north (*ascending node*) or from north to south (*descending node*). The line joining these two points is known as the *line of nodes.*

Nova: A star which suffers a sudden outburst, and flares up to many times its normal brightness for a few days, weeks or months before fading back to its former obscurity. The outburst affects only the outer layers of the star. Novæ have been seen in other galaxies as well as in our own. There is a probable link between novæ and irregular variable stars; there are even a few stars, termed *recurrent novæ,* which have been observed to suffer more than one outburst.

Nutation: A slight, slow "nodding" of the Earth's axis, due to the fact that the Moon is sometimes above and sometimes below the **ecliptic,** and therefore does not always pull on the Earth's equatorial bulge in the same direction as the Sun.
The result is that the position of the celestial pole seems to "nod" by about 9 seconds of arc to either side of its mean position, in a period of 18 years 220 days. Nutation is superimposed on the more regular shift of the celestial pole caused by **precession.**

O

Object-glass (objective): The main lens of a refracting telescope (see **Refractor**).

Obliquity of the ecliptic: The angle between the **ecliptic** and the celestial equator. Its value is 23°26'54". It may also be defined as the angle by which the Earth's axis is tilted from the perpendicular to the orbital plane.

Occultation: The covering up of one celestial body by another. Thus the Moon may pass in front of a star or (occasionally) a planet; a planet may occult a star; and there have been cases when one planet has occulted another – for instance, Venus occulted Mars in 1590. Strictly speaking, a solar **eclipse** is an occultation of the Sun by the Moon.

Opposition: The position of a planet when it is exactly opposite the Sun in the sky, and so lies due south at midnight.
At opposition, the Sun, the Earth and the planet are approximately aligned, with the Earth in the mid position. Obviously, the **inferior planets** (Mercury and Venus) can never come to opposition.

Orbit: The path of an artificial or natural celestial body. See also **Transfer orbit.**

Ozone: Triatomic oxygen (O_3). The ozone layer in the Earth's upper atmosphere absorbs many of the lethal short-wavelength radiations coming from space. Were there no ozone layer, it is unlikely that life on Earth could ever have developed.

P

Parallax, trigonometrical: The apparent shift of a body when observed from two different directions. The separation of the two observing sites is called the baseline. The Earth's orbit provides a baseline 186 million miles long (since the radius of the orbit is 93 million miles); therefore, a nearby star observed at a six-monthly interval will show a definite parallax shift relative to the more distant stars. It was in this way that Bessel, in 1838, made the first measurement of the distance of a star (61 Cygni). The method is useful out to about 300 light-years, beyond which the parallax shifts become too small to be detectable.

Parsec: The distance at which a star would show a parallax of one second of arc It is equal to 3·26 **light-years,** 206,265 astronomical units, or 19,150,000,000,000 miles. (Apart from the Sun, no star lies within one parsec of us.)

Penumbra: (1) The outer, relatively light parts of a **sunspot.**
(2) The area of partial shadow lying to either side of the main cone of shadow cast by the Earth. During a lunar **eclipse,** the Moon must move through the penumbra before reaching the main shadow (**umbra**). Some lunar eclipses are penumbral only.

Periastron: The point of the orbit of a member of a **binary** system in which the stars are at their closest to each other. The most distant point is termed **apastron.**

Perigee: The point in the orbit of the Moon or an artificial satellite at which the body is closest to the Earth. The most distant point is the **apogee.**

Perihelion: The point in the orbit of a member of the **Solar System** in which the body is at its closest to the Sun. The most distant point is the **aphelion.** The Earth reaches perihelion in early January.

Periodic time: See **Sidereal time.**

Perturbations: The disturbances in the orbit of a celestial body produced by the gravitational pulls of others.

Phases: The apparent changes in shape of the Moon and some planets depending upon the amount of the sunlit hemisphere turned toward us. The Moon, Mercury and Venus show complete phases, from new (invisible) to full. Mars can show an appreciable phase, since at times less than 90 per cent of its sunlit face is turned in our direction. The phases of the outer planets are insignificant.

Photometry: The measurement of the intensity of light. The device now used for accurate determinations of star magnitudes is the *photoelectric photometer,* which consists of a photoelectric cell used together with a **telescope.** (A photoelectric cell is an electronic device. Light falls upon the cell, and produces an electric current; the strength of the current depends on the intensity of the light.)

Photosphere: The bright surface of the Sun.

Planet: A non-luminous body moving round a star. There are nine known planets in our **Solar System,** some of which are attended by **satellites.** There is every reason to suppose that other stars, too, have families of planets. With a few relatively near stars, irregularities in their proper motion have led to the detection of associated invisible bodies which seem insufficiently massive to be stars, and which are probably planets. Barnard's Star, at 6 light-years from us, is an example of this. Unfortunately, a planet moving round another star will be much too faint to be observable with our present equipment.

Planetarium: An instrument used to show an artificial sky on the inner surface of a large dome, and to reproduce celestial phenomena of all kinds. A planetarium projector is extremely complicated, and is very accurate. The planetarium is an educational device, and has become very popular in recent years. Planetaria have been set up in many large cities all over the world, and are also used in schools and colleges.

Planetary nebula: A faint star surrounded by an immense "shell" of tenuous gas. More than 300 are known in our Galaxy. They are so called because their telescopic appearance under low magnification is similar to that of a planet.

Plasma: A gas consisting of ionized atoms (see **Ion**) and free electrons, together with some neutral particles. Taken as a whole, it is electrically neutral, and is a good conductor of electricity.

Poles, celestial: The north and south points of the **celestial sphere.**

Populations, stellar: Two main types of star regions. *Population I* areas contain a great deal of interstellar material, and the brightest stars are hot and white; it is assumed that star formation is still in progress. The brightest stars in *Population II* areas are red giants, well advanced in their evolutionary cycle; there are almost no hot, **white giant** stars, and there is little interstellar material, so that star formation has apparently ceased. Though no rigid boundaries can be laid down, it may be said that the arms of spiral galaxies are mainly of Population I; the central parts of spirals, as well as elliptical galaxies and globular clusters, are mainly of Population II.

Position angle: The apparent direction of one object with reference to another, measured from the north point of the main object through east (90°), south (180°) and west (270°).

Precession: The apparent slow movement of the celestial poles. It is caused by the pull of the Moon and the Sun upon the Earth's equatorial bulge. The Earth behaves rather in the manner of a top which is running down and starting to topple, but the movement is very gradual; the pole describes a circle on the celestial sphere, centred on the pole of the ecliptic, which is 47° in diameter and takes 25,800 years to complete. Because of precession, the celestial equator also moves, and this in turn affects the position of the **First Point of Aries** (vernal equinox), which shifts eastward along the **ecliptic** by 50 seconds of arc each year. Since ancient times, this motion has taken the vernal equinox out of Aries into the adjacent constellation of Pisces (the Fishes). Our present Pole Star will not retain its title indefinitely. In A.D.12,000, the north polar star will be the brilliant Vega, in Lyra.

Prism: A glass block having flat surfaces inclined to one another. Light passing through a prism will be split up, since different colours are refracted by different amounts.

Prominences: Masses of glowing gas, chiefly hydrogen, above the Sun's bright surface. They are visible with the naked eye only during a total solar **eclipse,** but modern equipment allows them to be studied at any time. They are of two main types, *eruptive* and *quiescent.*

Proper motion: The individual motion of a star on the **celestial sphere.** Because the stars are so remote, their proper motions are slight. The greatest known is that of Barnard's Star (a red dwarf at a distance of 6 light-years); this amounts to one minute of arc every 6 years, so that it will take 180 years to move by an amount equal to the apparent diameter of the Moon. The proper motions of remote stars are too slight to be measured at all.

Proton: A fundamental particle with unit positive electrical charge. The nucleus of a hydrogen atom consists of one proton. See also **Neutron.**

Pulsar: A radio source which does not emit continuously, but in rapid, very regular pulses. Their periods are short (often much lsss than one second). In the light of recent evidence it now seems certain that they are **neutron stars.**

Purkinje effect: An effect inherent in the human eye, which makes it less sensitive to light of longer wavelength when the general level of intensity is low. Consider two lights, one red and one blue, which are of equal intensity. If now both are reduced by equal amounts, the blue light will appear to be the brighter of the two.

Q

Quadrature: The position of the Moon or a planet when at right angles to the Sun as seen from Earth. Thus the Moon is in quadrature when it is at half-phase.

Quantum: The smallest amount of light-energy which can be transmitted at any given wavelength.

Quasar: A very remote, almost incredibly luminous object which emits strong radio radiation as well as visible light. Quasars were first identified in 1963; their nature is still very much of a mystery. The tremendous red shifts in their spectra are interpreted as **Doppler effects,** in which case some quasars must be at least 6000 million light-years away. To appear as bright as they actually do, they must shine as brilliantly as perhaps 200 complete galaxies! They are also receding at high velocities, and some of them are variable at both optical and radio wavelengths. A few authorities believe that the spectral red shifts of quasars are not cosmological, in which case the quasars would be less remote and less luminous than is generally thought; at the moment, however, this is the minority view.
The relationship between quasars and galaxies is uncertain, though it has been suggested that quasars and **Seyfert galaxies** represent two stages in the evolution of the same class of object. Neither have we any plausible explanation of the way in which quasars produce their tremendous amounts of energy. In some cases the energy output appears to be equivalent to the total annihilation of 100 million suns occurring in volumes of space only a few light-years in diameter.

R

Radar astronomy: A branch of astronomy developed since the end of World War II. It depends on obtaining radar echoes from various celestial bodies. Among objects contacted by radar are the Moon, Venus, Mars, Mercury and the Sun; indeed, radar methods have given us our only positive information about the topography of Venus.
The distances of heavier celestial bodies

Glossary

can be determined by measuring the length of time required for a radar pulse to return from these bodies. Radar measurements have enabled a more accurate value of the distance of Venus, and hence of the astronomical unit, to be determined. Meteors can also be studied by radar, since, when a meteor burns up in the atmosphere, it leaves a trail of ionization, and the free electrons are efficient scatterers of the radar waves. Radar methods are, of course, all-important in the Apollo project of landing men on the Moon, since the final descent of the lunar module depends upon accurate radar measures.

Radial velocity: The toward-or-away movement of a celestial body, measured by the **Doppler effect** in its spectrum. If the spectral lines are red-shifted, the object is receding; if the shift is to the blue, the object is approaching. Conventionally, radial velocity is said to be positive with a receding body, negative with an approaching body.

Radiant: The point in the sky from which the **meteors** of any particular shower appear to radiate (for instance, the August shower has its radiant in Perseus, so that the meteors are known as the Perseids). The meteors in a shower are really moving through space in parallel paths, so that the radiant effect is due merely to perspective.

Radio astronomy: Astronomical studies carried out in the long-wavelength region of the **electromagnetic spectrum**. The main instruments used are known as *radio telescopes;* they are of many kinds, ranging from "dishes", such as the 250-foot paraboloid at Jodrell Bank (Cheshire), to long lines of aerials.

Radio galaxies: Galaxies which are extremely powerful emitters of radio radiation. The source of this energy is not known; the old theory of colliding galaxies has had to be rejected, and it now seems more likely that the radio emission is associated with some violent disruptive event in the nucleus of the galaxy.

Red shift: The **Doppler** displacement of spectral lines toward the red or long-wave end of the spectrum, indicating a velocity of recession. Apart from the members of the **Local Group**, all galaxies show red shifts in their spectra.

Reflector: A **telescope** in which the light is collected by means of a mirror.

Refraction: The change in direction of a ray of light when passing from one transparent substance into another.

Refractor: A **telescope** which collects its light by means of a lens. The light passes through this lens (**object-glass**) and is brought to focus; the image is then magnified by an eyepiece.

Resolving power: The ability of a **telescope** to separate objects which are close together; the larger the telescope the greater is resolving power. Radio telescopes (see **Radio astronomy**) have poor resolving power compared with optical telescopes.

Retardation: The difference in the time of moonrise between one night and the next. It may exceed one hour, or it may be as little as a quarter of an hour.

Retrograde motion: In the Solar System, movement in a sense opposite to that of the Earth in its orbit; some comets, notably Halley's, have retrograde motion. The term is also used with regard to the apparent movements of planets in the sky; when the apparent motion is from east to west, relative to the fixed stars, the direction is retrograde. The term may be applied to the rotations of planets. Since Uranus has an axial inclination of more than a right angle, its rotation is technically retrograde; Venus also seems to have retrograde axial rotation.

Reversing layer: The gaseous layer above the bright surface or **photosphere** of the Sun. Shining on its own, the gases would yield bright spectral lines; but as the photosphere makes up the background, the lines are reversed, and appear as dark absorption or **Fraunhofer lines.** Strictly speaking, the whole of the Sun's **chromosphere** is a reversing layer.

Right ascension: The right ascension of a celestial body is the time which elapses between the culmination of the **First Point of Aries** and the culmination

of the body concerned. For instance, Aldebaran in Taurus culminates 4h.33m. after the First Point of Aries has done so; therefore the right ascension of Aldebaran is 4h.33m.
The right ascensions of bodies in the **Solar System** change quickly. However, the right ascensions of stars do not change, apart from the slow cumulative effect of **precession.**

Roche limit: The distance from the centre of a planet, or other body, within which a second body would be broken up by gravitational distortion. This applies only to an orbiting body which has no appreciable structural cohesion, so that strong, solid objects, such as artificial satellites, may move safely well within the Roche limit for the Earth. The Roche limit lies at 2·44 times the radius of the planet from the centre of the globe, so that for the Earth it is about 5700 miles above ground-level. All known planetary natural satellites lie outside the Roche limits for their primaries; however, Saturn's ring system lies within the Roche limit for Saturn.

RR Lyræ variables: Regular **variable stars,** whose periods are very short (between about 1¼ hours and about 30 hours). They seem to be fairly uniform in luminosity, each is around 100 times as luminous as the Sun. They can therefore be used for distance measures, in the same way as **Cepheids.** Many of them are found in star-clusters, and they were formerly known as cluster-Cepheids. No RR Lyræ variable appears bright enough to be seen with the naked eye.

S

Saros: A period of 18 years 11·3 days, after which the Earth, Moon, and Sun return to almost the same relative positions. Therefore, an **eclipse** of the Sun or Moon is liable to be followed by a similar eclipse 18 years 11·3 days later. The period is not exact, but is good enough for predictions to be made – as was done in ancient times by the Greek philosophers.

Satellite: A secondary body moving round a planet. Jupiter has 12 satellites, Saturn 10, Uranus 5, and Neptune and Mars 2 each. The Moon appears too large to be a normal satellite of the Earth, and it is probably better to regard the Earth–Moon system as a double planet.

Schmidt telescope (or Schmidt camera): A type of **telescope** which uses a spherical mirror and a special glass correcting plate. With it, relatively wide areas of the sky may be photographed with a single exposure; definition is good all over the plate. In its original form, the Schmidt telescope can be used only photographically. The largest Schmidt in use is the 48-in. instrument at Palomar.

Scintillation: Twinkling of stars. It is due entirely to the effects of the Earth's atmosphere; a star will scintillate most violently when it is low over the horizon, so that its light is passing through a thick layer of atmosphere. A planet, which shows up as a small disk rather than a point, will generally twinkle much less than a star.

Seasons: Effects on the climate due to the inclination of the Earth's axis. The fact that the Earth's distance from the Sun is not constant has only a minor effect upon our seasons.

Seismometer: An earthquake recorder. Very sensitive seismometers were taken to the Moon by the Apollo astronauts, and provided interesting information about seismic conditions there.

Secular acceleration: Because of friction produced by the tides, the Earth's rotation is gradually slowing down; the "day" is becoming longer. The average daily lengthening is only 0·00000002 seconds, but over a sufficiently long period the effect becomes detectable. The lengthening of terrestrial time periods gives rise to an apparent speeding-up of the periods of the Sun, Moon and planets. Another result of these tidal phenomena is that the Moon is receding from the Earth by about 4 inches per month.

Seeing: The quality of the steadiness and clarity of a star's image. It depends upon conditions in the Earth's atmosphere. From the Moon, or from space, the "seeing" is always perfect.

Selenography: The study of the Moon's surface.

Sextant: An instrument used for measuring the altitude of a celestial body above the horizon.

Seyfert galaxies: Galaxies with small, bright nuclei. Many of them are radio sources, and show evidence of violent disturbances in their nuclei.

Shooting-star: The luminous appearance caused by a **meteor** falling through the Earth's atmosphere.

Sidereal period: The time taken for a planet or other body to make one journey round the Sun (365·2 days in the case of the Earth). The term is also used for a satellite in orbit round a planet. It is also known as *periodic time.*

Sidereal time: The local time reckoned according to the apparent rotation of the celestial sphere. It is 0 hours when the **First Point of Aries** crosses the observer's meridian. The sidereal time for any observer is equal to the right ascension of an object which lies on the meridian at that time. Greenwich sidereal time is used as the world standard (this is, of course, merely the local sidereal time at Greenwich Observatory).

Solar apex: The point on the celestial sphere toward which the Sun is apparently travelling. It lies in the constellation Hercules; the Sun's velocity toward the apex is 12 miles per second. The point directly opposite in the sky to the solar apex is termed the *solar antapex.* This motion is distinct from the Sun's rotation around the centre of the Galaxy, which amounts to about 200 miles per second.

Solar constant: The unit for measuring the amount of energy received on the Earth's surface by solar radiation. It is equal to 1·94 calories per minute per square centimetre. (A calorie is the amount of heat needed to raise the temperature of 1 gram of water by 1 °C.)

Solar flares: See **Flares, solar.**

Solar parallax: The trigonometrical parallax of the Sun. It is equal to 8·79 seconds of arc.

Solar System: The system made up of the Sun, the planets, satellites, comets, asteroids, meteoroids, and interplanetary dust and gas.

Solar time, apparent: The local time reckoned according to the Sun. Noon occurs when the Sun crosses the observer's meridian, and is therefore at its highest in the sky.

Solar wind: A steady flow of atomic particles streaming out from the Sun in all directions. It was detected by means of space-probes, many of which carry instruments to study it. Its velocity in the neighbourhood of the Earth exceeds 600 miles per second. The intensity of solar wind is enhanced during solar storms.

Solstices: The times when the Sun is at its northernmost point in the sky (declination 23½ °N, around 22 June) or southernmost point (23½ °S, around 22 December). The actual dates of the solstices vary somewhat, because of the calendar irregularities due to leap years.

Space medicine: A new field of medical science, concerned with the effects of space operations upon human subjects. Once beyond the atmosphere, an astronaut is exposed to all manner of radiations which can never reach the surface of the Earth; he is also existing under conditions of weightlessness, and in every way the environment is unfamiliar. Not many years ago it was thought unlikely that the human body could endure space conditions. However, astronauts who have been as far afield as the Moon have shown no ill-effects, and it is now certain that men can endure the conditions easily over limited periods of time. Whether this will also apply to longer periods – as, for instance, a journey to Mars – remains to be seen.
Medical studies in space research have also provided valuable general information, and there can be no doubt that a medical research base set up on the surface of the Moon will be of great benefit to mankind.

Space-suit (or pressure suit): Clothing able to give adequate protection to an astronaut who is in space, outside his space-craft, or who is on the surface of

the Moon. The space-suits worn by Apollo astronauts are reasonably flexible, and are very elaborate. It now seems certain that full space-suiting will be needed for an astronaut who walks about on the surface of Mars, where the atmospheric pressure is extremely low. On the other hand, Apollo-type space-suits would be of no use on the surface of Venus, where the pressure is so great.

Specific gravity: The density of any substance compared with that of an equal volume of water.

Spectroheliograph: An instrument used for photographing the Sun in the light of one particular wavelength only. If adapted for visual use, it is known as a *spectrohelioscope.*

Spectroscope: An instrument used to analyse the light from a star or other luminous object. Astronomical spectroscopes are used in conjunction with **telescopes.** Without them, our knowledge of the nature of the universe would still be very rudimentary.

Spectroscopic binary: A **binary star** whose components are too close together to be seen separately, but whose relative motions cause opposite **Doppler** shifts which are detectable spectroscopically.

Speculum: The main mirror of a reflecting telescope (see **Reflector**). Older mirrors were made of speculum metal; modern ones are generally of glass.

Spherical aberration: The blurred appearance of an image as seen in a telescope, due to the fact that the lens or mirror does not bring the rays falling on its edge and on its centre to exactly the same focus. If the spherical aberration is noticeable, then the lens or mirror is of poor quality, and should be corrected.

Spicules: Jets up to 10,000 miles in diameter, in the solar chromosphere. Each lasts for 4–5 minutes.

Spiral nebula: A now obsolete term for a spiral **galaxy.**

Star: A self-luminous gaseous body. The Sun is a typical star.

Steady-state theory: A theory of the universe according to which the universe has always existed, and will exist indefinitely; new matter is being spontaneously created at a rate which is sufficient to form new galaxies to take the place of those which are moving over our observational horizon because of the expansion of the universe.

Stratosphere: The layer in the Earth's atmosphere lying above the **troposphere.** It extends from about 7 to about 40 miles above sea-level.

Sublimation: The change of a solid body to the gaseous state without passing through a liquid condition. (This may well apply to the polar caps on Mars.)

Sundial: An instrument used to show the time, by using an inclined style, or gnomon, to cast a shadow on to a graduated dial. The gnomon points to the celestial pole. A sundial gives apparent time; to obtain mean time, the value shown on the dial must be corrected by applying the equation of time.

Sunspots: Darker patches on the solar **photosphere;** their temperature is around 4000 °C (as against about 6000 °C for the general photosphere), so that they are dark only by contrast; if they could be seen shining on their own, their surface brilliance would be greater than that of an arc-light. A large sunspot consists of a central darkish area or **umbra,** surrounded by a lighter area of **penumbra,** which may be very extensive and irregular. Sunspots tend to appear in groups, and are associated with strong magnetic fields; they are also associated with **faculæ** and with solar **flares.** They are most common at the time of solar maximum (approximately every 11 years). No sunspot lasts for more than a few months at most.

Sun-grazers: **Comets** which at perihelion make very close approaches to the Sun. All the sun-grazers are brilliant comets with extremely long periods.

Supergiant stars: Stars of exceptionally low density and great luminosity. Betelgeux in Orion is a typical supergiant.

Superior planets: The planets beyond the orbit of the Earth in the **Solar System:**

that is to say, all the principal planets apart from Mercury and Venus.

Superior conjunction: The position of a planet when it is on the far side of the Sun as seen from Earth.

Supernova: A star which suffers a cataclysmic explosion, sending much of its material away into space and never reverting to its old form. Supernovæ are of two main types. One class attains a maximum luminosity of 100 million suns; the other attains a luminosity of 10 million suns. Four supernovæ have been observed in our Galaxy during recorded times; many have been recorded in external galaxies.
Supernova remnants are strong radio sources, and it is thought that many of the radio sources in our Galaxy are of this nature. On average each major galaxy seems to produce one supernova every 600 years or so.

Synchrotron radiation: Radiation emitted by charged particles moving at relativistic velocities in a strong magnetic field. Much of the radio radiation coming from the Crab Nebula is of this type.

Synchronous satellite: An artificial satellite moving in a west-to-east equatorial orbit in a period equal to that of the Earth's axial rotation (approximately 24 hours); as seen from Earth, the satellite appears to remain stationary and is of great value as a communications relay. Various synchronous satellites are now in orbit.

Synodic period: The interval between successive **oppositions** of a **superior planet.** For an **inferior planet,** the term is taken to mean the interval between successive conjunctions with the Sun.

Syzygy: The position of the Moon in its orbit when at new or full phase.

T

Tektites: Small, glassy objects found in a few restricted areas on the Earth, notably in Australia. They are aero-dynamically shaped, and seem to have been heated twice. Their origin remains very much of a mystery. Most authorities, though not all, consider that they have come from space.

Telemetry: The technique of transmitting the results of measurements and observations made from instruments in inaccessible positions (such as unmanned probes in orbit) to a point where they can be used and analysed.

Telescope: The main instrument used to collect the light from celestial bodies, thereby producing an image which can be magnified. There are two main types, the **reflector** and the **refractor.** All the world's largest telescopes are reflectors, because a mirror can be supported by its back, whereas a lens has to be supported round its edge — and if it is extremely large, it will inevitably sag and distort under its own weight, thereby rendering itself useless.

Terminator: The boundary between the day and night hemispheres of the Moon or a planet. Since the lunar surface is mountainous, the terminator is rough and jagged, and isolated peaks may even appear to be detached from the main body of the Moon. Mercury and Venus, which also show lunar-type phases, seem to have almost smooth terminators, but this is probably because we cannot see them in such detail (at least in the case of Mercury, whose surface is likely to be as mountainous as that of the Moon). Mars also shows a smooth terminator, though it is now known that the surface of the planet is far from being smooth and level. Photographs of the Earth taken from space or from the Moon show a smooth terminator which appears much "softer" than that of the Moon, because of the presence of atmosphere.

Thermocouple: An instrument used for measuring very small quantities of heat. When used in conjunction with a large telescope, it is capable of detecting remarkably feeble heat-sources.

Tides: The regular rise and fall of the ocean waters, due to the gravitational pulls of the Moon and (to a lesser extent) the Sun.

Time dilation effect: According to relativity theory, the "time" experienced by two observers in motion compared with each other will not be the same. To an observer moving at near the velocity of light, time will slow down; also, the observer's mass will increase, until at the actual velocity of light time will stand still and mass will become infinite! The time and mass effects are entirely negligible except for very high velocities, and at the speeds of modern rockets they may be ignored completely.

Transfer orbit (or Hohmann orbit): The most economical **orbit** for a space-craft which is sent to another planet. To carry out the journey by the shortest possible route would mean continuous expenditure of fuel, which is a practical impossibility. What has to be done is to put the probe into an orbit which will swing it inward or outward to the orbit of the target planet. To reach Mars, the probe is speeded up relative to the Earth, so that it moves outward in an elliptical orbit; calculations are made so that the probe will reach the orbit of Mars and rendezvous with the planet. To reach Venus, the probe must initially be slowed down relative to the Earth, so that it will swing inward toward the orbit of Venus. With a probe moving in a transfer orbit, almost all the journey is carried out in free fall, so that no propellent is being used. On the other hand, it means that the distances covered are increased, so that the time taken for the journey is also increased.

Transit: (1) The passage of a celestial body, or a point on the celestial sphere, across the observer's meridian; thus the **First Point of Aries** must transit at 0 hours **sidereal time.**
(2) Mercury and Venus are said to be in transit when they are seen against the disk of the Sun at inferior conjunction. Transits of Mercury are not very infrequent (the next will be in 1973 and in 1986), but the next transit of Venus will not occur until 2004; the last took place in 1882. Similarly, a satellite of a planet is said to be in transit when it is seen against the planet's disk. Transits of the four large satellites of Jupiter may be seen with small telescopes; also visible are shadow transits of these satellites, when the shadows cast by the satellites are seen as black spots on the face of Jupiter.

Transit instrument: A **telescope** which is specially mounted; it can move only in elevation, and always points to the meridian. Its sole use is to time the moments when stars cross the meridian, so providing a means of checking the time. The transit instrument set up at Greenwich Observatory by Sir George Airy, in the 19th century, is taken to mark the Earth's prime meridian (longitude 0°). Transit instruments are still in common use, though it is likely that they will become obsolete before long.

Trojans: Asteroids which move round the Sun at a mean distance equal to that of Jupiter. One group of Trojans keeps well ahead of Jupiter and the other group well behind, so that there is no danger of collision. More than a dozen Trojans are now known.

Troposphere: The lowest part of the Earth's atmosphere, reaching to an average height of about 7 miles above sea-level. It includes most of the mass of the atmosphere, and all normal clouds lie within it. Above, separating the troposphere from the **stratosphere,** is the **tropopause.**

Twilight, astronomical: The state of illumination of the sky when the Sun is below the horizon, but by less than 18°.

Twinkling: The common term for **scintillation.**

U

Ultra-violet radiation: Electromagnetic radiation which has a wavelength shorter than that of violet light, and so cannot be seen with the naked eye. The ultra-violet region of the **electromagnetic spectrum** lies between visible light and X-radiation. The Sun is a very powerful source of ultra-violet, but most of this radiation is blocked out by layers in the Earth's upper atmosphere — which is fortunate for us, since in large quantities ultra-violet radiation is lethal. Studies of the ultra-violet radiations emitted by the stars have to be carried out by means of instruments sent up in rockets or artificial satellites.

Umbra: (1) The dark inner portion of a sunspot.
(2) The main cone of shadow cast by a planet or the Moon.

Universal time: The same as **Greenwich Mean Time.**

Universe, origin of: It is best to admit that we have no knowledge whatsoever about the origin of the universe. We do not know how matter came into existence; neither does it help to claim that the universe has always existed, so that there was no actual moment of creation. All we can do is to try to trace the evolution of the universe from its early stages. According to the evolutionary theory, all the matter in the universe came into existence at one moment, more than 10,000 million years ago. The "primæval atom", as it has been called, exploded and sent material outward in all directions; the universal expansion is still going on. On the **steady-state theory,** the universe has always existed, and new material is being spontaneously created all the time. The oscillating or cyclic theory seems to be gaining popularity at present; according to this idea, the universe alternately expands and contracts, so that the present stage of expansion will not continue indefinitely. A new cycle begins every 60,000 million years or so (though this figure may be much too large; some authorities prefer a mere 25,000 million years or so).

V

Van Allen Zones (or Van Allen Belts): Zones around the Earth in which electrically charged particles are trapped and accelerated by the Earth's magnetic field. They were detected by J. Van Allen and his colleagues in 1958, from results obtained with the first successful U.S. artificial satellite, Explorer I. Apparently there are two main belts. The outer, made up mainly of **electrons,** is very variable, since it is strongly affected by events taking place in the Sun; the inner zone, composed chiefly of **protons,** is more stable. On the other hand, it may be misleading to talk of two separate zones; it may be that there is one general belt whose characteristics vary according to distance from the Earth. The Van Allen radiation is of great importance in all geophysical research, and probably represents the major discovery of the first years of practical astronautics.

Variable stars: Stars which fluctuate in brightness over relatively short periods. They are of different kinds. *Eclipsing variables,* such as Algol and Beta Lyræ, are not truly variable at all, and are better termed eclipsing **binaries;** but all other types are intrinsically variable. **Cepheids** have short periods (up to a few weeks), and are very regular; their periods are linked with their real luminosities, so that their distances may be determined by simple observation. The **RR Lyræ variables** are of shorter period, and seem to be relatively uniform in luminosity. *Long-period variables,* of which the most famous example is Mira in Cetus, are usually (though not always) red giants; they have large magnitude-ranges, and periods of several months or, in some cases, well over a year. They are not perfectly regular, and there is no period-luminosity law as with the Cepheids. *Irregular variables* are not predictable; here, too, there are many different kinds — for instance, there are the U Geminorum or SS Cygni stars, nicknamed "dwarf novæ" because they suffer mild outbursts quite frequently, and the R Coronæ Borealis stars, which remain at maximum brightness for most of the time but which undergo sudden, unpredictable drops of many magnitudes. Variable stars are, in general, unstable; most of them are well advanced in their evolutionary careers, though this is by no means a firm rule. Studies of them make up an important part of modern astrophysical research.

Variation: (1) An inequality in the motion of the Moon, due to the fact that the Sun's pull on it throughout its orbit is not constant in strength.
(2) Magnetic variation: the difference, in degrees, between magnetic north and true north. It is not the same for all places on the Earth's surface, and it changes slightly from year to year because of the wandering of the magnetic pole.

Vernal Equinox: See **First Point of Aries.**

Vulcan: The name given to a hypothetical planet believed to move round the Sun at a distance less than that of Mercury. It is now virtually certain that Vulcan does not exist.

W

White dwarf: A very small, extremely dense star. The atoms in it have been broken up, and the various parts packed tightly together with almost no waste space, so that the density rises to millions of times that of water; a spoonful of white dwarf material would weigh many tons! Evidently a white dwarf has used up all its nuclear "fuel"; it is in the last stages of its active career, and has been aptly described as a bankrupt star. **Neutron stars** may be even smaller and denser than white dwarfs.

Widmanstätten patterns: If an iron **meteorite** is cut, polished and then etched with acid, characteristic figures of the iron crystals appear. These are known as Widmanstätten patterns. They are never found except in meteorites.

Wolf–Rayet stars: Exceptionally hot, greenish-white stars whose spectra contain bright emission lines as well as the usual dark absorption lines. Their surface temperatures may approach 100,000°C, and they seem to be surrounded by rapidly expanding envelopes of gas. Attention was first drawn to them in 1867 by the astronomers Wolf and Rayet, after whom the class is named. Recently, it has been found that many of the Wolf–Rayet stars are **spectroscopic binaries.**

X

X-ray astronomy: X-rays are very short electromagnetic radiations, with wavelengths of from 0·1 to 100 Ångströms. Since X-rays from space are blocked by the Earth's atmosphere, astronomical researches have to be carried out by means of instruments taken up in rockets. The Sun is a source of X-rays; the intensity of the X-radiation is greatly enhanced by solar **flares.** Sources of X-rays outside the **Solar System** were first found in 1962 by American astronomers, who located two sources, one in Scorpius and the other in Taurus; the latter has now been identified with the Crab Nebula. Since then, various other X-ray sources have been discovered, some of which are variable.

Y

Year: The time taken for the Earth to go once round the Sun; in everyday life it is taken to be 365 days (366 days in Leap Year).
(1) *Sidereal year:* The true revolution period of the Earth: 365·26 days, or 365 days 6 hours 9 minutes 10 seconds.
(2) *Tropical year:* The interval between successive passages of the Sun across the **First Point of Aries.** It is equal to 365·24 days, or 365 days 5 hours 48 minutes 45 seconds. The tropical year is about 20 minutes shorter than the sidereal year because of the effects of **precession,** which cause a shift in the position of the First Point of Aries.
(3) *Anomalistic year:* The interval between successive **perihelion** passages of the Earth. It is equal to 365·26 days, or 365 days 6 hours 13 minutes 53 seconds. It is slightly longer than the sidereal year because the position of the perihelion point moves by about 11 seconds of arc annually.
(4) *Calendar year:* The mean length of the year according to the Gregorian calendar. It is equal to 365·24 days, or 365 days 5 hours 49 minutes 12 seconds.

Z

Zenith: The observer's overhead point (altitude 90°).

Zenith distance: The angular distance of a celestial body from the observer's **zenith.**

Zodiac: A belt stretching right round the sky, 8° to either side of the **ecliptic,** in which the Sun, Moon and bright planets are always to be found. It passes through 13 constellations, the twelve commonly known as the Zodiacal groups plus a small part of Ophiuchus (the Serpent-bearer).

Zodiacal light: A cone of light rising from the horizon and stretching along the **ecliptic.** It is visible only when the Sun is a little way below the horizon, and is best seen on clear, moonless evenings or mornings. It is thought to be due to small particles scattered near the main plane of the Solar System. A still fainter extension along the ecliptic is known as the *Zodiacal Band.*

Acknowledgements

The original *Atlas of the Universe* and the new, completely updated, *Concise Atlas of the Universe* owe their publication to teamwork nationally and internationally. We have as a team been enormously helped and encouraged throughout by Professor Sir Bernard Lovell's advice and comments; and to him, therefore, first and foremost our most sincere and grateful thanks. His influence on our work has been immensely constructive. Mr Harold Fullard, Cartographic Director of George Philip in London, has also given crucial help on the planning and design of the cartographic element and to him also warm thanks.

Both editions of the *Atlas* owe much to international research and teamwork; and indeed it has been support from the international publishing community which has made them economically viable. I must, therefore, next pay a tribute to those men and organizations upon whose generosity and goodwill the scientific community of the world has grown to rely, and perhaps to take for granted all too often. And first I want to thank our friends in America. American technological endeavour was responsible for the building of the great 200-in. reflector at Palomar, a remarkable achievement spanning the 15 years from 1932 to 1947. It is through this, the Hale Telescope, that so much of what we now know of the nature, structure and beauty of heavenly objects has been revealed. In the last decade it has been another American body, the National Aeronautics and Space Administration, which has led the world in the new sciences of manned and unmanned space exploration. Both these bodies I thank, especially for their help with these publications, particularly NASA and its Administrator, Dr Thomas O. Paine, and the late Dr Arch Gerlach, Chief Geographer at the United States Geological Survey and geographical adviser to the NASA/USGS Earth Resources Studies programme, whose help and comments, particularly on the "Atlas of the Earth from Space" section, were most gratefully appreciated. I cannot find space to name all the many other members of NASA staff who have contributed so much, but I must especially single out for thanks Les Gaver of the audio-visual branch and Dr James Wray, consultant to the NASA/USGS Geography Program. I am also extremely grateful to Dr William Pickering and Frank Bristow of the Jet Propulsion Laboratory at Pasadena, California, for advice on the NASA photographs of Mars.

Then, too, I thank the professional astronomers, astronomical observatories and radio telescope laboratories across the world; the Nuffield Radio Astronomy Laboratory at Jodrell Bank (and particularly Dr Palmer and Miss Sanderson); the Royal Astronomical Society; the Royal Greenwich Observatory, Herstmonceux; the Astronomical Faculty of the University of London (especially its Lunar Department under Dr Fielder, and where Dr John Guest has been of great help); the Royal Observatory at Cape Town, South Africa; Dr W. S. Finsen, former Director of the Republic Observatory, also in South Africa; the Lick Observatory; the Lunar and Planetary Laboratory at Tucson, Arizona (and especially Dr G. P. Kuiper); the Lincoln Laboratory of the Massachusetts Institute of Technology (and particularly John Kessler); the U.S. Naval Observatory; the Lowell Observatory at Flagstaff, Arizona; the Dearborn Observatory of Northwestern University (and particularly the other Dr James Wray who has done so much to liaise on the content and presentation of astronomical information for the American editions); the Department of Astronomy of the University of Michigan (and especially Orren C. Mohler, Chairman); the Observatoire d'Astrophysique at Meudon, Paris; the Pic du Midi Observatory and the Nice Observatory, both in France; the Astronomical Society of Switzerland and the astronomers of the University of Berne; the Astronomical Observatory of Rome; the David Dunlap Observatory, and the Dominion Astrophysical Observatory, British Columbia, both in Canada; the Lund Observatory, and the Uppsala Astronomical Observatory, in Sweden; and the Berlin Observatory.

Throughout its history astronomy has perhaps benefited more than any other branch of science from the work of skilled amateur observers and enthusiasts, and I am proud that so much work from contemporary amateur observers has been found of sufficient importance to be included here. I thank especially Commander H. R. Hatfield, R.N., who has gone to great trouble on our behalf; Henry Brinton, Alan Williams, Peter Gill, W. J. Rippengale, Frank J. Acfield, P. Glaser, and my many friends of the British Astronomical Association (particularly W. M. Baxter, Director of the Solar Section).

I am very grateful to Kenneth Gatland for his original help in planning the *Atlas* and for his advice throughout on matters of space flight and space research and to Dr Anthony Michaelis for providing illustrative material. A. D. Andrews of Armagh Observatory has also been of great assistance in this respect, as have Miss Pat Outram, producer of the BBC television programme *The Sky at Night*, and the other members of the production team.

Among our home and overseas publishing partners I should especially like to thank Mr Paul Tiddens of Rand McNally, who has done a tremendous amount for the project, together with Mr Sanford Cobb, Mr Art Dubois, Mr Don Eldredge and Mr Dick Randall of that company, which generously made available the Moon maps which the members of the cartographic staff of George Philip Ltd (especially Mr Pointer) have kindly adapted for the *Atlas*. In Switzerland we have been much helped by the constructive comments and advice of Mr Werner Merkli and Dr Peter Meyer of Hallwag, who provided the magnificent star maps. In France I must thank the directors, translators and staff of the Club Français du Livre and Robert Laffont for their first-class co-operation, besides the staffs of Mondadori in Italy, Spectrum in Holland and Editorial Labor in Spain. A work of this magnitude has been made possible only by their willingness to work at great speed, and we are extremely grateful to them.

Patrick Moore,
Selsey, January 1974

Picture Credits

A great many individuals and organizations have provided photographs for use in *The Concise Atlas of the Universe*. To each and every one we wish to extend our thanks for their help and cooperation.

F. Acfield
W. M. Baxter
W. Bohenblust
H. Brinton
H. E. Dall
W. S. Finsen
P. Gill
P. Glaser
H. R. Hatfield
C. Hunt

G. P. Kuiper
E. M. Lindsay
A. R. Michaelis
A. H. Mikesell
T. Moseley
T. W. Rackham
W. Rippengale
A. Williams
F. E. Wright
W. Zünti

Aeronautical Chart and Information Centre
Associated Press
Georgetown University Observatory
Lick Observatory
Lincoln Laboratory, M.I.T.

National Aeronautics and Space Administration
Novosti
Observatoire du Pic du Midi
Picturepoint Ltd
Radcliffe Observatory
Ronan Picture Library
Royal Astronomical Society
Tass
The Hale Observatories (Mount Wilson and Mount Palomar)
United States Geological Survey
United States Information Service, London
Uppsala, Stockholm and Lund Observatories
U.S. Naval Observatory
Wilhelm-Förster Sternwarte
Yerkes Observatory

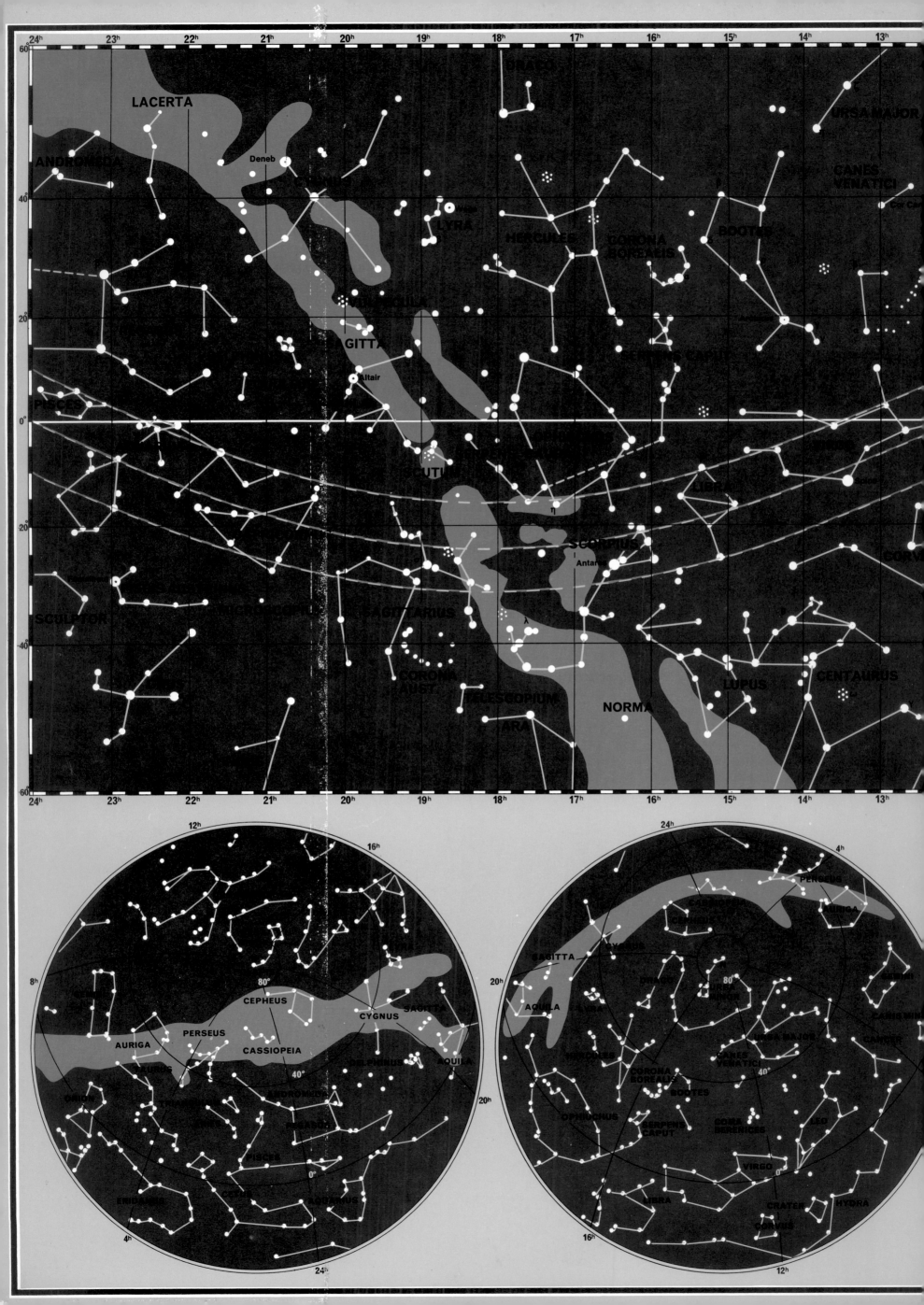